An Introdu
ACOUS

Robert H. Randall

DOVER PUBLICATIONS, INC.
Mineola, New York

Bibliographical Note

This Dover edition, first published in 2005, is an unabridged republication of the work originally published by Addison-Wesley Press Inc., Cambridge, Massachusetts, in 1951, as part of the Addison-Wesley Principles of Physics Series.

Library of Congress Cataloging-in-Publication Data

Randall, Robert H. (Robert Hulbert), 1903–1983.
An introduction to acoustics / Robert H. Randall.
p. cm.
Originally published: Cambridge, Mass. : Addison-Wesley Press, 1951, in series: Addison-Wesley principles of physics series.
Includes bibliographical references and index.
ISBN-13: 978-0-486-44251-8
ISBN-10: 0-486-44251-9
1. Sound. I. Title.

QC225.15.R36 2005
534—dc22

2004065693

www.doverpublications.com

PREFACE

Anyone who has thoughtfully taught the subject of acoustics for any length of time must surely be struck by the basic nature of the material, both in the fields of pure and of applied physics. For the student who has completed a general college course in physics there is hardly a better starting point for more advanced study. A course in acoustics very naturally begins with a study of vibrations, as preliminary to the introduction of the wave equations. It is impossible to overemphasize the importance of the two subjects — vibrations and waves — to all branches of physics and engineering. In addition, there are distinct advantages in first discussing waves of the mechanical type, rather than electromagnetic waves, with their more abstract nature and added subtleties.

Of growing importance during the last ten or twenty years is the very fruitful use of electrical analogs in acoustics. Electrical engineers are most aware of the extreme usefulness of the analog method, particularly in problems originating during World War II. In a book of this type no attempt can be made to give a complete treatment, even in the field of acoustics alone, of the use of analogs taken from electrical circuits. However, the author believes that so useful a tool in this and other branches of physics and engineering should be given more attention than is ordinarily afforded in an intermediate text.

In connection with these more quantitative aspects of the subject, it might be said that the great difficulty of setting down the features of most actual acoustical problems in precise mathematical form is of great instructive value to the physics student. Coming fresh from more elementary courses, where the problems supply just the necessary data to achieve the exact answer, he may be appalled at the extent to which approximations *must* be made to get any kind of an answer at all in acoustical problems. Experience of this kind is good preparation for the later practical use of, say, electromagnetic field equations which involve complicated boundary conditions, where the mathematical problems are very similar. A course in acoustics may incidentally serve to discourage a pure mathematician, to whom some of the approximations of physics are anathema, from entering upon a career unsuited to his temperament and point of view.

The average undergraduate is greatly interested in many of the more popular and applied features of the subject. Among these are the physics of musical instruments, peculiarities of hearing, the design of radio loudspeakers, some consideration of electronic devices as used in electroacoustical equipment, the acoustics of auditoriums, etc. As one whose interest in acoustics was originally aroused, in part, by a love of music, the author believes no text in acoustics should omit some reference to these subjects, which are as essential in their way as a consideration of the wave equations.

There are a number of elementary books on acoustics published in this country, of which Colby and Watson are good examples. Above this level there is quite a choice of specialized books on the engineering or graduate level. By far the most original and thoughtful general book on acoustics is Morse's *Vibration and Sound.* While of considerable use as a reference, this book is too difficult as a whole for undergraduate use. Chapter 5 has drawn generously upon certain parts of Morse. Mention should also be made of *Acoustic Measurements* by L. L. Beranek, an excellent survey of modern experimental techniques in acoustics. In Chapter 10 frequent reference is made to Beranek's work. There is practically no book available at the intermediate level except for the British imports, and it is hoped that the present book will help to fill the gap.

A year of college physics and a year of calculus constitute a minimum preparation for the subject as presented here. A previous knowledge of the complex notation, as used in a.c. circuit analysis, would be helpful, but Chapter 5 contains a summary of the essential material sufficient to the understanding of the text. While the book has been written mainly for undergraduates in physics, it is believed that engineering students who may later wish to specialize in communications and electroacoustics would greatly profit from a basic course using this kind of book.

The author wishes to thank Professor Francis W. Sears for his kind interest in this project and to express his gratitude to Professor A. Wilson Nolle of the Department of Physics, University of Texas, for his careful and critical reading of the manuscript and his many helpful suggestions on matters of precision and clarity.

ROBERT H. RANDALL
The City College of New York

April, 1951

TABLE OF CONTENTS

INTRODUCTION

There is no branch of classical physics that is older in its origins and yet more modern in its applications than that of acoustics. As long ago as the time of Galileo, quantitative experiments were performed on the vibrations of strings and the sound that is so produced. Boyle, Hooke, and Newton were interested in sound, and Newton undertook to compute, theoretically, its speed. Later on, the great mathematicians Laplace, Euler, d'Alembert, Bernoulli, Lagrange, and Poisson laid the bases for what was to become the general subject of hydrodynamics, although there was a great scarcity of experimental data with which to test their conclusions. In the nineteenth century, the results of the experiments of Doppler, Kundt, Kelvin, and others added to the growing body of the subject. Helmholtz, that Leonardo da Vinci of modern times, wrote his monumental work, the *Sensations of Tone*, largely from the physiological approach. Late in the nineteenth and during the early twentieth century, finishing touches to the already elegant formulation of the mechanics of sound propagation were added by Rayleigh and Lamb, whose writings on the subject have become "standard" treatises.

Along with this continuous scientific preoccupation with the problems of acoustics has gone a very lively interest, among laymen as well as among scientists, in the more qualitative aspects of the subject. Musicians are closer to science than they perhaps realize when they play musical instruments and wonder as to the quality of the sound flowing from them. Laymen of all kinds are interested in speech and song, music and noise. These are, it would appear, permanent interests which will probably persist, even with the competing glamour of the atom and its nucleus!

With the beginning of the twentieth century it would have been safe to say that the subject of acoustics was as nearly complete as it would ever be. Even were this so, a study of acoustics would still be a "must" for the proper understanding of the great body of related scientific literature. Vibrations, whether connected with strings and diaphragms or with subatomic oscillators radiating electromagnetic waves, are all of a kind, and to understand the one type is a great help towards understanding the other. In addition, the "fields" of sound and the "fields" of electromagnetic radiation are kindred in more ways than one, with the former a preferred starting point from the standpoint of concreteness and simplicity.

Two developments in the field of applied acoustics have given impetus, in recent years, to further study and growth of the subject. The first is the rise of a whole new industry, devoted to the realistic reproduction of speech and music through the mediums of the radio and the phonograph.

The second, less beneficent in nature, arose as the result of war needs, both in the field of undersea signaling and in connection with problems in aeronautics. As so often occurs when interest in a subject revives, other fields, like those of medicine and pure physics, have been stimulated to make use of new tools and new refinements of the older theoretical work. Thoughtful comparison between acoustics and other branches of physics and engineering has brought to light little realized interrelations, of great use to all fields concerned. The electrical circuit analogs discussed in Chapter 5 are a good example of this.

As an introduction to a logical presentation of the subject, a broad outline of the scope of acoustics, together with a certain definition of terms, will be helpful.

I–1 Sound vs acoustics. In the strict sense, the word *sound* should be used only in connection with effects directly perceivable by the human ear. These effects are ordinarily due to the wave motion set up in air by the vibration of material bodies, the frequencies which are audible to the ear being in the approximate range of 30 to 15,000 cycles/sec. In this book we shall consider the word *sound* to cover the entire wave phenomena in air of this frequency range and we shall use it as a qualifying adjective in connection with such wave properties as "particle displacement," "excess pressure," and the like. Whenever the frequencies are well outside the above range, we shall call the disturbance a longitudinal *wave*, rather than a sound. Waves set up in media other than air we shall also not call sound, since the ear is not ordinarily capable of responding to this type of energy directly. Waves set up within solid rods, crystals, etc., are of this type.

For no very good reason, the word *acoustics*, originally associated with the sound properties of rooms, auditoriums, etc., has been broadened to include almost the whole field of mechanical vibration and waves, whether of audible frequencies or not, and without regard to the medium. While the emphasis is still on what can be heard, many of the most interesting recent applications in acoustics are concerned with a range of frequencies well outside the audible range, particularly in the ultrasonic (high-frequency) region. Some of these applications will be discussed later in this book.

I–2 Vibrating bodies. Before there can be sound waves in air, there must be vibration of some material body. The character of the sound is so dependent upon the nature of this vibration that a careful study of the possible kinds of vibration is imperative. The simplest type of vibration to discuss is that of an idealized particle. Under certain special conditions, as will be seen, actual sound sources may be discussed as if they were particles. More often than not, due to the complexity of shape and motion of actual sound sources, such a simple picture is inadequate. Nevertheless, a consideration of particle vibration theory is basic to the understanding of the more complicated motions of extended bodies such as strings, bars, plates, etc., to be considered later.

I–3 Frequency. The *frequency* of a vibrating source of sound is the repetition rate of its periodic motion, assuming this to be simple harmonic. It is usually specified in cycles per unit time. In the wave phenomenon set up in the air, frequency refers to the vibration rate of layers of air, and is to be distinguished from *pitch*, a word used to describe the subjective sensation perceived by the listener. The sensation of pitch is a psychophysiological matter and is only imperfectly understood. As we shall see in Chapter 9, the relation between frequency and pitch is a complicated one. The range of frequencies to which a young, healthy ear will respond is enormous, from possibly as low as 15 cycles/sec to as high as 20,000 cycles/sec. The ear is by no means of equal sensitivity over this frequency range, but in studying the complex thing called musical sound and in designing modern electrical and electromechanical apparatus to reproduce this sound, we must cover the extremes of the frequency range of the ear. The design of such equipment is difficult, as we shall see, and it is only recently that any considerable success has been achieved.

I–4 Amplitude. The *amplitude* of any vibratory motion has the usual meaning associated with simple harmonic motion, i.e., the maximum excursion from the mean central position. Such amplitudes may refer to the motion of the source, the motion of the receiver of the sound, or the motion of the layers of air where the wave exists. Everyone knows how a motion of small amplitude over a sufficiently large area may give rise to tremendous sound disturbances. At the receiving end, whether it be at the ear or at a microphone, amplitudes may be unbelievably small. An amplitude of motion of the air of 10^{-8} cm is by no means the least to which the ear will respond.

I–5 Waves. It is one peculiarity of a fluid like air, with little or no resistance to shear, that only longitudinal waves may be propagated. All disturbances of any other nature will tend to disappear at a small distance from the source. A consideration of the elastic and inertial properties of the medium leads to a beautiful and complete theory of longitudinal wave propagation which is useful as well as elegant. The great difficulty with the differential equations for sound waves is in obtaining all the details of particular solutions to practical problems. Sound sources are rarely simple or symmetrical in shape, and the irregularities in contour introduce serious trouble. Useful solutions *may* be obtained, if one is willing to accept certain approximations. As always, approximations are dangerous and must be made with the utmost care, keeping in mind the essential physics of the problem. The results of this process might appear to be crude in many cases, but the student should appreciate that the ear itself is, fortunately for the analyst, a rather crude device, incapable under ordinary conditions of detecting discrepancies of less than 10% to 20%.

I–6 Wavelength. Frequency in the wave. For disturbances of a simple harmonic nature, the *wavelength* is the distance, at any one instant, between adjacent wave crests. The frequency, within the body of the wave disturbance, may be defined as the number of crests passing any one point in space per unit time, and is ordinarily the same as the frequency of vibration of the source of the wave disturbance. If the source is not stationary with respect to the medium, the frequency in the wave is not the same as that of the source. This is a situation that is one cause of the well-known Doppler effect.

I–7 The principle of superposition. It is a general property of many mechanical systems that when two different types of motion are impressed simultaneously, the resultant total motion may be described as the sum effect of the two motions considered independently. This is one statement of the Superposition Theorem. It is a very broad principle in physics. The student of elementary physics has seen the general principle applied many times in connection with such subjects as the composition of force vectors, the summing up of assorted emf's in electrical circuits, the interference effects in light, etc. It will be a correct principle to use whenever the system is "linear," that is, whenever its behavior may be accurately described by a linear differential equation. The vibrations of material bodies and of the particles in a deformable medium like air obey such equations, provided the amplitudes of motion are small. Fortunately, this is usually so in acoustics. We shall make frequent use of the Superposition Theorem throughout this book.

I–8 Energy density. Intensity in the wave. The average energy per unit volume in the medium, due to the presence of a wave, is called the *energy density*. The *intensity* in the wave is defined as the energy flow, per unit time and per unit area, across an area taken normally with respect to the direction of wave propagation. Energy density and intensity are simply related through the velocity of wave propagation. Both these quantities may be computed from measurements made with suitable laboratory instruments, whose operation depends in no way upon the properties of the ear. The student is cautioned not to use the word "loudness" as synonymous with "intensity." The loudness of a sound, in the language of acoustics today, is a measure of the purely subjective sensation arising when a sound wave strikes the ear. The exact relationship between loudness and intensity is difficult to determine, as one would expect; the student is referred to Chapter 9 for a further discussion of this matter. (We are using "loudness" here in the purely qualitative sense. We shall later refer to the loudness *level*, a numerical measure of loudness which is defined directly in terms of the *pressure* in the wave disturbance, rather than the intensity.)

The familiar unit, the *decibel*, is fundamentally a quantitative measure of *relative* (not absolute) *intensity*, and is used to compare one sound intensity with another. The decibel scale is defined in a logarithmic manner, as will be seen in Chapter 2, to conform to the approximately logarithmic behavior of the ear. Its exact meaning and use will be made clear when it is needed.

I–9 Sound "quality." The *quality* of a musical note, as played on some instrument, or coming from a singer's throat, is a most important characteristic, connected, in part, with the physiological, the psychic, and the aesthetic in the listener. From a purely objective point of view, it has been common to explain quality as due solely to the number and prominence of the steady-state harmonic overtones. There are other factors to be considered, however. Recent studies by Fletcher have revealed the importance of the *transient* period of vibration, the time during which the instrument and sound vibrations are building up or dying down. There is even evidence that it is during the transient period of "attack," for instance, that a violin is recognized as such, rather than as, say, a cello. The ear will apparently tend to confuse the two instruments when a sustained note is being played.

I–10 The use of electrical analogs. While it is somewhat in the nature of a digression in the logical development of the subject, the discussion of sound waves along classical lines will be followed by a brief introduction to the electrical analog method as applied to acoustics, with chief emphasis upon the concept of "acoustic radiation impedance." Applied with equal success in the subject of electromagnetic radiation, this idea, borrowed from a.c. circuit theory, is of especial aid in predicting the total radiation of power from a given sound source. It is of considerable assistance in the design of aperiodic radiators, like radio loudspeakers, where the problem is too difficult for complete analysis by means of the classical wave equations.

I–11 Waves in solids. Plane longitudinal waves set up in solids are very similar to such waves in air, with, of course, different elastic and inertial factors. Unlike gases and liquids, solids, with their resistance to shear, can sustain transverse vibrations. The simplest of all transverse vibrations for an extended body are those of the ideal flexible string, whose standing wave characteristics are so important to all stringed instruments. In fact, a discussion of string vibrations leads quite naturally to a consideration of some design features of the violin, the piano, etc. In only a few cases, in particular for the piano, is the mathematics capable of predicting the intensity of some of the more important harmonics that are so essential to the quality of the emitted sound. The great difficulty in precisely describing the initial conditions, when the string is struck, plucked, or bowed, as the case may be, is the main stumbling block to exact analysis. When it is realized that not only the string properties but also the shape and complex characteristics of the body of the instrument greatly determine the nature of the radiated sound, one is ready to accept the fact that the design of a high quality musical instrument is as much a matter of art as of science.

The problems of the vibration of membranes, bars, and plates become progressively more complicated. The more important general features of such motions will be discussed in Chapter 7.

I–12 Experimental technique. Sound measurements are some of the more difficult in experimental physics. While sensitive linear microphones and associated electronic amplifiers are now available, there are always two major difficulties with their use in a "field" of sound. First, there is the precise, absolute calibration of the equipment over a wide range of sound frequencies and intensities. Second, there is the disturbing effect that any detection device whose dimensions are comparable to the wavelength of the sound introduces upon the field of sound itself. The errors involved are somewhat similar to the potential errors encountered in the use of a voltmeter; one would like to measure the potentials existing *before* connecting the meter! In addition, the standing wave patterns set up in any ordinary room make impossible any accurate measurement of the true radiation properties of the source itself. One is then driven either to outdoor experiments or to building very elaborate and expensive sound rooms with especially treated wall surfaces and complicated structural supports. These and other difficulties will be discussed in the chapter on experimental methods.

I–13 Applied acoustics. Much of the renewed interest in acoustics has come from the applied field. Music has long felt itself an art to be insulated as far as possible from the mechanics of science. Yet the advent of "canned" music, deplored by so many musicians, has stimulated the scientific study of the quality of sound to the point where it is deemed possible to create new instruments having tonal qualities undreamed of by the old masters. It is true that thus far the instruments born of modern science, such as the electronic organ and the like, have aped the older traditional instruments. But as has been pointed out by Fletcher and others, the possibilities of sound synthesis have hardly been tapped, and entirely new instruments without prototype will undoubtedly be evolved.

Acoustics plays an important part in the reproduction of speech and music through the radio and the phonograph. With the refinements achieved in the electrical circuit and electronic fields, the importance of improving the acoustical features of such reproducing systems has become more and more apparent. As a result, much careful study of loudspeaker design has been made in recent years. As the musical sophistication of the general public rises, the results of this study will undoubtedly be realized in home radios and phonographs of higher acoustical quality.

Other applications of acoustics will be considered in Chapter 12. Vibrations and waves of ultrasonic frequency were first studied in detail by Wood and Loomis, and also by G. W. Pierce. Since those early experiments, much quantitative work has added to the knowledge. The general scope of the war work in connection with underwater signaling is well known. Industry has found many uses for ultrasonic waves in the testing of materials. In the realm of pure physics the use of high frequency longitudinal waves has become a valuable means for the study of interatomic forces in solids, both at normal temperatures and near the temperature of absolute zero.

Interest in acoustics has stimulated further study along physiological lines. In Chapter 9 some of the established facts will be reviewed. While no attempt can be made in a book of this kind to deal with this aspect of acoustics exhaustively, enough will be said to impress the student with the essential unity of science, and the importance of considering related fields whenever they have some bearing upon the subject at hand.

I-14 Systems of units. The cgs system is universally employed in all the important acoustical literature of the past, and it is still generally used in the current writing. Acoustics is concerned primarily with the mechanics of fluids, and for mechanics the cgs system is thoroughly self-consistent. In addition, the centimeter, the gram, the dyne, etc., are units well adapted to the small scale of acoustical phenomena. With the discussion of electroacoustical devices, however, a hybrid cgs system becomes necessary. Since the mks system is far better suited to systems containing both electrical and mechanical features, and since the use of this system is rapidly becoming common in various branches of physics and engineering, it seems unwise to ignore it completely in acoustics. The cgs system will be used, generally, in this book, but occasional reference will be made to the mks system as well.

CHAPTER 1

FUNDAMENTAL PARTICLE VIBRATION THEORY

The production of sound always involves some vibrating source. Such a source is often of irregular shape, and rarely do all parts of the vibrating surface move as a unit. It is the very complexity of the vibration of a sound source that makes it necessary to consider first the simplest vibrating body, the *particle*. The motion of actual sources may approximate that of a particle, particularly at low frequencies. Whenever this approximation may not be made, the vibrating surface may be broken up into smaller areas, infinitesimal if desired, the sum effect of which is equivalent to that of the total surface area of the actual source. The mathematics of this summation may be extremely complicated, but approximations will often lead to useful results.

1-1 Simple harmonic motion of a particle. Simple harmonic motion originates, in mechanics, because of the existence of some kind of unbalanced elastic force. With such a force, Newton's second law becomes, for a particle of mass m, free to move along the x-axis,

$$m\ddot{x} = -Kx. \qquad (1\text{–}1)$$

In the expression on the right for the force, K is called the elastic constant, and the negative sign indicates that the restoring force always acts towards the origin. Equation (1–1) may also be written

$$\ddot{x} = -\omega^2 x, \qquad (1\text{–}2)$$

where $\omega^2 = K/m$. This differential equation completely defines the type of motion and from it all other properties of simple harmonic motion may be obtained. By integrating Eq. (1–2) twice, the displacement equation may be shown to be of the form

$$x = x_m \cos(\omega t + \alpha), \qquad (1\text{–}3)$$

where x_m is the amplitude of the motion and α is called the phase angle. The quantities x_m and α are essentially constants of integration, whose values depend upon the mathematical boundary conditions. They may easily be determined, for instance, if one knows the value of x and of the velocity, $\dot{x}$, at either the time $t = 0$, or at any other specific value of the time. Whether the cosine or the sine function appears in Eq. (1–3) is dependent upon these boundary conditions. If, for instance, α turns out to be $\pm\pi/2$, Eq. (1–3) may be written in the sine form. The angular frequency, ω, is equal to $2\pi f$, where f is the repetition rate in cycles per unit time.

Besides the displacement equation, two similar equations for the velocity, $\dot{x}$, and the acceleration, $\ddot{x}$, are important:

$$\dot{x} = -\omega x_m \sin(\omega t + \alpha), \tag{1-4}$$

$$\ddot{x} = -\omega^2 x_m \cos(\omega t + \alpha). \tag{1-5}$$

These are obtained by a simple differentiation of Eq. (1-3). All three equations can also be obtained by considering the projection, on a diameter of a circle, of the motion of a particle moving around the circle with a constant speed, as is usually shown in elementary physics. The phase relationship is apparent from Eqs. (1-3), (1-4), and (1-5). The velocity and displacement bear a 90° relationship, while acceleration and displacement are 180° apart. The 90° relationship which always results from differentiating a sine or cosine function will be an important feature of our discussion of sound waves in air, as will be seen later.

1-2 Energy in SHM. In sound, we are always dealing with the vibration of material bodies, or media having the property of mass, and since the particle being considered is moving, it will, in general, have a kinetic energy equal to $\frac{1}{2}m(\dot{x})^2$. This energy varies with the velocity, being zero at the ends of the motion, where $x = x_m$, and a maximum when the particle is passing through the position $x = 0$. Since no dissipative force is being considered, the total energy of the system must remain constant. Therefore when the kinetic energy decreases, as the particle approaches $x = x_m$, the potential energy must increase. Clearly, the maximum potential energy must equal the maximum kinetic energy. The maximum potential energy, $(E_p)_m = \int_0^{xm} Kx\,dx = \frac{1}{2}Kx_m^2$. It is easy to show that this energy is equal to the maximum kinetic energy, $(E_k)_m$, possessed by the particle when it is moving through the central position. For, if $\dot{x}_m$ is the maximum velocity,

$$(E_p)_m = \tfrac{1}{2}\,Kx_m^2 = \tfrac{1}{2}\,K\,\frac{(\dot{x}_m)^2}{\omega^2} = \tfrac{1}{2}\,m(\dot{x}_m)^2 = (E_k)_m. \tag{1-6}$$

At positions other than the central one and the extreme end points, the energy is partly kinetic and partly potential. The total energy of the system may obviously be taken as either the maximum potential energy or the maximum kinetic energy. Using the latter,

$$E_{\text{total}} = \tfrac{1}{2}m(\dot{x}_m)^2 = \tfrac{1}{2}m\omega^2x_m^2 = \tfrac{1}{2}m(4\pi^2)f^2x_m^2. \tag{1-7}$$

It is interesting to note that for particles of equal mass executing simple harmonic motions of the same energy but of different frequencies, the amplitudes must be inversely proportional to the frequency. The paper cone of a radio loudspeaker, fed with the same energy at a variety of frequencies, will have imperceptible amplitudes at the high audible frequencies, whereas at low frequencies, visible amplitudes of as much as a millimeter or two may easily occur.

1-3 Combinations of SHM's along the same straight line. The combination of several collinear simple harmonic vibrations may be discussed either analytically or, more conveniently, by use of the graphical method commonly employed in a.c. circuit theory. This method is fundamentally based on the rectilinear projection of uniform circular motion, so often used in elementary physics to introduce SHM. In Fig. 1-1 the length of the vector represents the amplitude of the motion, x_m. The vector is conventionally assumed to rotate counterclockwise at the angular rate, ω (in radians per second). It is clear that the expression for the instantaneous projection of this vector, i.e., $x_m \cos(\omega t + \alpha)$, where α is the starting angle at $t = 0$, is identical with the displacement equation for SHM, Eq. (1-3).

Suppose, now, that we wish to represent the simultaneous execution, by a particle, of several SHM's along x, of differing amplitude, frequency, and phase angle. Each of these separate motions may be represented as the projection of an appropriate rotating vector. The simplest case to consider is when the frequencies are the same. The total displacement of the particle is

$$x_r = x_1 + x_2 + \cdots + x_n,$$

where x_1, x_2, etc., represent the separate displacements. Since all angular frequencies are the same, the relative angles between the different amplitude vectors are maintained at all times. Therefore it is possible at any time, such as at time $t = 0$, to sum up vectorially the several amplitude vectors and to consider the total motion, x, to be simply the projection of this resultant upon the x-axis. In Fig. 1-2 two amplitude vectors $(x_m)_1$ and $(x_m)_2$ are drawn for the time $t = 0$. The magnitude of the resultant

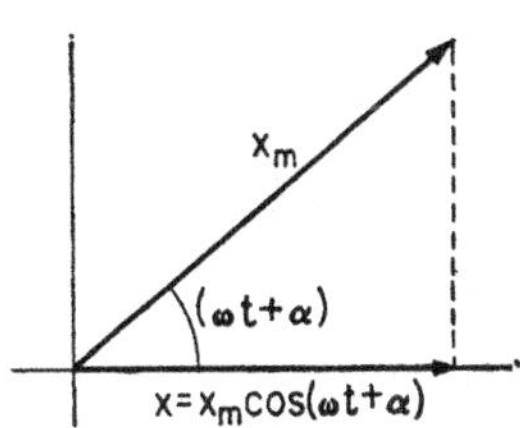

Fig. 1-1. Polar representation of SHM.

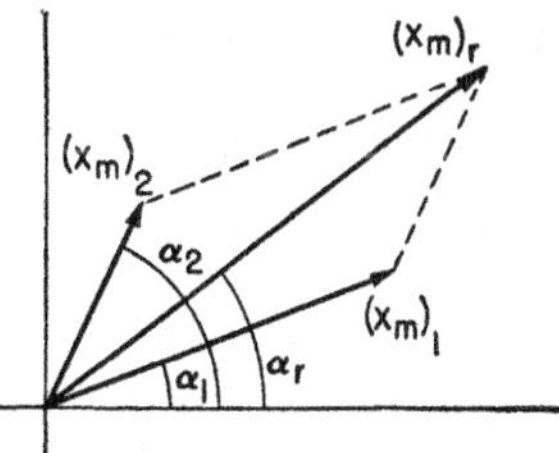

Fig. 1-2. Amplitude summation for two SHM's of the same frequency.

vector, $(x_m)_r$, may be obtained most simply by finding its x- and y-components, as is done in mechanics with force vectors:

$$(x_m)_r = \sqrt{[\Sigma(x_m)_x]^2 + [\Sigma(x_m)_y]^2}. \tag{1-8}$$

Also,

$$\tan \alpha_r = \frac{\Sigma(x_m)_y}{\Sigma(x_m)_x},$$

where $\Sigma(x_m)_x$ and $\Sigma(x_m)_y$ are the sums of the x- and y-components of the separate amplitude vectors at the time $t = 0$. The total motion, x_r, may then be written:

$$x_r = (x_m)_r \cos(\omega t + \alpha_r). \tag{1-9}$$

It is seen that such a combination of SHM's is always equivalent to a single pure SHM. This is a fact of fundamental practical importance in the production of music. In the first violin section of an orchestra, for instance, while at a given instant all violins are presumably playing at the same frequency and with approximately the same amplitudes, the relative phases are quite randomly related. Since these relative phases undoubtedly are shifting continuously due to slight frequency variations, the phase of the sum effect at the ear is also changing. As we shall see later, the ear ordinarily is insensitive to phase effects in music, and in the case of the violinists, only a single note of the common approximate frequency is heard.

This vector method of summing up SHM's of the same frequency but of differing phase will prove very useful in Chapter 4 in the consideration of sound diffraction.

Example. Reduce the following two collinear SHM's to a single equivalent vibration, finding the amplitude and the phase angle.

$$x_1 = 5 \cos(\omega t + 65°),$$
$$x_2 = 7 \cos(\omega t + 30°).$$

The two amplitude vectors are located at the time $t = 0$, as in Fig. 1-3. Making use of the cosine law, the resultant amplitude, $(x_m)_r$, may be found directly:

$$(x_m)_r = \sqrt{(5)^2 + (7)^2 + 2(5)(7)\cos 35°} = 11.4.$$

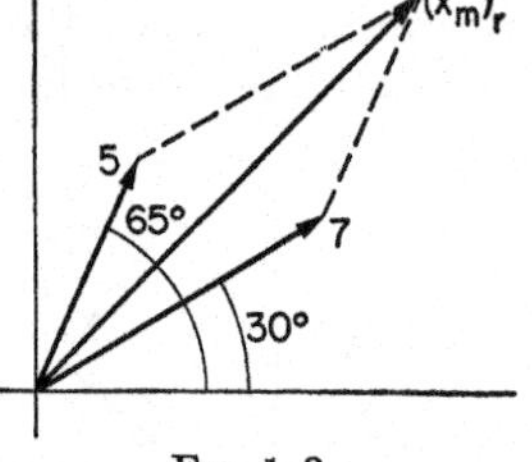

FIG. 1-3.

Or, using the x- and y-components:

$$\Sigma(x_m)_x = 5 \cos 65° + 7 \cos 30° = 8.18,$$

$$\Sigma(x_m)_y = 5 \sin 65° + 7 \sin 30° = 8.03,$$

$$(x_m)_r = \sqrt{[\Sigma(x_m)_x]^2 + [\Sigma(x_m)_y]^2} = \sqrt{(8.18)^2 + (8.03)^2} = 11.4.$$

The phase angle of the resultant vibration is arctan $\Sigma(x_m)_y/\Sigma(x_m)_x$ = arctan (0.982) = 44° 30′. Therefore the equation for x_r is

$$x_r = 11.4 \cos(\omega t + 44° \, 30').$$

1-4 Two collinear SHM's whose frequencies differ by a small amount. Beats. The phenomenon of beats, in sound, is a familiar one. As it is commonly observed, it is the slow, audible "throbbing," or variation in intensity, associated with two sounds of nearly the same frequency which alternately reinforce and partially or completely cancel each other. In Fig. 1-4a are shown two SHM's of slightly different frequency, the ordinate being the displacement and the abscissa, time. In the presence of two such sound waves, a layer of air (equivalent to the particle under discussion) will execute a motion which is the graphical sum of the two separate motions. In Fig. 1-4b is drawn the graphical sum of the curves of Fig. 1-4a. The periodic variation in amplitude, in the case of the sum curve, is to be expected, in view of the effects observed aurally.

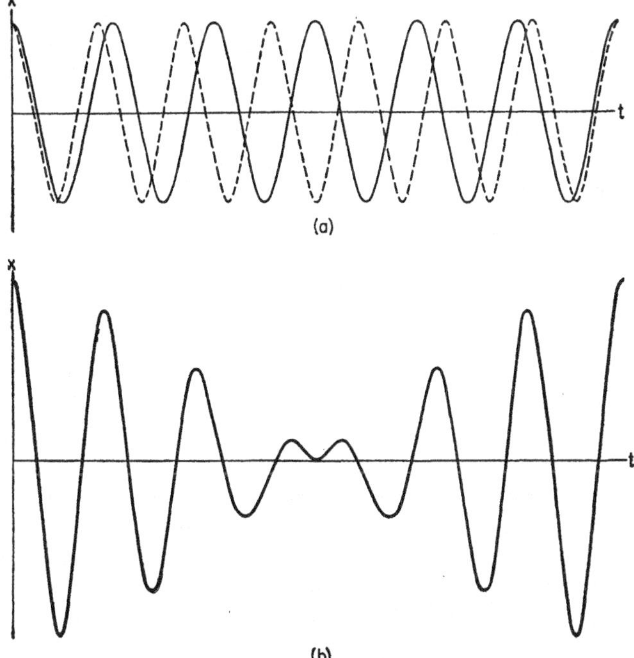

FIG. 1-4. Beats.

The two separate SHM'S may be represented analytically as

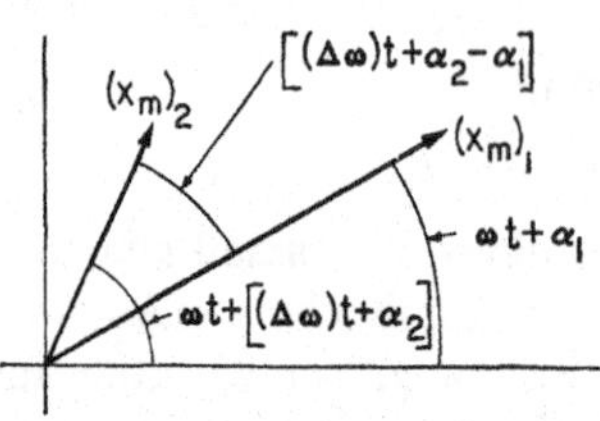

Fig. 1–5.

$$x_1 = (x_m)_1 \cos [\omega t + \alpha_1] \tag{1–10}$$

and

$$x_2 = (x_m)_2 \cos [(\omega + \Delta\omega)t + \alpha_2], \tag{1–11}$$

where $\Delta\omega$ is small compared with ω. Equation (1–11) may be rewritten:

$$x_2 = (x_m)_2 \cos \{\omega t + [(\Delta\omega)t + \alpha_2]\}. \tag{1–12}$$

Both x_1 and x_2 may be thought of as projections of rotating vectors, as discussed in Section 1–3. Since the two amplitude vectors, $(x_m)_1$ and $(x_m)_2$, rotate with nearly the same angular velocity, the term $[(\Delta\omega)t + \alpha_2]$ in Eq. (1–12) may be considered as a slowly changing phase angle. When the two vectors are in the positions shown in Fig. 1–5, the resultant vector, $(x_m)_r$, may be computed by means of the cosine law:

$$(x_m)_r = \sqrt{(x_m)_1^2 + (x_m)_2^2 + 2(x_m)_1(x_m)_2 \cos [(\Delta\omega)t + \alpha_2 - \alpha_1]}. \tag{1–13}$$

The magnitude of $(x_m)_r$ will slowly change, as time goes on, due to the variation of the cosine function in Eq. (1–13) with the time. The maximum and minimum values of $(x_m)_r$ will occur when the cosine function is equal to $+1$ and -1, respectively. The corresponding values for $(x_m)_r$ will be $(x_m)_1 + (x_m)_2$ and $(x_m)_1 - (x_m)_2$, assuming $(x_m)_1 > (x_m)_2$. The frequency, f_b, of this cyclic change in $(x_m)_r$ is plainly $\Delta\omega/2\pi$. Since $\Delta\omega$ is the difference between the angular frequencies for the two vibrations, ω_1 and ω_2, f_b will equal the difference between the vibration rates, f_1 and f_2.

If f_1 is nearly equal to f_2, what has been said in the preceding paragraph regarding the variations in $(x_m)_r$ will closely describe the variation in the amplitude of the motion along x, which is the *projection* of $(x_m)_r$. It is the projection of $(x_m)_r$, of course, which represents the instantaneous sum of x_1 and x_2, and which directly describes the beat phenomenon. The maximum value of $x = x_1 + x_2$ will vary periodically, at a frequency very close to the beat frequency, f_b, between limits which are very nearly $(x_m)_1 + (x_m)_2$ and $(x_m)_1 - (x_m)_2$. These are not exact statements, since in general $(x_m)_1$ and $(x_m)_2$ will not become coincident when in the horizontal position. However, since the two amplitude vectors are rotating with nearly the same angular velocity, it is clear that at whatever angle to the x-axis coincidence occurs, the two vectors will have only slight relative displacement by the time they *do* reach the horizontal, and the above statements are, for all practical purposes, valid. If $\Delta\omega$ is quite large compared with ω (not the case in ordinary sound beats) this method of interpretation has little meaning.

1-5 Mathematical vs audible beats. There is an interesting distinction between what might be called "mathematical" and "audible" beats. It can be shown that unless the two angular frequencies ω_1 and ω_2 are commensurate, that is, unless the ω's and therefore the two actual frequencies f_1 and f_2 bear a whole number ratio, the sum motion will *never* repeat exactly. Therefore no recurring beat phenomenon, in the strict mathematical sense, will exist. In addition, if the whole number relation *does* exist, each separate vibration must execute some integral number of cycles before a repetition of the sum motion can occur. To take numerical examples, suppose the two frequencies are 406 and 404 cycles/sec, respectively. The two vibrations will be in phase twice each second. This can readily be seen by reducing the frequency ratio to the smallest whole number ratio, i.e., $\frac{203}{202}$. If the two frequencies start in phase, after $\frac{1}{2}$ second, when they have executed 203 and 202 cycles respectively, they will be in phase again. The two beats each second obtained in this way would indicate that the beat frequency is always $f_1 - f_2$. This, however, is not invariably so, for if the two frequencies were 407 and 404 the difference frequency would indicate three beats a second, whereas $\frac{407}{404}$ being already the smallest whole number ratio, there is a mathematical repetition only *once* a second.

The above statements can be easily checked by consideration of the rotating vector example.

The *audible* effect of beats contains none of the subtleties discussed above. If two sources initially emit sound waves of the same frequency and then one frequency is gradually raised, the beat effect begins to occur smoothly and continuously, with no gaps occurring at discrete frequencies. This is because the ear is sensitive only to the envelope of the sum function, as in Fig. 1-4b, and an absence of an exact mathematical repetition *within* the envelope goes unnoticed.

When the simple difference frequency becomes greater than about ten per second, the alternation in intensity is no longer observed and, instead, one receives the impression of a steady sound which is either harmonious or discordant, depending on the frequency interval. (This will be discussed later in Chapter 9 in connection with consonance and dissonance.) With ordinary sound intensities a real difference frequency is never observed, that is, a third musical note is never evident. (With very large sound intensities, it is another matter. See Chapter 9.) This is not surprising, since there are really only two SHM's involved. The true beat effect is merely the alternation in intensity of what appears to be one frequency.

1-6 Combinations of more than two SHM's of different frequencies. From the discussion just concluded, a mixture of frequencies not bearing a whole number relation is equivalent to no repetitive steady state vibration. This situation is not often encountered in problems in sound — at least one does not usually attempt to analyze problems in which it *does* arise. Most musical instruments, fortunately, vibrate in such a way as to give rise to a "fundamental" tone and "overtones," all of which bear whole number ratios to one another, and consequently the over-all vibration is a repeating function. There is a theorem, due to Fourier, so powerful in its ability to analyze such a repeating function into its separate component frequencies that it deserves considerable attention in any discussion of vibration and sound.

1-7 Fourier's theorem. Stated briefly, this theorem asserts that any single-valued periodic and continuous function may be expressed as a summation of simple harmonic terms, finite or infinite in number (depending on the form of the function), whose frequencies are integral multiples of the repetition rate of the given function.* The restrictions that the function be single-valued and continuous are easily met in the case of the vibrations of material bodies, and the theorem is therefore of the greatest use in acoustics.

The most useful analytic expression for the harmonic series for periodic functions of the time is as follows:

$$\begin{aligned} x = f(t) = A_0 &+ A_1 \sin \omega t + A_2 \sin 2\omega t + \cdots + A_n \sin (n\omega t) + \cdots \\ &+ B_1 \cos \omega t + B_2 \cos 2\omega t + \cdots + B_n \cos (n\omega t) + \cdots, \end{aligned} \tag{1-14}$$

where the A's and B's are constants, to be determined.

Every term in this series may not always be present, depending on the nature of the function to be expanded. This will be made clear presently by an illustrative example. The presence or absence of a term will be known when one determines the constants A_0, A_n, and B_n. Formulas for this determination are obtained quite easily.

1-8 Determination of the Fourier coefficients. The constant term, A_0, is obtained by multiplying both sides of Eq. (1-14) by dt and then integrating over the time $t = T$, where T is the period ($T = 2\pi/\omega$) of the first term of *lowest* frequency. With this integration, all sine and cosine terms will disappear, since the area under any integral number of sine or cosine cycles is zero. Only the constant term will remain, and solving for A_0,

$$A_0 = \frac{1}{T}\int_0^T x\,dt. \tag{1-15}$$

---

* There are a number of additional mathematical restrictions placed upon the form of the function. The theorem fully applies to all functions encountered in problems in acoustics.

To evaluate A_0 it is necessary, of course, to have the expression for x as a function of time.

To obtain a typical coefficient, A_n, for the sine series, both sides of Eq. (1–14) are multiplied by $\sin(n\omega t)\,dt$ and again integrated from $t = 0$ to $t = T$. On the right-hand side, all but one of the integrations will involve products of the type $\sin(n\omega t)\sin(n'\omega t)\,dt$, where n and n' are different integers. Since

$$\sin(n\omega t)\sin(n'\omega t) = \frac{\cos[(n-n')\omega t] - \cos[(n+n')\omega t]}{2},$$

and since the integration will always be over an integral number of cycles, the result of all integrations on the right-hand side of Eq. (1–14) will be zero, except in the case where $n = n'$. For this latter case, the integration becomes

$$A_n \int_0^T \sin^2(n\omega t)\,dt = A_n \frac{T}{2}.$$

Therefore, integration of both sides of Eq. (1–14) yields

$$\int_0^T x\sin(n\omega t)\,dt = A_n \frac{T}{2}.$$

Solving for A_n, we obtain

$$A_n = \frac{2}{T}\int_0^T x\sin(n\omega t)\,dt. \tag{1–16}$$

In a similar way, by multiplying each term in (1–14) by $\cos(n\omega t)\,dt$ and integrating, term by term, from $t = 0$ to $t = T$, one may obtain the expression for B_n, the coefficient of a typical cosine term in the series:

$$B_n = \frac{2}{T}\int_0^T x\cos(n\omega t)\,dt. \tag{1–17}$$

Whether or not the integrations represented by Eqs. (1–15), (1–16), and (1–17) are feasible will, of course, depend on the nature and complexity of the function, $x = f(t)$, to be expanded. In addition, while the harmonic series can be shown always to be convergent, so that the coefficients A_n and B_n become progressively smaller as the frequency of the term rises, this rate of convergence may be slow in the case of certain functions. In these cases, it may be necessary to include a large number of harmonic terms in order to achieve a reasonably good equivalence to the original function. In problems in sound the convergence is frequently fairly rapid. In addition, to the average ear, the over-all effect due to a complex sound vibration is often only slightly modified if the very high harmonics are removed or ignored.

In a function which exactly represents the combination of a finite number of pure sine or cosine variations, the series obtained by analysis of the sum function will contain a finite, not an infinite, number of terms. Analysis, for instance, of the vibration effect known as beats will yield only the two frequencies present. Similarly, the complex sound constituting the sum of three pure musical notes will analyze into those three frequencies alone.

Example. To illustrate the application of the formulas developed above for the series coefficients, an analysis of the function represented graphically by the so-called "saw-toothed" wave will suffice. This function, shown graphically in Fig. 1–6, may be defined analytically as

$$f(t) = b\left(\frac{1}{2} - \frac{t}{T}\right)$$

for the time interval $t = 0$ to $t = T$. After this time the function repeats with a fundamental period, T ($1/T$ is then the frequency of the first sine or cosine term). Then

$$A_0 = \frac{1}{T}\int_0^T x\,dt = \frac{b}{T}\int_0^T \left(\frac{1}{2} - \frac{t}{T}\right) dt = 0.$$

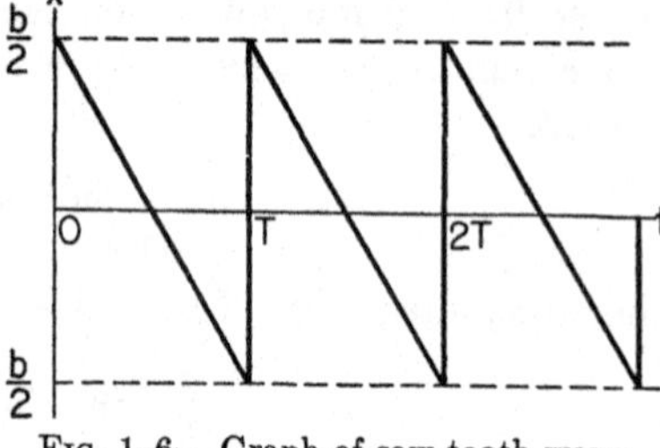

FIG. 1–6. Graph of saw-tooth wave.

It should be noted that A_0 is here zero because of the complete symmetry of the graph about the time axis. Wherever this symmetry is lacking, the constant term will not be zero.

The coefficient of a typical sine term becomes, in this problem,

$$A_n = \frac{2b}{T}\int_0^T \left(\frac{1}{2} - \frac{t}{T}\right) \sin(n\omega t)\, dt = \frac{2b}{n\pi}.$$

The amplitudes of the successive terms are then

$$\frac{2b}{\pi}, \frac{2b}{2\pi} \cdots \frac{2b}{n\pi}.$$

The cosine series is, in this problem, completely absent, since

$$B_n = \frac{2b}{T}\int_0^T \left(\frac{1}{2} - \frac{t}{T}\right) \cos(n\omega t)\, dt = 0,$$

regardless of the value of n. The complete series equivalent to the saw-tooth wave is therefore

$$x = f(t) = \frac{2b}{\pi}\left(\sin \omega t + \frac{1}{2}\sin 2\omega t + \cdots + \frac{1}{n}\sin(n\omega t) + \cdots\right).$$

1–9 Even and odd functions. In general, the absence of all the sine terms, or of all the cosine terms, depends on whether the original repeating function is "even" or "odd." An even function is one such that $f(t) =$

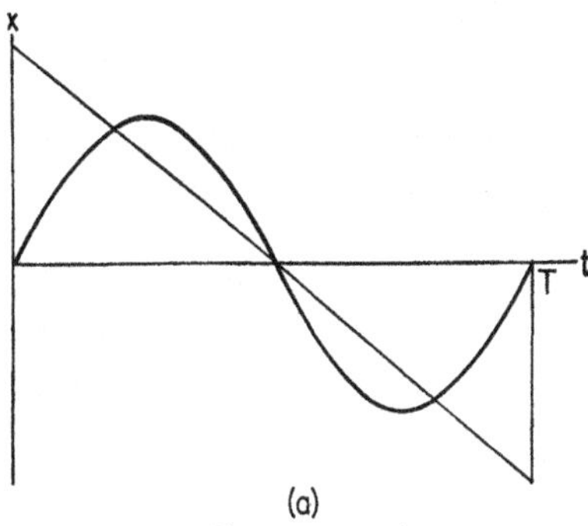

(a)
First term only

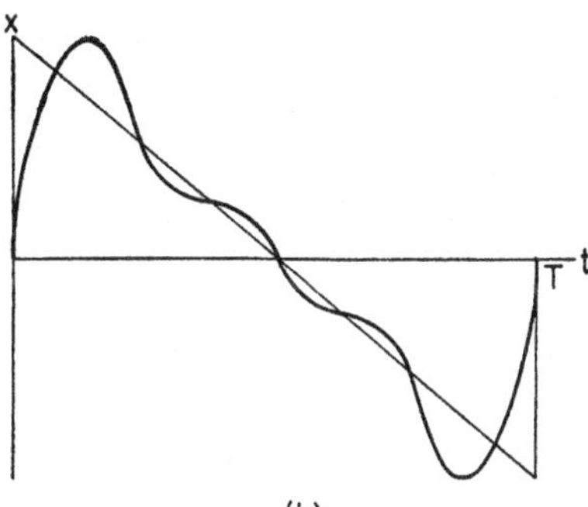

(b)
First three terms

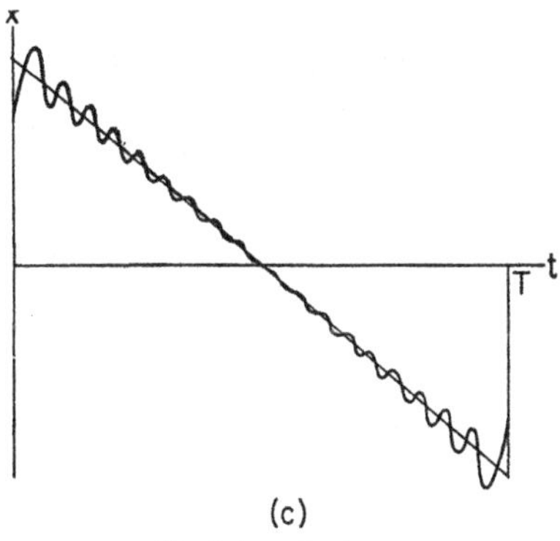

(c)
First twenty terms

Fig. 1–7. The effect of including additional terms in the Fourier series.

$f(-t)$. An odd function, on the other hand, is one where $f(t) = -f(-t)$. If the saw-tooth graph of Fig. 1–6 is repeated to the left of the origin, where t is negative, it may be verified that in this problem the conditions for an odd function are satisfied. Therefore the equivalent series contains only the sine terms. For the function to be even there must plainly be a mirror symmetry around the y-axis. This symmetry obtains, for instance, in the case of a simple cosine curve, there being, in this case, no sine terms. In the interests of saving computing labor, it will pay to first classify the given function as either even or odd. There are many functions, of course, which are neither even nor odd, in which case there will be both sine and cosine terms.

1–10 Convergence. It is clear in the problem just discussed that the harmonic terms become of smaller and smaller amplitude as the frequency rises. The complete infinite series must be considered for a complete equivalence. In Fig. 1–7 one can see the approach to the saw-tooth wave as more and more terms are added. The precision desired determines how far the computation is carried out. In general, it is near points of abrupt changes of slope that the "fit" is poorest, when using a finite number of terms.

The example above will suffice to show the general method of computing the Fourier coefficients. Other practice problems of a similar nature will be found at the end of the chapter.

1–11 Application of the Fourier analysis to empirical functions. Because one starts the analysis already knowing the analytic expression for the function, it may appear that problems of the above type are very artificial. Experimentally, the motions of vibrating bodies and the vibrations of air itself are usually picked up by electrical or electromagnetic means and are studied by means of a recording galvanometer or an oscilloscope, and we therefore have a graph to analyze, not an analytic function. From the principles of the Fourier analysis just discussed, graphical methods may be developed whereby, through the use of selected ordinates, the amplitude of the various harmonic terms may be determined with any desired precision. (This material may be found in many texts on electrical engineering.) In recent years many so-called harmonic analyzers have been built which, by mechanical or electronic means or a combination of both, perform the desired analysis with great saving of labor and with the highest precision. In Chapter 10 there is described an acoustical equivalent to the optical diffraction grating that may be used to determine very quickly the approximate harmonic content in any complex sound.

1–12 Damped vibrations of a particle. So far no force other than an elastic restraining force has been assumed to act upon the particle (or upon the sound source treated as equivalent to a particle). No such mechanical system exists in nature (at least in the large scale or macroscopic world!), since some sort of friction or dissipative force is always present. It will be assumed that the dissipative force acting upon the particle is in the nature of fluid friction and is of the form $F = -r\dot{x}$. The constant r is the force per unit velocity. The negative sign is necessary to show that the force is always opposite in direction to the velocity. In general, fluid friction is a function of the velocity raised to some power. The first power is used here as a first approximation. If the velocity is not too great, this approximation is reasonably good and, in addition, the use of the first power greatly simplifies the differential equation.

For a particle moving under the action of an elastic force and also of a viscous force of the above type, Newton's second law may be written

$$m\ddot{x} = -Kx - r\dot{x}$$

or, after transposing all terms to the left,

$$m\ddot{x} + r\dot{x} + Kx = 0. \tag{1–18}$$

This linear differential equation arises many times in different branches of physics. The student of electricity, for instance, will encounter an equation of exactly this form when he studies the transient behavior of an *L-R-C* circuit. This analogy will be discussed in more detail in Chapter 5.

The solution to Eq. (1-18) may be obtained in a number of ways. In Chapter 5, when the use of complex quantities is introduced, a more general method of solving this and related equations will be discussed. At this point, a change of variable will yield results more quickly. Let

$$x = y\epsilon^{-bt}, \tag{1-19}$$

where b is an arbitrary constant. Differentiating Eq. (1-19) and substituting in Eq. (1-18), we obtain

$$\ddot{y} + \left(\frac{r}{m} - 2b\right)\dot{y} + \left(\frac{K}{m} + b^2 - \frac{r}{m}b\right)y = 0. \tag{1-20}$$

In this new equation in y, the constants m, r, and K are fixed by the nature of the system being considered, but the constant b which appears first in the change of variable equation, Eq. (1-19), may be selected quite arbitrarily. If b is chosen equal to $r/2m$, the second differential term in Eq. (1-20) will vanish and the whole equation will take the much simpler form

$$\ddot{y} + (\omega_u^2 - b^2)y = 0, \tag{1-21}$$

where ω_u^2 has been substituted for K/m. The values of y which are solutions to Eq. (1-21) can be obtained quite simply. Then, according to Eq. (1-19), x, the actual particle displacement, may be obtained by simply multiplying the value of y by ϵ^{-bt}.

There are three important types of solutions to Eq. (1-21), whose form depends on the values of the system parameters, m, r, and K.

1-13 Case I. $\omega_u^2 < b^2 \left(\text{or } \frac{K}{m} < \frac{r^2}{4m^2}\right)$. Large frictional force.

When the system constants are such that ω_u^2 is less than b^2, the algebraic sign of the coefficient of y in Eq. (1-21) is negative. The solution to the differential equation can then readily be shown to be

$$y = A_1\epsilon^{\sqrt{b^2-\omega_u^2}\,t} + A_2\epsilon^{-\sqrt{b^2-\omega_u^2}\,t}, \tag{1-22}$$

A_1 and A_2 being integration constants. Therefore, using Eq. (1-19), we find that

$$x = A_1\epsilon^{-(b-\sqrt{b^2-\omega_u^2})\,t} + A_2\epsilon^{-(b+\sqrt{b^2-\omega_u^2})\,t}. \tag{1-23}$$

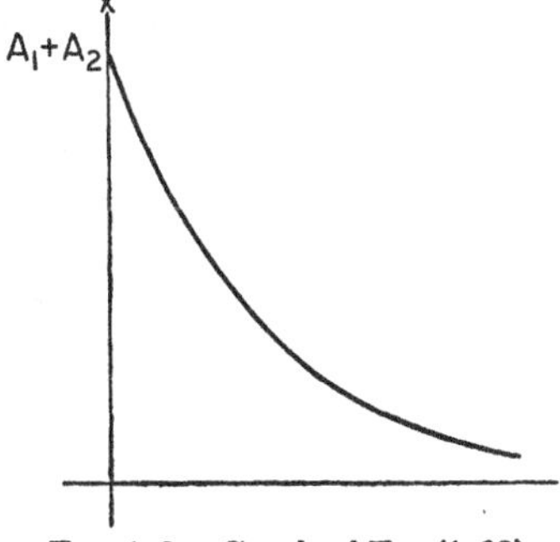

Fig. 1-8. Graph of Eq. (1-23).

The values of the integration constants A_1 and A_2 may be determined if the initial or other time conditions of the problem are known. Since $b > \sqrt{b^2 - \omega_u^2}$, both exponents are intrinsically negative, and the particle, once displaced, will always return to the position $x = 0$ asymptotically with time. The rate of this approach to zero will depend on the values of ω_u and b. The graph in Fig. 1–8 shows the subsequent motion after $t = 0$, in the case where both A_1 and A_2 are positive. Two initial conditions, such as displacement and velocity at the time $t = 0$, may be used to determine the constants A_1 and A_2.

1–14 Case II. $\omega_u^2 > b^2 \left(\text{or } \dfrac{K}{m} > \dfrac{r^2}{4m^2}\right)$. **Small frictional force.** In this case the coefficient of y in Eq. (1–21) changes sign, i.e., $(\omega_u^2 - b^2)$ is positive, and the equation is readily recognized as in the form for SHM. Clearly, then, the solution is

$$y = y_m \cos(\omega' t + \alpha),$$

and therefore

$$x = y_m \epsilon^{-bt} \cos(\omega' t + \alpha). \tag{1–24}$$

where

$$\omega' = \sqrt{\omega_u^2 - b^2}.$$

Equation (1–24) describes a *damped* harmonic motion, whose effective amplitude, $x_m = y_m \epsilon^{-bt}$, dies out exponentially with the time. The initial amplitude and phase angle are, respectively, y_m and α. The constant $b(= r/2m)$ determines the time rate of damping. The envelope of the curve represented by Eq. (1–24) is, effectively, the exponential curve $x = y_m \epsilon^{-bt}$, as shown in Fig. 1–9.

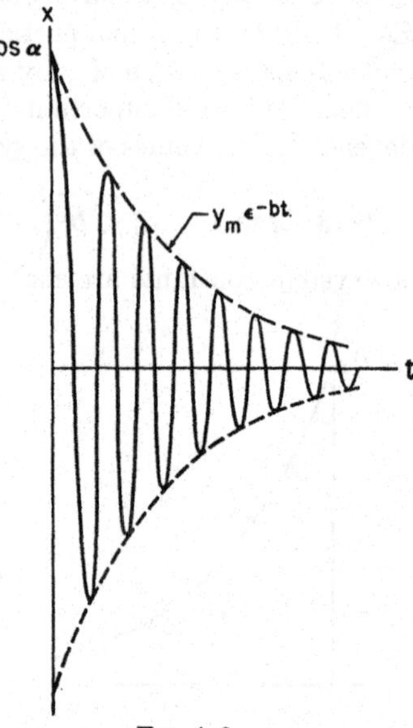

Fig. 1–9.

With no damping, i.e., when $b = 0$, the frequency of the motion is $\omega_u/2\pi$. Where damping exists, the natural frequency is always lowered, since the frequency is $\omega'/2\pi$ and ω' is always less than ω_u. In fact, as the value of b is increased (say, by keeping m constant and increasing the frictional coefficient, r), the oscillation frequency approaches zero as b approaches ω_u. Practical sound sources are usually so lightly damped that the damping factor, b, does not greatly affect the frequency. This is especially true because of the quadratic relation between ω', ω_u, and b.

The length of time required for an oscillation to die out is of practical importance in sound. The time for x to become zero is, of course, infinite from the mathematical point of view, but in the case of sound waves an amplitude below a certain minimum will be inaudible to the ear, and some quantitative measure of the rate at which the amplitude diminishes is desirable. A commonly used quantity for this measure is the modulus of decay, $1/b$, often called the *time constant*. This is the time for the amplitude of the cosine function in Eq. (1-24) to drop to the fraction $1/\epsilon$ of its initial value. Since $b = r/2m$, it will be seen that a large frictional coefficient, r, and a small mass, m, will make the time constant small. A small time constant implies a rapid rate of decay. It will be seen in Chapter 7 that the moduli of decay of the different harmonic frequencies generated by musical instruments are of considerable importance in determining the quality of the sound produced.

1-15 Case III. $\omega_u^2 = b^2 \left(\text{or } \frac{K}{m} = \frac{r^2}{4m^2}\right)$. Critical damping.

This is a case of more importance in scientific instrument design than in the behavior of sound sources. When $\omega_u^2 = b^2$, Eq. (1-21) becomes simply

$$\ddot{y} = 0. \tag{1-25}$$

The solution to this equation is a straight line, of the form

$$y = A_1 t + A_2,$$

where A_1 and A_2 are again constants of integration. The expression for x then becomes

$$x = \epsilon^{-bt}(A_1 t + A_2). \tag{1-26}$$

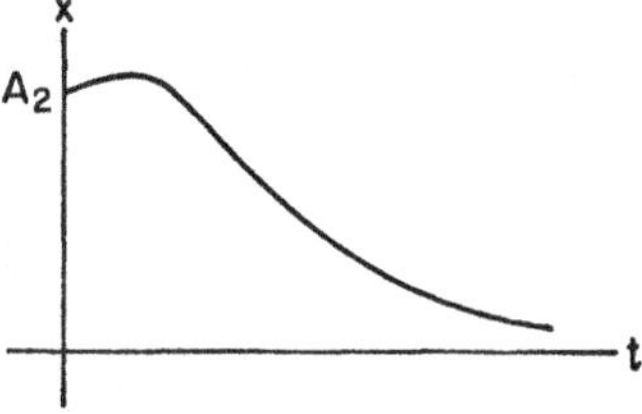

Fig. 1-10. Graph of Eq. (1-26).

Plotted, this equation does not look greatly different from the solution for the case where b^2 is greater than ω_u^2 (friction large). Figure 1-10 represents the plot of Eq. (1-26) for the case where A_1 is large and where both A_1 and A_2 are positive (this is, of course, not necessarily so). If A_1 is numerically large, x will increase at first, but eventually the exponential coefficient will bring about a reversal of slope and x will approach zero as time progresses. Rarely are actual sound sources so critically damped.

Example. An example will show how the physical constants of the vibrating system are used and also the technique for the evaluation of the constants of integration.

A particle of mass 3 gm is subject to an elastic force of 27 dyne-cm^{-1} and a damping force of 6 dyne-cm^{-1}-sec. It is displaced a distance of $+1.0$ cm from its equilibrium position and released. It is required to determine whether or not the motion is oscillatory and, if so, to find its period; also the complete equation for x as a function of time is to be obtained, with the numerical values of the amplitude x_m and the phase angle α.

From the data given, $\omega_u = \sqrt{K/m} = 3$ sec^{-1} and $b = r/2m = 1.0$ sec^{-1}. Since $\omega_u > b$, the solution is oscillatory, of the form:

$$x = y_m \epsilon^{-bt} \cos\left(\sqrt{\omega_u^2 - b^2}\, t + \alpha\right).$$

The period is $2\pi/\sqrt{\omega_u^2 - b^2} = 2.23$ sec. To find the initial amplitude y_m and the phase angle α (the integration constants), the initial position and velocity may be used. Differentiating x, we obtain

$$\dot{x} = -y_m \epsilon^{-bt}\left[\sqrt{\omega_u^2 - b^2} \sin\left(\sqrt{\omega_u^2 - b^2}\, t + \alpha\right) + b \cos\left(\sqrt{\omega_u^2 - b^2}\, t + \alpha\right)\right].$$

Setting $t = 0$ and inserting the values $x = 1.0$ and $\dot{x} = 0$, two equations may be obtained for the determination of α and y_m, that is:

$$\tan \alpha = -\frac{b}{\sqrt{\omega_u^2 - b^2}}$$

and

$$y_m = \frac{1.0}{\cos \alpha}.$$

Solving for α and y_m, we obtain

$$\tan \alpha = -0.353; \qquad \alpha = -19°\ 30'; \qquad y_m = 1.06 \text{ cm}.$$

Therefore the complete expression for x is

$$x = 1.06 \epsilon^{-1.0t} \cos(2.82t - 19°\ 30').$$

1-16 Forced vibrations. All sound sources are set into vibration by some external source of energy, capable of supplying some kind of periodic force. Sometimes the mechanism of this energy transfer is quite complicated, as, for instance, in the excitation of a violin string or in the sounding of an organ pipe. A simpler example to consider is the setting into motion of a pendulum by the application of an external force of a periodic nature.

In practice, the periodic driving force is rarely a simple harmonic variation of a single frequency. The cone of a radio loudspeaker which is reproducing music, for instance, is being driven by a variable force equivalent to a mixture of periodic forces of assorted frequencies. If, however, we can discover how the particle will behave under the action of a driving force of one particular frequency, we are ready, by means of the superposition principle, to understand its motion when there are many frequencies.

1-17 The differential equation. Let the instantaneous driving force be represented by $F = F_m \cos \omega t$, where, as before, $\omega = 2\pi f$. Writing Newton's second law for a particle subject to an elastic and a damping force, we obtain

$$m\ddot{x} + r\dot{x} + Kx = F_m \cos \omega t. \tag{1-27}$$

The general solution to this equation is made up of two parts, mathematically speaking. The first part is the *complementary function*, which is the solution to Eq. (1-27), with the right-hand side set equal to zero. Since this modified differential equation is exactly the one just discussed under the heading of "damped vibrations of a particle," it is clear that the complementary function may actually take one of three forms, depending on the factors m, r, and K.

The complete solution to Eq. (1-27) must contain, besides the complementary function, a second part, which constitutes a *particular solution* to Eq. (1-27), with the right-hand side $\neq 0$. This particular solution must satisfy the complete differential equation for all values of the time t. It will be remembered that in the presence of damping all solutions to the simpler differential equation (where the right-hand side of (1-27) is set equal to zero) are of a form such that x approaches zero with the passage of time. This part of the general solution to Eq. (1-27) (i.e., the complementary function) is therefore called the *transient* part. With physical vibrations it can usually be neglected after a short time. The remaining part of the solution, the *particular* or *steady state solution* referred to, will then be the only significant part, for later times. It is this important steady state solution that we shall now consider.

1-18 The steady state solution for forced vibrations. To obtain the steady state part of the solution to Eq. (1-27), it is most convenient to compare the differential equation with an exactly similar one arising in electricity. If an emf, varying in a simple harmonic manner, is impressed upon a series circuit with inductance L, resistance R, and capacitance C, we may write the equation

$$L\ddot{q} + R\dot{q} + \frac{1}{C}q = E_m \cos \omega t, \tag{1-28}$$

where E_m is the maximum value of the impressed emf and q is the charge on the capacitor at any instant. Since for an electrical circuit the current i is equal to dq/dt, we may write Eq. (1-28) in terms of the current:

$$L\frac{di}{dt} + Ri + \frac{1}{C}\int i\,dt = E_m \cos \omega t. \tag{1-29}$$

A comparison of Eq. (1–27) with (1–28) will show the mathematical form to be identical. In addition, there is an equation for the mechanical system in terms of the *velocity* $\dot{x}$, which is the exact counterpart of Eq. (1–29) for the current $i, = \dot{q}$. This means that if the electrical equations have been solved, the equations for the mechanical system have also been solved. Writing down the solutions to the electrical equations, we have only to replace the electrical parameters with those of the particle system and to insert the variable x instead of the variable q to obtain the solutions to the mechanical equations.

The steady state solution to Eq. (1–29), the electrical equation in terms of the current, is the ordinary expression for the instantaneous value of the alternating current in an L–R–C circuit, familiar to most students of elementary electricity. The expression for this current, i, is

$$i = \frac{E_m}{\sqrt{R^2 + \left(\omega L - \frac{1}{\omega C}\right)^2}} \cos(\omega t - \alpha), \tag{1–30}$$

where

$$\tan \alpha = \frac{\omega L - \frac{1}{\omega C}}{R}. \tag{1–31}$$

In this equation the angle α represents the phase relationship between the impressed potential and the current. Not so familiar is the expression for the charge q. The equation for q may be easily obtained by integrating Eq. (1–30) with respect to time. (Note that the constant of integration must be zero, since there is no d.c. component to the impressed potential.)

$$q = \frac{E_m}{\omega\sqrt{R^2 + \left(\omega L - \frac{1}{\omega C}\right)^2}} \sin(\omega t - \alpha). \tag{1–32}$$

It will be remembered that the expression in the denominator of Eq. (1–30) is called the total electrical *impedance* of the circuit, while the collection of terms, $\left(\omega L - \frac{1}{\omega C}\right)$, is called the circuit *reactance*, commonly represented by the symbol X.

We can now write the analogous equations for the mechanical system, where the displacement x replaces charge, and the velocity $\dot{x}$ replaces current:

$$x = \frac{F_m}{\omega\sqrt{r^2 + \left(\omega m - \frac{K}{\omega}\right)^2}} \sin(\omega t - \alpha) \tag{1–33}$$

and

$$\dot{x} = \frac{F_m}{\sqrt{r^2 + \left(\omega m - \frac{K}{\omega}\right)^2}} \cos(\omega t - \alpha). \tag{1–34}$$

So exact is the parallel between the mechanical and the electrical problem that it is common to use for the mechanical system such expressions as "mechanical impedance," "mechanical resistance," and "mechanical reactance." (This use of the concept of impedance, as applied to a particle or its equivalent, is not to be confused with the idea of "radiation impedance," to be introduced in Chapter 5. This latter concept is used only in connection with wave propagation and involves a quite different use of the word impedance.) Note that in the comparison of the mechanical and the electrical parameters, r is analogous to R, m to L, and $1/K$ to C. $1/K$ is called the "compliance" of the system, since it is the reciprocal of K, the elastic "stiffness" constant. More will be said about the use of analogies in Chapter 5.

1–19 Velocity and displacement resonance. In the electrical equation, (1–30), so-called series resonance occurs when the current is in phase with the applied potential or, from (1–31), when the reactance is zero, i.e., $\omega L = 1/\omega C$. Under these conditions, since the impedance is a minimum, the value of the current, I_m, will be a maximum, and so will the "root mean square" current, I_{rms}. For the mechanical system, this means that the criterion for *velocity* resonance is that $\omega m = K/\omega$. If this condition is brought about by the variation in the angular driving frequency ω, other parameters remaining constant, $\dot{x}_m$ will then be a maximum. This corresponds to the maximum current observed in the circuit.

Of more interest in the mechanical than in the electrical problem is another kind of resonance, *displacement* resonance. Again considering ω as the variable, this resonance may be said to occur when the amplitude of x, i.e., x_m, in Eq. (1–33) is a maximum. Since ω appears outside the radical in the denominator, as well as inside, it is necessary to differentiate with respect to ω the coefficient of the sine expression on the right and set the result equal to zero, in order to determine the exact criterion for resonance. The necessary condition may be stated as follows:

$$\omega^2 = \omega_u^2 - 2b^2, \tag{1–35}$$

where ω_u^2, as earlier, $= K/m$ and $b = r/2m$.

It should be noted that if the frictional coefficient, r, is small, so that $2b^2$ is much less than ω_u^2, then the condition for amplitude resonance is very nearly that $\omega^2 = \omega_u^2$. Since $\omega_u^2 = K/m$, this condition is seen to be identical with that for velocity resonance. It is worth noting that with low damping the frequency f at which amplitude resonance occurs is identical with the natural frequency of vibration of the particle under the action of an elastic force only, i.e., $f = \omega/2\pi = \frac{1}{2\pi}\sqrt{\frac{K}{m}}$. (With velocity resonance this is *always* true, regardless of the degree of the damping.) When the damping is large, however, so that the term $2b^2$ in Eq. (1–35) becomes important, the frequency for amplitude resonance is lowered. Indeed, if the damping is so large that $2b^2$ is greater than ω_u^2, there is then no true resonance at all, since in Eq. (1–35) ω is then imaginary.

In Fig. 1–11 are shown a number of curves for different degrees of damping, each curve being a plot of the amplitude x_m against the angular frequency of the driving force, ω. With low damping it is seen that resonance virtually occurs when $\omega = \omega_u$. When the damping is increased, the position of the maximum shifts to the left. Curve 4 represents the transition case such that with any increased damping, no true maximum occurs.

1–20 The amplitude at resonance. It is clear from Fig. 1–11 that the maximum value of x_m at resonance is a function of the degree of damping. The exact value of this maximum ordinate, $(x_m)_{res}$, is determined by inserting the condition given by Eq. (1–35) into the expression

$$(x_m)_{res} = \frac{F_m}{\omega\sqrt{r^2 + \left(\omega m - \frac{K}{\omega}\right)^2}}. \tag{1–36}$$

A simpler, approximate expression for $(x_m)_{res}$ may be readily obtained if the damping is low (usually the case in acoustics). In this case the condition for amplitude resonance is practically that for velocity resonance, i.e., that the mechanical reactance $X = (\omega m - K/\omega) = 0$. We then have

$$(x_m)_{res} = \frac{F_m}{\omega r} = \frac{F_m}{\omega_u r}. \tag{1–37}$$

For low damping, it is seen that the amplitude at resonance is inversely proportional to the frictional coefficient, r, becoming very large as the

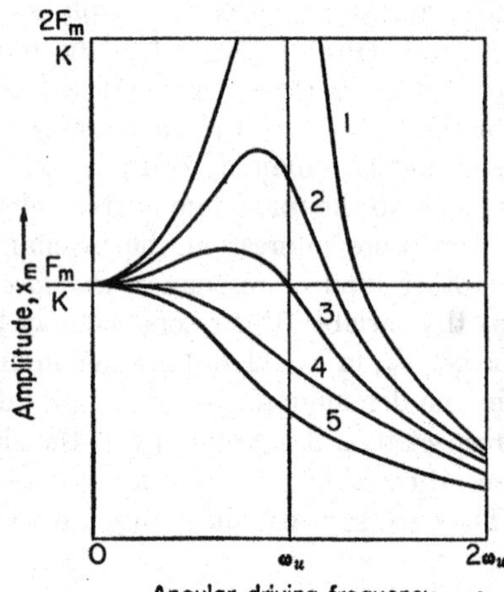

FIG. 1–11.

damping factor approaches zero. As in the case of electrical resonance, it is near resonance that the amplitude is affected most markedly by the value of the dissipative element. Well off resonance, it is the mechanical reactance, $\left(\omega m - \frac{K}{\omega}\right)$, that mainly determines the amplitude.

1-21 Phase relationships. In general, varied phase relations will obtain between particle displacement and driving force, and between particle velocity and driving force. In the latter case, the phase angle relationship should be familiar from alternating current circuit theory; at velocity resonance, $\dot{x}$ and F are in phase. At frequencies above resonance the effect of the mechanical mass reactance, ωm, predominates; $\dot{x}$ lags F by a greater and greater angle, approaching $\pi/2$ for values of ωm large compared with K/ω and r. Below resonance, the angle is a lead, since it is the term K/ω, containing the compliance, that is important, and the angle approaches $\pi/2$ for large values of K/ω. In the case of the displacement x the angles are different, since the displacement is 90° out of phase with the velocity. Figure 1-12 is a graph of the phase angle between the displacement x and the driving force, $-(\alpha + \pi/2)$, plotted against angular driving frequency for various values of b. When $\omega = \omega_u$, regardless of the value of b, the phase angle is $\pi/2$ and is a lag. At very low frequencies the angle approaches zero; for very high frequencies the lag approaches π. For low damping, where b is small, the phase angle shifts rather abruptly as the driving frequency is varied from a little below the value $\omega_u/2\pi$ to a little above. With greater damping the change is more gradual.

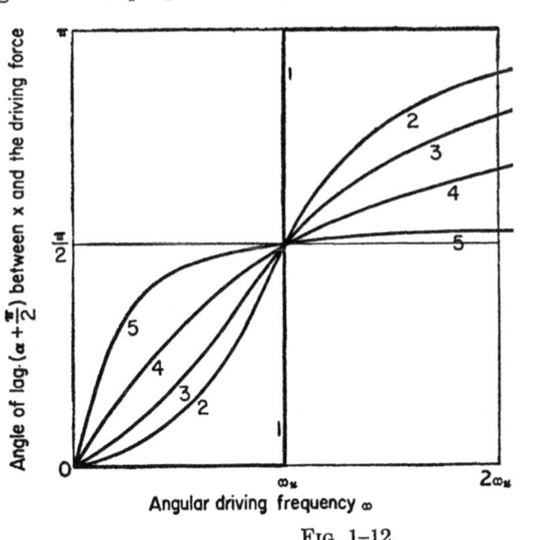

Fig. 1-12.

A simple demonstration of the above phase relationships can be set up as in Fig. 1–13. A heavy and a light plumb bob, M and m respectively, are suspended from a somewhat flexible common support, such as a horizontally stretched string. The two pendulum lengths are adjusted to be slightly different. If the heavy pendulum is set swinging, the lighter one will soon begin to oscillate also, due to the coupling at the support. The amplitude of this induced motion will alternately build up and die down as energy flows back and forth between the two pendulums in this coupled system (the heavier one, having the larger mass and energy, will not be appreciably affected). During the peaks of the induced oscillations, the above phase relations may be clearly seen. The driven system here is one of very low damping. Therefore if the heavier pendulum is longer and the driving frequency consequently lower than that of the driven system, the phase angle will be almost zero. On the other hand, if the heavier pendulum is shortened so that the driving frequency is higher than that of the driven system, the two pendulums will be almost 180° out of phase, being at opposite ends of their motions at the same time. When the two pendulums are of the same length ($\omega = \omega_u$), the 90° relationship can also be clearly seen.

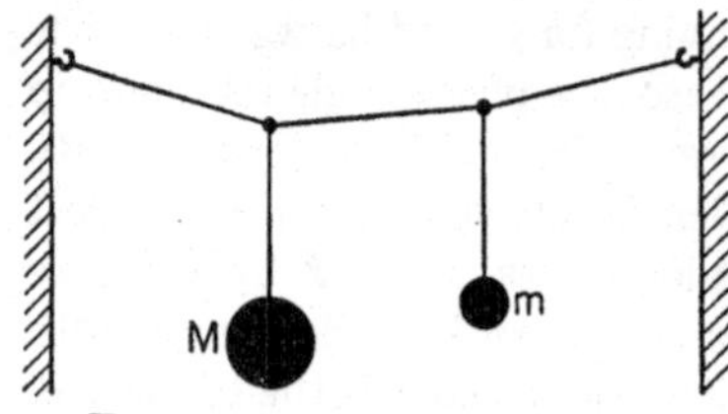

FIG. 1–13. Forced oscillations.

1–22 Energy transfer in forced oscillations. Unless the two pendulums in the above experiment are of nearly the same length, very little energy will be transferred. This is in line with common experience and can readily be shown with equations. The instantaneous power delivered to the particle system is $F\dot{x}$. This is the analog of electrical power, ei. In both the electrical and the mechanical case it is the time average of this product over a large number of cycles which constitutes the real power delivered. In the electrical case the time average of the product $ei(= E_m \cos \omega t\ I_m \cos (\omega t - \alpha))$ becomes $E_{rms}I_{rms} \cos \alpha = I^2_{rms}R$, where E_{rms} and I_{rms} are root mean square values. The angle α is the angle between current and applied potential, and R is the circuit resistance. Analogously, for the particle, since F and $\dot{x}$ are periodic functions of the time, just as are e and i, real average mechanical power may be written $F_{rms}\dot{x}_{rms} \cos \alpha$ or $(\dot{x}_{rms})^2r$. The expression $(\dot{x}_{rms})^2r$ shows that with a system having fixed damping characteristics the power delivered will be a maximum whenever the velocity, $\dot{x}_{rms}$, is a maximum. $\dot{x}_{rms}$ is itself a function of r and at resonance $= F_{rms}/r$. Therefore real power, Ω_{res}, at resonance may be written

$$\Omega_{res} = (\dot{x}_{rms})^2_{res} r = \frac{(F_{rms})^2}{r}. \tag{1-38}$$

(Compare with the electrical equivalent, $(E^2_{rms})/R$.) The lower the damping, obviously, the greater will be the delivered power.

Example. A particle has a mass of 2 gm. It is free to vibrate under the action of an elastic force of 128 dyne-cm^{-1} and a damping force of 8 dyne-cm^{-1}-sec. A periodically varying outside force of maximum value 256 dynes is applied to the particle. It is required to find the frequency $(f_{res})_d$ for displacement resonance and also the frequency $(f_{res})_v$ for velocity resonance, and the approximate amplitude at displacement resonance.

In this case $\omega_u^2 = K/m = 64$ sec^{-2} and $b^2 = r^2/4m^2 = 4$ sec^{-2}. For displacement resonance, $\omega^2 = \omega_u^2 - 2b^2$. Therefore the required frequency is

$$(f_{res})_d = \frac{\omega}{2\pi} = \frac{1}{2\pi}\sqrt{\omega_u^2 - 2b^2} = 1.19 \text{ sec}^{-1}.$$

For velocity resonance, $\omega m = K/\omega$, or

$$(f_{res})_v = \frac{1}{2\pi}\sqrt{\frac{K}{m}} = 1.27 \text{ sec}^{-1}.$$

Since $2b^2$ is considerably less than ω_u^2, we may use the approximate expression for the amplitude at displacement resonance:

$$(x_m)_{res} \cong \frac{F_m}{\omega r} = 4.0 \text{ cm}.$$

1-23 Some applications of the theory of forced vibrations. From the graphs of Fig. 1-11 several useful conclusions can be drawn. If we are interested in transferring the maximum energy at a *single* frequency to a system capable of vibration, it is obvious from the graphs and from the previous discussion of energy that the damping factor of the system should be as small as possible and that the driving frequency should be near the natural frequency of the system. The crystal vibrators used in the production of ultrasonic waves are good examples of low-damped systems. In addition, the smaller the damping factor, the longer the persistence of any sound energy set up after the driving force has ceased. The vibrations of musical instruments persist for an appreciable time after energy ceases to be supplied. There are two kinds of damping involved in the decay of these vibrations. First, there are the internal frictions set up within the sound source (string, bar, or plate, as the case may be). This type of friction is undesirable from an energy point of view, as it results in the degeneration of vibrational energy into thermal energy. The second kind of damping is due to the presence of the surrounding air, and constitutes the only mechanism by which sound energy is radiated into space. For high radiation efficiency this type of damping should be large compared with the damping due to internal friction. In Chapter 5 this aspect of vibration will be discussed more fully in connection with "radiation resistance."

Ever since the advent of the phonograph and the radio set there has been a need for a source of sound reproduction which is capable of vibration at *all* audible frequencies, with no partiality to any one such frequency. For many different reasons the ideal source has not been found; some of the difficulties will be discussed later. A study of the curves of Fig. 1–11 will suggest one possible solution to the problem. By designing the system (treated as a particle) so as to have rather high damping, and by placing the resonant frequency above audibility, we may obtain a virtually aperiodic response to a driving force over a wide, useful frequency range.

In Fig. 1–11, Curve 3 shows this approximately aperiodic property for values of $\omega < \omega_u$. Unfortunately, in order for the system to have this type of response, the damping must be quite large. If, in a radio loudspeaker, the damping could be mainly that due to the air load, this would be all to the good, for the sound radiating efficiency would then be high. Unless the area of the vibrating source is impossibly large, as will be shown, the damping due to the air is likely to be much smaller than is necessary to approach critical damping. The required damping must then be obtained by artificially increasing the internal losses, which will result in very low over-all sound efficiency. Fortunately for efficiency, it is actually undesirable for such a sound source to have strictly aperiodic properties. Sound sources are usually poor radiators at low frequencies, for reasons not connected with their own intrinsic vibration properties. By reducing the damping well below the critical and placing the loudspeaker resonance near the lower end of the audible spectrum, the increased amplitude near resonance will make the output more uniform.

There is another interesting use that may be made of the phase angle graphs of Fig. 1–12. In general, the motion of a particle undergoing forced oscillations, with or without damping, will lag the driving force by some small time which will depend, in a rather complicated way, upon the driving frequency. This means that when a series of frequencies of particular relative phases are impressed upon the particle (constituting a complex forced vibration), the resulting particle motion will not be a complete replica of the variation in the driving force because of the assorted phase lags. If, however, referring to Fig. 1–12, a damping factor is so chosen as to make the phase angle approximately linear with driving frequency (such as with $b \cong .75\omega$), the original phase relationship will be maintained.

This can be seen from Eqs. (1-33) and (1-34). If $\alpha \propto \omega$ or $\alpha = A\omega$, then the angle on the right-hand side may be written $[\omega(t - A)]$, showing what amounts to a simple shift of the time axis, all frequencies being shifted together by the amount A. In actual practice, it is usually unnecessary to worry about phase shifts in sound, since the ear, at least for stimuli of the usual steady state type, is unaware of the phase relations in a complex sound wave. This may not be true, however, in the case of short-duration transients.

1-24 The importance of the transient response. A word may be said here about the transient response of a system equivalent to a particle, undergoing forced vibrations. The transient part of the solution to Eq. (1-27), while of short duration, may have considerable effect on the quality of a musical instrument and in some cases may distort or even mask the desired steady state frequency. The difference in the quality of a violin during the rapid playing of scales as compared with the sound of long, sustained notes is quite apparent. It is only in the latter case that the transient vibrations have had time to die out. The characteristic sound of a drum is due entirely to a transient, the driving force being of very short duration. Consider again the radio loudspeaker, whose purpose it is to transform into sound all driving frequencies applied to it. Whenever a new driving force is applied, there may be an important transient response amounting to some 10 to 20 vibrations or more at the natural frequency of the diaphragm, this frequency having nothing whatever to do with the driving frequency. As a result, all sounds which are abruptly cut off appear to have a "tail" or "hangover." Short duration sounds, like those originating from the drum, appear to have about the same monotonous frequency (i.e., that of the *speaker* resonance). These effects are minimized by an increase in the damping factor, but for most present-day radio reproducers this damping is not sufficient to overcome these effects.

1-25 Superposition of SHM's mutually perpendicular. This interesting case is important mainly because of the modern use of the oscilloscope in the study and measurement of sound. In this instrument, vertical and horizontal motions are imparted to an electron beam by means of vertical and horizontal field forces. If these forces vary sinusoidally with the time, the luminous spot on the screen will execute the motions to be described. The curves traced are called *Lissajous' figures*. Lissajous himself obtained these figures originally by observing the rectilinear vibrations of a particle while sighting through a microscope, itself mounted upon the prong of a tuning fork, free to oscillate at right angles to the particle motion.

There is almost no limit to the variety of the curves that may be obtained, depending on the amplitudes, frequencies, and relative phases of the two motions. If the frequencies are the same, but the amplitudes and phase angles are different, the equations for the vertical and horizontal motions may be written:

$$x = x_m \cos(\omega t + \alpha_1) \tag{1-39}$$

and

$$y = y_m \cos(\omega t + \alpha_2). \tag{1-40}$$

If the time is eliminated between these two equations, the following equation is obtained:

$$\frac{x^2}{x_m^2} + \frac{y^2}{y_m^2} - \frac{2xy}{x_m y_m}\cos(\alpha_1 - \alpha_2) - \sin^2(\alpha_1 - \alpha_2) = 0. \tag{1-41}$$

This represents an ellipse whose eccentricity and inclination depend upon the phase relations and the amplitudes. If the relative phase angle $(\alpha_1 - \alpha_2)$ happens to be $\pi/2$, the principal axes of the ellipse are vertical and horizontal, since the term containing the product xy is absent. If, in addition, $x_m = y_m$, the ellipse becomes a circle.

If the relative phase angle is zero, the equation degenerates into two identical straight lines, given by

$$y = \frac{y_m}{x_m}x. \tag{1-42}$$

These curves can be made the basis of an exceedingly sensitive test for frequency measurement. If the vertical motion is of unknown frequency and if the frequency of the horizontal motion can be controlled with a calibrated variable frequency electrical oscillator, it is only necessary to adjust the oscillator until a stationary ellipse, circle, or straight line appears, and then read off the unknown frequency.

If the frequencies of the vertical and horizontal motions are not the same, no steady pattern will appear upon an oscilloscope unless the two frequencies bear a whole number relationship, as indicated earlier in con-

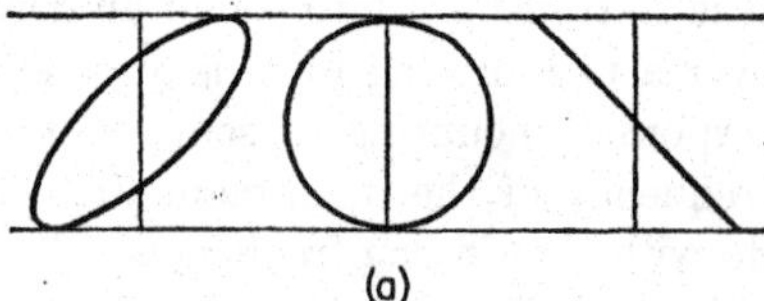

(a)
Frequency ratio 1:1

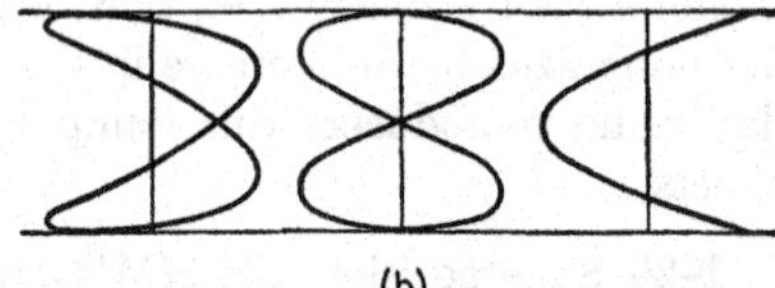

(b)
Frequency ratio 1:2

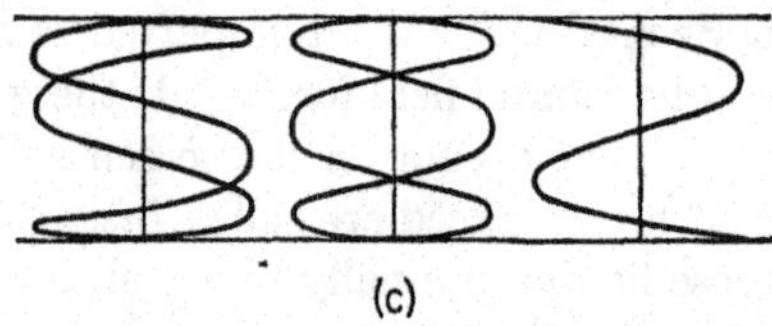

(c)
Frequency ratio 1:3

FIG. 1–14. Lissajous figures.

nection with beats. The steady patterns are all closed curves representing higher degree equations. A few of the simpler ones are illustrated in Fig. 1–14. Some of the patterns may be used, practically, to determine the ratio of a known to an unknown frequency, provided that the whole number ratio of the two frequencies does not involve integers which are too large. In this latter case, the patterns are too crowded to interpret easily.

PROBLEMS

1. Using a single pair of rectangular axes, draw three graphs to represent, for simple harmonic motion, the displacement x, the velocity $\dot{x}$, and the acceleration $\ddot{x}$, each as a function of the time. Besides showing the relative phases, indicate the maximum values of the three variables in terms of the proper constants.

2. (a) For simple harmonic motion, find the displacement x as a function of the time, by integrating the equation $m\ddot{x} = -Kx$. (b) Show that the period of the motion is given by $T = 2\pi\sqrt{m/K}$.

3. (a) Find the displacement x as a function of the time, if the differential equation for the motion is $m\ddot{x} = +Ax$, where A is a constant. Assume that the initial velocity is not zero, but has some value v_0. (Why is this necessary?) (b) Is the resulting motion periodic? Give a physical description of the motion.

4. A perfectly elastic ball is bouncing on a rigid floor. If the constant height to which it rebounds is h, find the period of the motion. Is the motion simple harmonic?

5. Two collinear harmonic motions of the same frequency have amplitudes of 2 cm and 3 cm respectively, and corresponding phase angles of +10° and +30°. Find by the "method of components" used in mechanics (a) the amplitude, and (b) the phase angle of the sum vibration.

6. Two collinear simple harmonic motions are given by

$$x_1 = (x_m)_1 \cos (2\pi ft + \alpha_1)$$

and

$$x_2 = (x_m)_2 \cos (2\pi ft + \alpha_2).$$

By expanding the cosines of the sums of angles and adding, show that the resultant displacement x so obtained is equivalent to that obtained by the purely vector method.

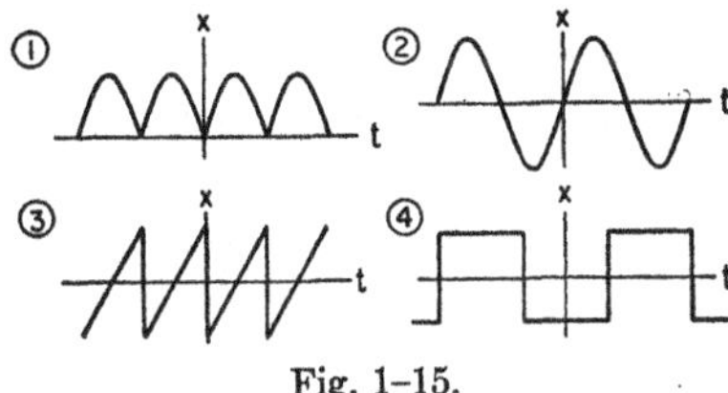

Fig. 1–15.

7. Two collinear simple harmonic motions have frequencies of 1024 and 1021 cycles-sec^{-1} respectively. (a) What is the number of "mathematical" beats per second? Of audible beats? (b) Answer the same questions if the two frequencies are 1024 and 1022 cycles-sec^{-1}.

8. (a) Which of the graphs of Fig. 1–15 represent even functions and which represent odd functions? (b) In which cases will a Fourier expansion involve a constant term?

9. Find the first few terms of the Fourier series equivalent to the square wave specified by $x = a$, from $t = 0$ to $t = T/2$, and $x = -a$, from $t = T/2$ to $t = T$.

10. Show graphically how close to the square wave is the sum of the first three periodic terms in the solution to problem 10.

11. The current in a circuit with a half-wave rectifier is given by $i = I_m \sin (2\pi ft)$ from $t = 0$ to $t = T/2$, and $i = 0$ from $t = T/2$ to $t = T$. Find the first few terms of the equivalent Fourier series.

12. A triangular wave is represented by the analytical expressions $x = 2at/T$ from $t = 0$ to $t = T/2$, and $x = 2a\ (1 - t/T)$ from $t = T/2$ to $t = T$. Find the first few terms of the Fourier expansion.

13. A telephone receiver diaphragm is considered as a particle of mass 1.0 gm. When displaced a distance 1.0 mm from its equilibrium position, the elastic restoring force is 10^6 dynes. The frictional force opposing its motion is 4.0×10^3 dynes per unit velocity (in cm-sec^{-1}). (a) If the diaphragm is displaced and then released, will its subsequent motion be oscillatory or not? (b) Find its natural frequency both with and without the presence of the damping force.

14. The diaphragm in problem 13 is driven by a force $F = 10^5 \cos (2\pi ft)$ dynes. (a) Plot a curve of velocity amplitude vs the driving frequency, from $f = 0$ to values of f beyond the resonance frequency. (b) Compute the frequency for displacement resonance, and compare with the frequency for velocity resonance.

15. It is desired to halve the free-oscillation resonance frequency (with damping) of the diaphragm of problem 13. If this is to be done by changing the mass alone, what will the new mass be?

CHAPTER 2

PLANE WAVES IN AIR

2–1 Introduction. The displacement of a single particle held in an elastic suspension gives rise to a simple harmonic vibration around a fixed point. On the other hand, the displacement of a portion of an extended medium, having the properties of distributed mass and elasticity, always results in *waves*, traveling out from the disturbed region. It is the mass or inertial property of the medium which keeps the propagation speed from becoming infinite, and in general the greater the specific mass (i.e., the density), the lower this speed will be. Conversely, the "stiffer" the medium, the greater will be the unbalanced force upon a portion of the medium adjacent to the disturbed region and the greater the resulting acceleration. A "stiff" medium will therefore make for a high propagation speed.

For the study of plane waves the general physical approach will be as follows. We will assume a deformable medium, having both elastic and inertial properties. A particular type of deformation will be assumed to exist at a certain location in space, at a particular time. Then, in view of the physical properties of the medium, we shall see that in the region being considered the degree of deformation changes with time and, in addition, new deformations appear in adjacent regions. This may be stated quite simply in terms of the rates of change of the degree of deformation with both position and time, i.e., in the form of a differential equation. It is the solution to this differential equation that describes completely the wave phenomena.

The student of electricity and magnetism should compare this procedure with the method of demonstrating the necessity, under certain conditions, for the existence of electromagnetic waves. In the case of electromagnetic waves we start, not with the mechanical properties of a material medium, but with Coulomb's law and the principle of electromagnetic induction. Two "fields," the electric and the magnetic, are assumed to be locally distorted, in the sense that they have local values differing from those existing in the surrounding regions. Just as the mechanical deformations in air then change with time and position, so one can also express the rates of change of the two fields with time and space coordinates. It is not strange that both the differential equation and the integral equation obtained in this manner are similar in many respects to the equations for sound waves.

As a preliminary to a more formal description of wave phenomena, let us clearly state the properties assumed for the medium. We describe it as a continuous, isotropic medium, of uniform density and having the property of perfect elasticity. As long as sound sources and receivers have dimensions large compared with the mean spacing between molecules, air is, to all intents and purposes, continuous. (More will be said about the molecular point of view of sound propagation in Chapter 6.) The assumption that air is perfectly elastic deserves additional attention. Sound attenuation actually *does* occur, due to the presence of dissipative factors. As sound energy is projected through limited regions of the air, viscous stresses in the nature of shear appear near the lateral boundaries of the disturbance, and tend to dissipate the energy associated with the wave motion. Away from the boundaries these effects are of negligible importance. A second possible means of wave energy dissipation is by the process of heat flow between adjacent regions of compression and rarefaction. Such a heat flow, because of its irreversible nature, would result in a constant degradation of the wave-motion energy into the energy of uncoordinated thermal motions. (This decrease in wave amplitude is not to be confused with the operation of the ordinary inverse square law in the case of spherical waves, where the same total wave energy simply spreads into a larger and larger volume.) The heat conductivity of gases is low, and therefore over the audible spectrum the deformation process can be described quite accurately as adiabatic, and heat flow is not a significant dissipative factor. More will be said on this matter in Chapter 6.

One other assumption will be made in the course of setting up the differential equations for waves, namely, that the disturbance in the normal mass distribution for the medium will always remain small. This is true in any ordinary sound wave and this assumption will greatly simplify the mathematics. Moreover, as a consequence of this assumption, there will appear certain important physical features of wave propagation characteristic of small amplitude waves only.

A few definitions will be of use in setting up the wave equations.

2–2 Dilatation and condensation. Let V_0 be the volume occupied by any fixed mass of air with no wave disturbance present. Similarly, ρ_0 is the density of the air under the same conditions. Then, if there is some small deformation of the medium, so that V_0 is increased by a small amount v, and ρ_0 is changed similarly by a small amount ρ_d, we may state that

$$\left.\begin{aligned} \text{Dilatation} &= \delta = v/V_0 \\ \text{and} \qquad & \\ \text{Condensation} &= s = \rho_d/\rho_0. \end{aligned}\right\} \tag{2–1}$$

These dimensionless ratios describe the instantaneous fractional change in volume and density at a point in a field of sound. They are small, but not truly differential quantities, and they vary in value both with position in space and with time. In the manner of physics, dilatation and condensation are sometimes treated as true differentials when their magnitudes are small compared with other variables.

In view of Eqs. (2-1), we may write for the volume V of the chosen mass of air and for the density ρ in the presence of the distortion in the air,

$$V = V_0(1 + \delta)$$

and

$$\rho = \rho_0(1 + s). \tag{2-2}$$

Therefore

$$(1 + \delta)(1 + s) = 1, \tag{2-3}$$

since ρV is constant, for constant mass.

If s and δ are small,

$$s \cong -\delta, \tag{2-4}$$

neglecting the product, $s\delta$, in comparison with s or δ. The values of s and δ rarely exceed 10^{-3} for ordinary sound waves, so that the error in this assumption is negligible. For the large amplitude waves which accompany explosions, the simple relation of Eq. (2-4) can no longer be assumed and the exact expression of Eq. (2-3) must be used. This greatly complicates the mathematics, as will be seen in Chapter 6.

2-3 Bulk modulus. One other definition, from elasticity, will be useful. For an elastic, isotropic medium, the bulk modulus is

$$\mathcal{B} = -V\frac{dP}{dV}, \tag{2-5}$$

where P and V represent the pressure and volume respectively of a given mass. With this definition the constant $\mathcal{B}$ is always positive, since the volume will decrease when the pressure increases and vice versa.

For a perfect gas there are two such moduli, the adiabatic modulus, $\mathcal{B}_a$, and the isothermal modulus, $\mathcal{B}_i$. The ratio $\mathcal{B}_a/\mathcal{B}_i = \gamma$, where γ is the ratio of the specific heat of the gas at constant pressure to the specific heat at constant volume (= 1.4 for air). Since the variations involved in sound propagation in air are closely adiabatic in nature, we will be concerned only with $\mathcal{B}_a$. Used without a subscript, $\mathcal{B}$ will be assumed, therefore, to be $\mathcal{B}_a$.

The relationship between pressure and volume for a gas is not a linear one, so that in general the value of $\mathcal{B}$ does not remain constant, for a given mass of gas, when the total pressure and volume are varied. However, for sound propagated in the ordinary open air, any variations due to changes in atmospheric conditions and also due to the presence of the sound wave itself are quite small. Therefore there is little error in assuming $\mathcal{B}$ constant. (For normal open air conditions, $\mathcal{B}$ is of the order of 1.4×10^6 dynes/cm^2, or 1.4×10^5 newtons/m^2.)

While Eq. (2–5) is the precise definition of $\mathcal{B}$, we may substitute small finite changes in pressure and volume for the differential changes without introducing any serious error. Let p and v be such small variations in the total pressure P and the volume V, due to the presence of the sound wave. If P_0 and V_0 are the normal undisturbed values for a given mass of gas, we may write

$$\mathcal{B} = -\frac{p}{v/V_0}$$

or

$$p = -\mathcal{B}\delta = \mathcal{B}s. \tag{2-6}$$

This is a most useful relation between the small "excess pressure" p and the condensation s. At any one position in space, both quantities will vary periodically as the wave passes by, and since they are linearly related through the bulk modulus, they will always be in phase.

2–4 Significant variables in the field of sound. The state of the air through which sound waves are traveling may be discussed in terms of any one of several physical variables. We have defined and related three such variables, the dilatation δ, the condensation s, and the excess pressure p. In setting up the wave equation in the next section, we shall introduce a fourth important variable, the "particle displacement" ξ, together with its time and space derivatives. The three quantities already discussed are related by quite simple equations and are also simply related to ξ. It is therefore equally correct to describe the wave as a traveling variation in the pressure, the condensation, the dilatation, or the particle displacement.

In modern experimental acoustics, microphones of an electrical type are used almost exclusively to detect sound waves. Such microphones respond primarily to the pressure variable in the wave. In addition, the "particle velocity" $\dot{\xi}$, which represents the time derivative of the particle displacement, presently to be introduced, will be a particularly convenient variable when we come to the use of electrical analogs. In our later discussion we shall make more use of "sound pressure" and "particle velocity" than of any other of the variables so far introduced. These other quantities will, however, be useful in setting up the wave equations and in addition they are important to a complete understanding of the physical nature of a longitudinal wave.

2–5 The differential equation for plane waves. The problem of the one-dimensional wave, where the deformations in the medium are a function of one cartesian space coordinate, is the simplest to analyze. Such a wave is called *plane* because conditions are uniform over the cartesian plane specified by the one space coordinate. Most sound waves are not plane, but at a considerable distance from sound sources of ordinary size and of any shape the curvature of the wave front is small, and the wave-front shape, for all practical purposes, becomes plane. At nearer points, where this cannot be assumed, we must make use of the more complicated three-dimensional equations developed in the next chapter.

In the following sections we shall make free use of partial derivatives. The integral equations in this chapter will often involve three or more variables, the two independent ones being the space coordinate x and the time t. The important physical parameters in the field of sound, p, ξ, $\dot{\xi}$, s, etc., will each be a function of both x and t. When we write $\partial\xi/\partial x$, we shall be assuming that time, t, is held constant, whereas when $\partial\xi/\partial t$ is used, it is understood that x is held constant.

Let the air be deformed, at a given instant of time, along the x-direction only (Fig. 2–1). Assume a layer of air, originally of thickness dx and of unit cross section, to be displaced along x in such a way that the face originally at x has moved a distance ξ, and the face at $x + dx$ has moved a distance $\xi + d\xi$. The increased thickness of the layer of air, due to the deformation, is plainly $d\xi$. Since $d\xi$ can be written as $\frac{\partial\xi}{\partial x}dx$, we can then evaluate the dilatation, at this instant, for the layer of unit area:

$$\delta = \frac{v}{V_0} = \frac{\frac{\partial\xi}{\partial x}\,dx}{dx} = \frac{\partial\xi}{\partial x}. \tag{2-7}$$

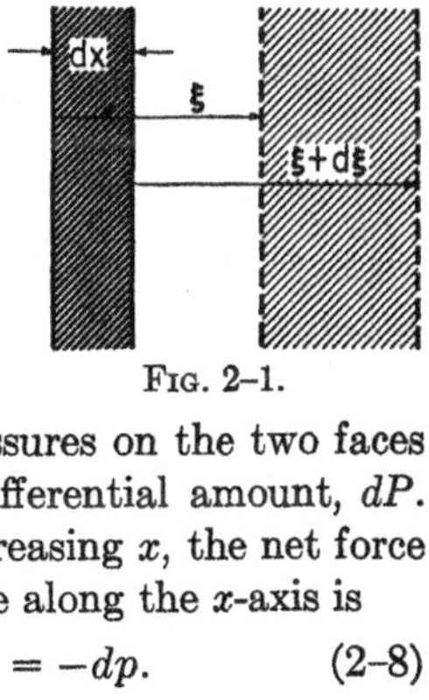

Fig. 2–1.

Due to the deformation of the medium, the pressures on the two faces of the layer will now be slightly different by a differential amount, dP. Assuming a positive increment in pressure with increasing x, the net force is to the left and therefore negative. This net force along the x-axis is

$$-P_{x+dx} + P_x = -dP = -d(P_0 + p) = -dp. \tag{2-8}$$

Therefore, writing Newton's second law for the matter within the layer, we have

$$-dp = \rho_0\,dx\,\frac{\partial^2\xi}{\partial t^2}, \tag{2-9}$$

where ρ_0 is the normal undisturbed density of the air. (This equation neglects the second order difference between the acceleration of the face displaced by an amount ξ, and that of the face displaced by an amount $\xi + d\xi$.) The acceleration is expressed as a partial derivative in recognition of the fact that ξ is a function of both x and t. Since the small change in the excess pressure dp is $\frac{\partial p}{\partial x}\,dx$, Eq. (2–9) may be written

$$-\frac{1}{\rho_0}\frac{\partial p}{\partial x} = \frac{\partial^2 \xi}{\partial t^2}. \tag{2-10}$$

This form of the wave equation involves four variables. By making use of Eq. (2–6), the number may be reduced to three. Differentiating (2–6) partially with respect to x, time being assumed constant, we obtain

$$\frac{\partial p}{\partial x} = -\mathcal{B}\frac{\partial \delta}{\partial x} = -\mathcal{B}\frac{\partial^2 \xi}{\partial x^2}. \tag{2-11}$$

Therefore Eq. (2–10) may be written

$$\frac{\partial^2 \xi}{\partial t^2} = \frac{\mathcal{B}}{\rho_0}\frac{\partial^2 \xi}{\partial x^2}$$

or, letting $\mathcal{B}/\rho_0$ equal c^2,

$$\frac{\partial^2 \xi}{\partial t^2} = c^2\frac{\partial^2 \xi}{\partial x^2}. \tag{2-12}$$

This is the most common form for the differential equation for plane waves, where ξ is the dependent variable and x and t are the independent variables. Equation (2–12) uniquely relates rates of change of ξ with respect to position and rates of change of ξ with respect to time. Before we discuss the solution of Eq. (2–12) and how it implies wave production, we should make clear the meaning of the variable ξ as applied to air.

2–6 Physical significance of the particle displacement, ξ. A gas like air is not, of course, made up of molecules having any fixed mean position in the medium, like the atoms of a solid. Even without the presence of a wave, gas molecules are in constant motion, with average velocities far in excess of any velocities associated with the wave motion (see Chapter 6). However, from a statistical point of view, a fluid, either gas or liquid, may be treated much as a solid because, when in the undisturbed state, molecules leaving a certain region as a result of their random motions are replaced by an exactly equivalent number of molecules, having exactly the same properties, thus keeping the macroscopic properties of the medium the same. Similarly, during the vibration cycle associated with the wave motion, the fact that a continuously changing group of molecules is involved rather than a fixed set is of no moment, so long as the average properties of the aggregate remain the same. In view of this equivalence, it is quite proper to speak of "particle" displacements, velocities, and accelerations for a fluid with much the same meaning as for a solid.

2–7 Solution of the wave equation. The most general solution of the differential equation, (2–12), can be shown to be of the form

$$\xi = f(x \pm ct), \tag{2-13}$$

where $c = \sqrt{\mathcal{B}/\rho_0}$. The exact nature of the function f is determined by the boundary conditions peculiar to a specific problem, in particular by the nature and behavior of the sound source. There is no mathematical restriction that the function be periodic, although in practical sound production this is usually the case.

If the reader has never encountered an equation of the type given by (2-13), he will see nothing in it to indicate a wave. A closer scrutiny of the term $(x \pm ct)$, however, will reveal its wave implications. Let us assume the negative sign for the term ct. At some specific time t_1 and at some specific position x_1, ξ will have some particular value ξ_1. If a small increment of time is added to the time t_1, so that t becomes $t_1 + \Delta t$, there will then be a slightly greater value of $x = x_1 + \Delta x$ such that the total value of $(x - ct)$ will remain the same (i.e., such that $(x_1 - ct_1) = [(x_1 + \Delta x) - c(t_1 + \Delta t)]$). Therefore ξ will still have the value ξ_1. Putting it as simply as possible, after a short time Δt, the same value of ξ will recur at a point a little farther along in the $+x$ direction. This is equivalent to describing a traveling disturbance, where the whole graphical representation of Eq. (2-13) moves along the x-axis from left to right. Figure 2-2 will help to clarify this interpretation of Eq. (2-13).

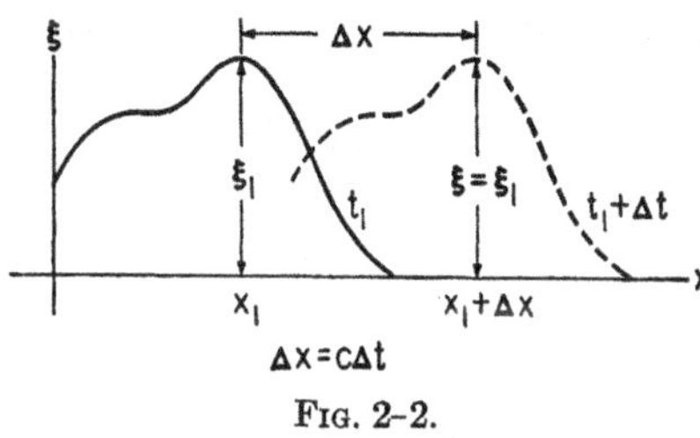

FIG. 2-2.

A similar consideration will show that with a positive sign in front of the term ct, the disturbance will move in the direction of $-x$. Whether the motion is in the positive or the negative direction, for identical values of the argument on the right of Eq. (2-13), and therefore for identical values of ξ, Δx must equal $c\,\Delta t$. The velocity of travel is therefore $\Delta x/\Delta t = c$. Since the quantity c, which is equal to $\sqrt{\mathcal{B}/\rho_0}$, is almost a constant, the velocity is independent of the nature of the function f. It should also be clear, from the above analysis, that for the small amplitude disturbances here considered, no change in graphical "shape" will occur during the propagation. If this were not true, the whole character of musical sound and speech would vary with the distance between the source and the observer!

Using the adiabatic bulk modulus and the density for dry air at normal atmospheric pressure and 0° C, the velocity of sound c becomes very nearly 33,100 cm/sec, or 331 m/sec. This is in close agreement with experiment. In Chapter 6 we shall consider some of the reasons for variations in this figure.

2–8 Disturbances of a periodic nature. The word *disturbance* is purposely used here instead of "wave" because *wave* generally implies a recurring pattern or, mathematically, a repeating function, and there are no such restrictions on the solution of Eq. (2–12). Much of the sound associated with music is transient in nature, with no true steady state frequencies. The transient component of the air disturbance travels with the same speed as does the regularly periodic portion, and plays an important part in determining the over-all effect on the hearer. For simplicity, most of our attention will be directed to disturbances of a steady state nature, originating from sustained vibrations at the source.

The following periodic expression for ξ satisfies the differential equation for plane waves:

$$\xi = \xi_m \cos \frac{2\pi}{\lambda}(x \pm ct), \tag{2–14}$$

since ξ is a function of $(x \pm ct)$. The quantities ξ_m and λ are constants. Equation (2–14) may also be written

$$\xi = \xi_m \cos\left[\pm \frac{2\pi}{\lambda}(ct \pm x)\right] = \xi_m \cos \frac{2\pi}{\lambda}(ct \pm x). \tag{2–15}$$

The student may check directly that any function $f(ct \pm x)$ is a solution of the differential equation, as well as $f(x \pm ct)$. Written either way, the use of the negative sign signifies a disturbance traveling in the $+x$ direction. We shall, for the most part, use the argument $(ct \pm x)$, since this form leads to the interpretation of phase in the conventional manner.

In physical problems, the solution of a differential equation must not only satisfy this equation but must also fit the boundary conditions. Suppose that the source of the plane waves being considered is one side of a rigid vibrating plate, the motion of every point of whose surface may be described by the equation

$$Q = Q_m \cos 2\pi ft, \tag{2–16}$$

Q being the instantaneous displacement of the surface of the plate. Such a source is often called an "acoustic piston." The air adjacent to the vibrating surface of the source must have a motion identical with that of the source itself. Let x in the wave equation (2–15) be measured from the source position. Then, provided that the constant $\lambda = c/f$, and if $\xi_m = Q_m$, it is seen that Eq. (2–15) becomes identical with (2–16) at $x = 0$. Thus the form of Eq. (2–15) is correct to fit this particular boundary condition.

2–9 The wavelength. The relation $\lambda = c/f$ will suggest that this constant is the *wavelength*, or the distance between adjacent crests in the traveling disturbance. That this interpretation is correct will be evident if Eq. (2–15) is rewritten as

$$\xi = \xi_m \cos 2\pi \left(\frac{c}{\lambda} t \pm \frac{x}{\lambda}\right). \tag{2-17}$$

Assuming time to be held constant, Eq. (2-17) becomes a relation between two variables only, ξ and x, and represents a sort of "frozen" picture of the various air layer displacements at a given instant. It is then seen that there is a spatial repetition of a given value of ξ every time x changes by an amount λ. This is the ordinary idea of wavelength. With x held constant, Eq. (2-17) becomes a relation between the two variables ξ and t. It then describes, as a function of time, and while the wave passes by, the vibration of a particular layer of air around its equilibrium position. The frequency of this vibration will be c/λ. In either case, the plot of ξ vs x or of ξ vs t is a sinusoid, whose position along the x- or t-axis, as the case may be, is determined by the particular value of x or t that is chosen.

2-10 Graphical representation. In the ordinary graphs of ξ vs t, the particle displacement ξ is plotted vertically, along the y-direction. It must be emphasized that since the wave is longitudinal, the actual physical direction of the displacement of a layer of air is parallel to the x-axis. In this connection it will be recalled that in setting up the original differential equation, a positive value of ξ was measured to the *right* of the equilibrium position, i.e., in the direction of $+x$. In Fig. 2-3 the graph is placed below the physical picture in order to clarify these relationships. The dashed lines represent the equilibrium positions for selected layers of air.

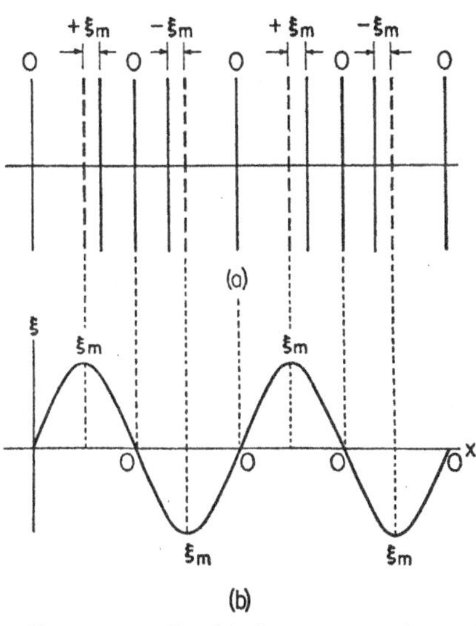

FIG. 2-3. Graphical representation of $\xi = f(x)$, time being held constant. Dashed lines in part (a) represent central position for each vibrating layer of air.

2-11 Waves containing more than one frequency component. If the vibrating source is simultaneously executing a number of simple harmonic motions of different frequencies, each of these motions will contribute a separate component displacement of the air. The total value of ξ, by the superposition principle, is the sum of these contributions. In general, then, we may write, for a wave traveling in the $+x$ direction,

$$\xi = (\xi_m)_1 \cos\left[\frac{2\pi}{\lambda_1}(ct - x) + \alpha_1\right] + (\xi_m)_2 \cos\left[\frac{2\pi}{\lambda_2}(ct - x) + \alpha_2\right] + \cdots + (\xi_m)_n \cos\left[\frac{2\pi}{\lambda_n}(ct - x) + \alpha_n\right], \quad (2\text{–}18)$$

where each of the λ's is associated with one of the particular frequencies present. Note that it is necessary to introduce a separate phase angle α into each component of the wave expression, the values of these angles being determined by the exact behavior of the source, where $x = 0$. The reader should recognize that the right-hand side of Eq. (2–17) is somewhat similar, in terms of the wave equation, to the harmonic series of a Fourier analysis, as discussed in Chapter 1.

As was brought out in Section 2–7, the speed of wave propagation, c, for small amplitude waves, is dependent only on the elastic and inertial properties of the medium. The various frequencies present in a complex wave, as represented by Eq. (2–18), all travel with the same speed. This means that the phenomenon of dispersion, so important in light, is almost nonexistent in sound. This is to be expected, in view of the purely mechanical nature of longitudinal waves. At very high audible frequencies and in the ultrasonic region, anomalous effects do occur (see Chapter 6), but not for the ordinary audible range. There is one interesting special case where dispersion *does* occur in air with ordinary sound frequencies and intensities, i.e., in the propagation of waves along an exponential horn. This will be mentioned again when this type of horn is discussed in Chapter 5.

2–12 Alternate forms for the steady state solution to the wave equation. By means of the relations given in Sections 2–2 and 2–3, it is possible to rewrite Eq. (2–15) in terms of any of the various field parameters. The results for a wave traveling in the $+x$ direction are summarized below, along with the original equation in terms of ξ.

$$\begin{aligned}
&\text{(a)}\ \ \xi = \xi_m \cos\frac{2\pi}{\lambda}(ct - x),\\
&\text{(b)}\ \ \dot{\xi} = -2\pi\frac{c}{\lambda}\xi_m \sin\frac{2\pi}{\lambda}(ct - x),\\
&\text{(c)}\ \ \delta = \frac{\partial\xi}{\partial x} = \frac{2\pi}{\lambda}\xi_m \sin\frac{2\pi}{\lambda}(ct - x),\\
&\text{(d)}\ \ s = -\frac{\partial\xi}{\partial x} = -\frac{2\pi}{\lambda}\xi_m \sin\frac{2\pi}{\lambda}(ct - x),\\
&\text{(e)}\ \ p = \mathcal{B}s = -\mathcal{B}\frac{\partial\xi}{\partial x} = -\mathcal{B}\frac{2\pi}{\lambda}\xi_m \sin\frac{2\pi}{\lambda}(ct - x).
\end{aligned} \qquad (2\text{–}19)$$

(The coefficient on the right of Eq. (2-19b) could also have been written as $2\pi f \xi_m$.)

It should be pointed out that the determination of any one of the group of field properties in a plane wave in free space uniquely determines all the others, since they are all simply related. This fact is important in connection with experimental acoustical measurements in the path of plane waves. A "pressure" detector may be used to measure indirectly all other important quantities as well.

2-13 Phase relationships. The phase relationships which appear in the above set of equations are worth some attention. The particle velocity $\dot{\xi}$, the condensation s, and the excess pressure p are all in phase. This means that the density and the pressure are a maximum when a layer of air is moving through its *central* position (where $\dot{\xi}$ is a maximum), *not* when it is at the extreme ends of its motions, as one might expect. The dilatation is, of course, 180° out of phase with the condensation and the excess pressure. The quantities $\dot{\xi}$; δ, s, and p are all 90° out of phase with the displacement.

The algebraic signs of these quantities introduce some subtleties in the phase relationships. In the first place, the variables ξ and $\dot{\xi}$ in Eqs. (2-19) represent *vector* quantities. As assumed in Section 2-5, $+\xi$ is in the $+x$ direction and $-\xi$ is in the $-x$ direction. This applies also to the particle velocity, $\dot{\xi}$. The variables δ and s may also be plus or minus, but since they represent *scalar* quantities, the algebraic sign simply indicates whether the volume or density change is an increase or a decrease. The excess pressure, p, is also a scalar, in this sense. This difference in the interpretation of sign for the vector and for the scalar equations in Eqs. (2-19) must be kept clearly in mind in connection with reflection phenomena.

For a wave traveling in the $-x$ direction, where $(ct - x)$ is replaced by $(ct + x)$, there will be a positive sign on the right-hand side of Eqs. (2-19d) and (2-19e), after the differentiation. No change of sign will take place, however, in Eq. (2-19b). This means that there will now be a 180° phase relationship between $\dot{\xi}$ and either s or p. When the particle velocity is to the right (positive), the density and total pressure will be *less* than the normal ρ_0 and P_0 respectively.

The vector interpretation of a positive or negative ξ or $\dot{\xi}$ is unaffected by the direction of wave propagation, since their sign is tied up with the sign convention associated with a fixed x-axis.

Example. A large flat plate is radiating plane sound waves from one side only. The amplitude of its motion is 0.01 mm and the frequency of vibration is 1000 cycles-sec^{-1}. For any point in the path of the waves, find the maximum values of ξ, $\dot{\xi}$, s, δ, and p.

Assume the velocity of sound c to be 331 m/sec and the bulk modulus, $\mathcal{B}$, to be 1.4×10^6 dynes-cm^{-2}. The amplitude of motion in the air is the same as that of the source. Therefore

$$\begin{aligned}
\xi_m &= 10^{-3}\text{ cm},\\
\dot{\xi}_m &= 2\pi f \xi_m = 2\pi(10^3)(10^{-3}) = 6.28\text{ cm-sec}^{-1},\\
s_m &= \delta_m = \frac{2\pi}{\lambda}\xi_m = \frac{2\pi}{33.1}10^{-3} = 1.9 \times 10^{-4},\\
p &= \mathcal{B}s = (1.4 \times 10^6)(1.9 \times 10^{-4}) = 2.66 \times 10^2\text{ dynes-cm}^{-2}.
\end{aligned}$$

These values are typical of sounds of high intensity.

2–14 Energy in the wave. For the simple harmonic motion of a mass particle, the energy was seen to be, on the average, half kinetic and half potential. One would therefore suspect that in a sound wave the energy is also so divided. While this turns out to be the case, there are certain features of energy storage which are peculiar to longitudinal waves and deserve some discussion.

2–15 Kinetic energy. Consider a longitudinal wave of sinusoidal form, progressing in the $+x$ direction. For a thin layer of air, of thickness dx and of unit cross section, moving with a velocity $\dot{\xi}$, the instantaneous kinetic energy is

$$dE_k = \frac{1}{2}\rho_0(\dot{\xi})^2\,dx. \tag{2–20}$$

The average kinetic energy density e_k in the medium may be obtained by integrating this expression with respect to x over an integral number of wavelengths $n\lambda$ (keeping the time constant), and then dividing the result by the volume this total energy occupies:

$$e_k = \frac{\frac{1}{2}\rho_0(\dot{\xi}_m)^2\int_0^{n\lambda}\sin^2\frac{2\pi}{\lambda}(ct - x)dx}{n\lambda} = \frac{1}{4}\rho_0(\dot{\xi}_m)^2 = \frac{1}{4}\rho_0(4\pi^2 f^2)\xi_m^2. \tag{2–21}$$

It will be noted that the result is identical with what one would expect for the time average of the kinetic energy of a particle whose mass is the mass per unit volume of the gas.

2–16 Potential energy. To obtain the corresponding *potential energy* density, we must consider the properties of a perfect gas. In Section 2–3 it was indicated that the volume and pressure changes that occur in air during the passage of longitudinal waves of audible frequencies are nearly adiabatic in character. The graph of Fig. 2–4 represents a small portion of a PV diagram for an adiabatic variation, using a fixed mass of gas. Let us suppose a volume V_0 to be reduced to a slightly smaller volume, V, the decrease being called v. As a result, the pressure will rise slightly from P_0 to P, the increase being called p. The relations between these quantities are shown in the graph. Assuming the curve to be straight for small changes, the work ΔW done upon the gas during the volume change, or the energy ΔE_p, stored potentially within the gas, will be the area under the curve between V and V_0:

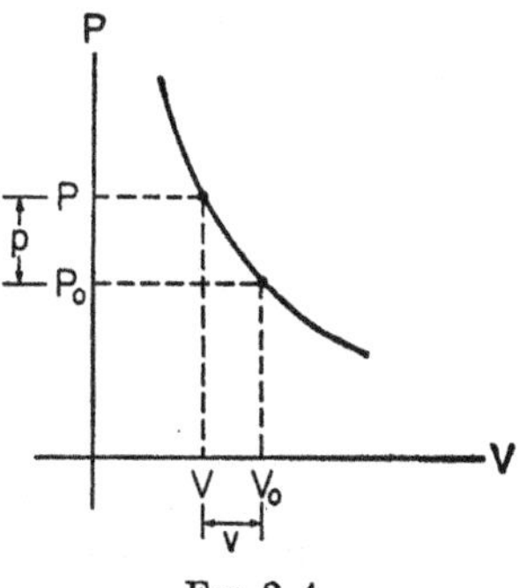

FIG. 2–4.

$$\Delta W = \Delta E_p = (P_0 + \frac{1}{2} p)v = P_0 v + \frac{1}{2} pv. \tag{2–22}$$

For simple harmonic variations around mean pressures and volumes, the v's and p's are alternately plus and minus. The average value of the first term on the right-hand side, $P_0 v$, over any integral number of cycles is therefore zero. In the second term, however, when v changes sign, so does p. The product sign is therefore always the same, showing potential energy to be stored in the gas whether there is a compression or a rarefaction. It is interesting to see that while the air is, of course, always "compressed" in the absolute sense, *total* pressures never changing sign, the medium nevertheless acts just like an unstressed spring which is alternately compressed and stretched.

The second term on the right-hand side of Eq. (2–22), $\frac{1}{2}pv$, which alone contributes to the average potential energy stored in the medium, may be written in terms of the other field parameters by means of the following transformations.

Since

$$p = -\mathcal{B}\delta \quad \text{and} \quad v = V_0\delta, \tag{2–23}$$

therefore

$$\frac{1}{2} pv = \frac{1}{2}\mathcal{B}\delta^2 V_0. \tag{2–24}$$

The minus sign of the first relation of Eq. (2–23) may be ignored, since a positive sign for p and a negative sign for δ both signify work done *on* the gas and therefore an increase in the stored potential energy. For a thin layer of air of unit cross section in the path of the wave, the potential energy becomes

$$dE_p = \frac{1}{2} \mathcal{B}\, \delta^2\, dx. \tag{2–25}$$

By partially differentiating the right-hand side of Eq. (2–13), first with respect to x and then with respect to t, we see that

$$\frac{\partial \xi}{\partial x} = \pm \frac{1}{c}\frac{\partial \xi}{\partial t}. \tag{2–26}$$

Inserting this value of $\partial\xi/\partial x(= \delta)$ into Eq. (2–25), we obtain

$$dE_p = \frac{1}{2}\frac{\mathcal{B}}{c^2}(\dot{\xi})^2\,dx = \frac{1}{2}\rho_0(\dot{\xi})^2\,dx. \tag{2–27}$$

This last expression is identical with Eq. (2–20), representing the instantaneous *kinetic* energy of a thin layer of air. It therefore follows that the *average potential energy density* in a region containing an integral number of wavelengths will also be $\frac{1}{4}\rho_0(\dot{\xi}_m)^2$.

2–17 Total energy density in the wave. The total average energy density in the wave will be the sum of the kinetic and the potential energies, or

$$e_{\text{total}} = \frac{1}{4}\rho_0(\dot{\xi}_m)^2 + \frac{1}{4}\rho_0(\dot{\xi}_m)^2 = \frac{1}{2}\rho_0(\dot{\xi}_m)^2. \tag{2–28}$$

One of the interesting things about the energy in a wave disturbance, not readily foreseen, is that the kinetic and the potential energies move along together in *identical regions*. Since the instantaneous energies can both be written in terms of the instantaneous particle velocity, they are each a maximum at the same position in space and also at other places and times are zero together. This is, of course, not true for the vibration of a single isolated mass particle. A plot of the total energy distribution in space for the wave, Fig. 2–5, shows a sort of pseudoquantum nature, regions of large total energy alternating with regions of little or no energy. In the case of traveling transverse waves on a stretched string, the above remarks do not apply. For these waves, as will be seen in Chapter 7, the kinetic energy maxima and the potential energy maxima are not coincident.

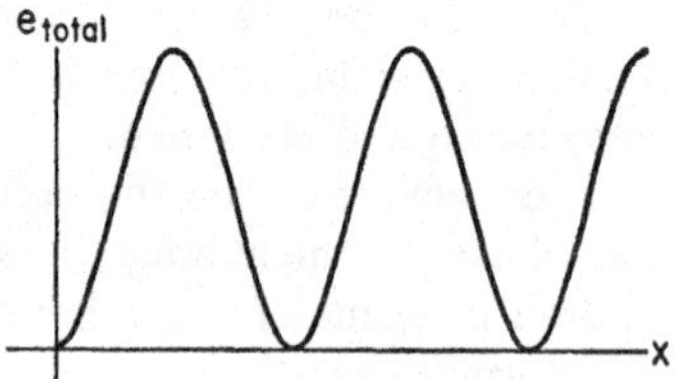

FIG. 2–5. Graph of energy distribution along x at a given instant of time.

2–18 Sound intensity. This important measure of sound wave amplitudes is defined as the energy flow across an area, per unit area and per unit time. This energy will plainly be equal to that contained in a column of unit cross section and of length c, the velocity of sound. Therefore the intensity I is the product of the total energy density, derived above, and the velocity c.

There are several different ways of writing the expression for the sound wave intensity, in view of the interrelations between all the important parameters. Several useful forms, whose validity are given below.

$$\text{(a)}\ \frac{1}{2}\,\rho_0 c(\xi_m)^2. \qquad \text{(d)}\ p_{\text{rms}}\dot{\xi}_{\text{rms}}.$$

$$\text{(b)}\ \frac{1}{2}\,\rho_0 c\, 4\pi^2 f^2(\xi_m)^2. \qquad \text{(e)}\ \frac{p^2_{\text{rms}}}{\rho_0 c}. \tag{2-29}$$

$$\text{(c)}\ \rho_0 c(\dot{\xi}_{\text{rms}})^2.$$

The form given in Eq. (2-29c) is algebraically similar to the expression for electrical power, RI^2, where $\rho_0 c$ replaces R and $\dot{\xi}$ takes the place of I. Extensive use of this analogy will be made in Chapter 5.

2-19 Units of intensity. Using cgs units, intensity is measured in ergs-cm^{-2}-sec^{-1}. For ordinary audible waves, intensities range from about 10^{-9} to about 10^{+3} cgs units. In mks units this corresponds to a range of from 10^{-12} to 1.0 joule-meter^{-2}-sec^{-1}. These numbers are an indication of how small are the energies associated with sound. The total energy coming from the throats of a crowd at a football game, in response to some spectacular play on the field, might perhaps be enough to heat a cup of coffee! Even the great crescendos of a large symphony orchestra involve very little sound wave energy. All of this is a tribute to the sensitivity of the human ear.

2-20 The decibel. For two sounds of intensities I_1 and I_2, one is said to be of a greater intensity than the other by a number of *decibels* (db), where

$$\text{Intensity difference in db} = 10 \log_{10} \frac{I_1}{I_2}. \tag{2-30}$$

The decibel is therefore not an absolute, but a comparative measure of intensity and is consequently a pure number. Without the factor 10, the comparison is in *bels*, a unit too large for most practical purposes.

It is because of the sensitivity of the ear, and the range of intensities to which it responds, that the decibel scale has been devised. The scale is based on the well-known observation that the human sensory response to a given increase in an objective stimulus is approximately proportional to the ratio of the increase in stimulus to the stimulus already present. To give a concrete example, the ear is capable of detecting a very small increase in sound intensity when the background intensity is low; with a great deal of background noise, a much larger increase of intensity is necessary to give to the ear the same sensations.

If h represents the sensation delivered to the brain, and g the objective stimulus, the proportion may be expressed mathematically as

$$h \propto \frac{\Delta g}{g}. \tag{2-31}$$

If this statement is essentially correct, as seems to be the case for sensations of sight, pain, etc., as well as for hearing, then with greater changes in stimuli Eq. (2–31) may be integrated between definite limits, to obtain

$$h_1 - h_2 = \log \frac{g_1}{g_2}. \tag{2-32}$$

It is upon this equation that the decibel scale is based.

The range of audible intensities mentioned above, 10^{+3} to 10^{-9} ergs-cm^{-2}-sec^{-1}, may be converted into a decibel comparison by inserting the ratio $10^3/10^{-9} = 10^{12}$ for I_1/I_2 in Eq. (2–30). There is then seen to be an approximate range of 120 db between very weak and very intense sounds (ranging from the so-called threshold of hearing to the threshold of feeling). Apart from the nature of the ear response, the numerical convenience of this compressed scale is obvious. The decibel is also a convenient sized unit to use because any intensity difference of the order of one decibel may usually be ignored as far as the ear is concerned. The average ear is unable, even under ideal laboratory conditions, to tell that two sounds differ in intensity when their difference, measured in power per unit area, is less than about 10%, and under ordinary listening conditions the difference must be much greater.

There is an important consequence of the ear's rather crude ability to differentiate among varying sound intensities. It was pointed out earlier that it is difficult to express the physics of actual sound problems in terms of precise mathematics, and often even more difficult to solve these approximate equations. Fortunately, a discrepancy of 10–15% between theory and experiment, as far as intensity is concerned, is of no significance to the ear. This is a great comfort to the designer of practical acoustical equipment.

2–21 Intensity "level"; pressure "level." In recent years there has been devised an absolute intensity scale, known as the *intensity level*. This scale is based on the arbitrary selection of a low reference intensity, I_0, with which other intensities are compared. The value generally used for I_0 for plane waves is 10^{-16} watt-cm^{-2} = 10^{-9} erg-cm^{-2}-sec^{-1}, an intensity which corresponds approximately to the average threshold of hearing, or the weakest sound which can be heard. With I_0 specified, the intensity level in decibels of some sound of intensity I is then computed by replacing the ratio I_1/I_2 in Eq. (2–30) by I/I_0. Using the reference intensity given above, the intensity level in a noisy machine shop might be 100 db.

Since virtually all modern sound detectors respond directly to the variations of *pressure* in the wave disturbance, rather than to the intensity itself, the so-called *pressure level* is considered more fundamental than the intensity level. By Eq. (2–29e) the intensity in a plane wave is seen to be proportional to p^2. Therefore we may write Eq. (2–30) in the form

$$\text{Intensity difference in db} = 10 \log_{10} \frac{p^2}{p_0^2} = 20 \log_{10} \frac{p}{p_0}. \qquad (2\text{–}33)$$

The rms value of p_0, the standard reference pressure, is commonly taken to be 0.0002 dyne-cm^{-2}. (This corresponds closely to the pressure in a wave whose intensity is the reference one given above, 10^{-16} watt-cm^{-2}.) The pressure level in a plane wave is therefore, by Eq. (2–33), 20 times the logarithm to the base 10 of the ratio of the pressure in the wave to the reference pressure, p_0.

Neither intensity level nor pressure level is the same as *loudness level*, which is a measure of subjective response that will be defined in Chapter 9.

PROBLEMS

1. Consider a plane wave traveling in the $+x$ direction. Using the same pair of rectangular axes, plot as a function of the time, for a fixed value of x, the particle displacement ξ, the particle velocity $\dot{\xi}$, the dilatation δ, and the excess pressure p. Besides showing the relative phases, indicate also the maximum values of each variable in terms of c, λ, etc.

2. Repeat problem 1 for a wave moving in the $-x$ direction.

3. Two plane waves are traveling along the x-axis. The particle displacements due to the separate waves are given by

$$\xi_1 = \xi_m \cos \frac{2\pi}{\lambda} (ct - x)$$

and

$$\xi_2 = -\xi_m \cos \frac{2\pi}{\lambda} (ct + x).$$

For the two waves, at the position $x = 0$, find the relative phase of (a) ξ_1 compared with ξ_2, (b) $\dot{\xi}_1$ compared with $\dot{\xi}_2$, and (c) p_1 compared with p_2.

4. For a certain plane wave traveling through air, the maximum value of the excess or acoustic pressure is 0.1 dyne-cm^{-2}. If the frequency is 1000 cycles-sec^{-1}, and the density of air is 1.29×10^{-3} gm-cm^{-3}, find (a) the maximum particle displacement, (b) the maximum particle velocity, (c) the maximum condensation, and (d) the maximum dilatation.

5. For the wave in problem 4, find (a) the average kinetic energy per unit volume, (b) the average potential energy per unit volume, and (c) the wave intensity, all in cgs units. (d) Also determine these values expressed in mks units.

6. A plane wave in air has an intensity of 40 erg-cm^{-2}-sec^{-1} and a frequency, f. A second wave, also in air, has an intensity of 10 erg-cm^{-2}-sec^{-1}, and a frequency $f/2$. For the two waves, find the ratio of (a) the maximum particle displacements, (b) the maximum particle velocities, and (c) the maximum acoustic pressures.

7. A plane wave in air and a plane wave in hydrogen have the same intensity and are of the same frequency. Find for the two waves the relative values of (a) the maximum particle displacements, (b) the maximum particle velocities, and (c) the maximum acoustic pressures. The density of hydrogen is 9×10^{-5} gm-cm^{-3} and the value of c is 1270 m-sec^{-1}.

8. The intensity level in a longitudinal wave in air is 20 db. Assuming a reference intensity of 10^{-9} erg-cm^{-2}-sec^{-1}, find the absolute intensity in the wave in cgs units. Assuming the rms reference pressure to be 2×10^{-4} dyne-cm^{-2}, find the rms acoustic pressure for the *pressure* level of 20 db.

9. Sound wave *B* has an intensity 10 db greater than wave *A*. Wave *C* has an intensity 10 db greater than wave *B*. (a) What is the intensity of wave *C* relative to wave *A*, in db? (b) Find the absolute ratio of the intensity of *B* to *A*, *C* to *B*, and *C* to *A*.

10. The pressure level in a sound wave in air is 30 db. (a) Find the absolute value of the acoustic pressure in the wave, in cgs units. (b) Also find the maximum value of the particle velocity. (Use the standard reference pressure given in problem 8.)

CHAPTER 3

WAVES IN THREE DIMENSIONS

Many problems in acoustics can be handled adequately by the methods discussed in Chapter 2. Whenever the vibrating surface is plane and rigid, or approximately so, the waves leaving the source are plane. The wave-front shape will be *maintained* as plane, provided that the surface area of the source has dimensions large compared with the wavelength in air (see Chapter 4). The waves leaving the mouth of an "ideal" horn, for instance, are approximately plane and remain so provided the perimeter of the mouth of the horn is large compared with the wavelength. All properties in the resulting field of sound may then be computed to within a fair degree of accuracy through the use of the plane wave equations just developed.

There are, however, so many important practical problems where the above state of affairs does not exist that a more general consideration of wave phenomena is desirable. Let us therefore consider the general problem of space waves.

3–1 Waves in three dimensions. The equation of continuity. Consider the medium to have the same isotropic properties assumed in setting up the differential equation for plane waves. In this case, however, we will consider a more general type of deformation. In Fig. 3–1 a differential volume element in cartesian coordinates, $dx\,dy\,dz$, is located at the position specified by x, y, z. Matter is pictured as in a general state of flux throughout the region, with velocity components at the point in question specified by u, v, and w, along the x-, y- and z-axes respectively. Positive vectors are assumed and u, v, and w are presumed to change with variations in x, y, and z. Positive increments in velocity accompany positive increments in position coordinates.

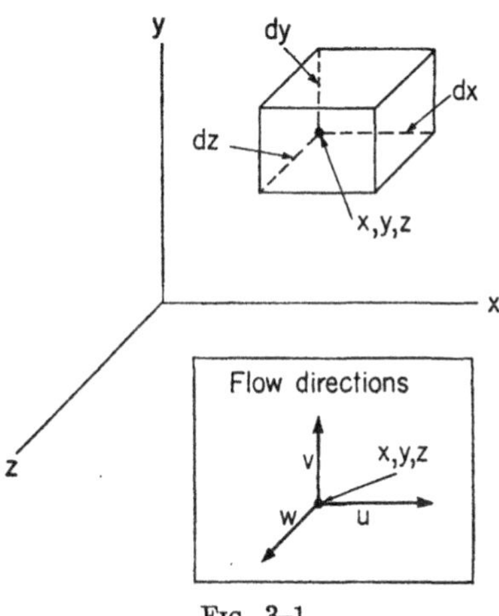

FIG. 3–1.

It is now possible to set up an equation which states, very simply, that the total time rate of mass flow *out* of the volume, dV, through all six faces of the cube, is equal to the rate of decrease of mass *within* this volume. This is called the *equation of continuity*, and it is set up as follows. Consider the face of area $dy\,dz$, lying in the yz plane, on the left-hand side of the cube. The mass of fluid entering this face per unit time is $\rho u\, dy\, dz$, where ρ is the fluid density at this face. (Note that without the factor $dx\, dy$, this expression represents the flow per *unit* area.) At the parallel face, located at $x + dx$, both u and ρ are assumed to be different, and greater, by a differential amount. We may therefore express the net rate of *efflux* of mass through the two parallel faces as

$$-\rho u\, dy\, dz + \left[\rho u + \frac{\partial(\rho u)}{\partial x}\, dx\right] dy\, dz = \frac{\partial(\rho u)}{\partial x}\, dx\, dy\, dz. \tag{3-1}$$

In a similar manner, the time rate of outward flow of mass through the other two sets of parallel faces may be shown to be $\frac{\partial(\rho v)}{\partial y}\, dy\, dx\, dz$ and $\frac{\partial(\rho w)}{\partial z}\, dz\, dx\, dy$ respectively. The sum of the three rates of flow represents the total rate at which matter is leaving the volume element, which in turn may be written as $-\frac{\partial \rho}{\partial t}\, dx\, dy\, dz$. Equating the two expressions for the efflux, we obtain

$$\frac{\partial(\rho u)}{\partial x} + \frac{\partial(\rho v)}{\partial y} + \frac{\partial(\rho w)}{\partial z} = -\frac{\partial \rho}{\partial t}. \tag{3-2}$$

This is the *equation of continuity* and is nothing more than a statement of the conservation of mass.

3-2 Application of Newton's second law. As in the case of the derivation for plane waves, we shall now make use of Newton's second law of mechanics. Along the x-direction there is an unbalanced force upon the matter within the differential cube. By an analysis similar to that just employed, it is easy to show that this unbalanced force, dF, is given by

$$dF = -dp\, dy\, dz = -\frac{\partial p}{\partial x}\, dx\, dy\, dz. \tag{3-3}$$

Here, as in the plane wave analysis, we may use the "excess" pressure, p, instead of the total pressure, P, since $dp = dP$. The negative sign appears because, with P increasing with positive increments along x, the net force is in the $-x$ direction. Expressing Newton's law in the general form involving momentum, we now have

$$-\frac{\partial p}{\partial x}\, dx\, dy\, dz = \frac{\partial(\rho u)}{\partial t}\, dx\, dy\, dz$$

or

$$-\frac{\partial p}{\partial x} = \frac{\partial(\rho u)}{\partial t}. \tag{3-4}$$

There are three symmetrical equations for the motions along y and along z.

Collecting the results for all three axes, we have

$$\begin{aligned} &\text{(a)} \quad -\frac{\partial p}{\partial x} = \frac{\partial(\rho u)}{\partial t}, \\ &\text{(b)} \quad -\frac{\partial p}{\partial y} = \frac{\partial(\rho v)}{\partial t}, \\ &\text{(c)} \quad -\frac{\partial p}{\partial z} = \frac{\partial(\rho w)}{\partial t}. \end{aligned} \tag{3–5}$$

3–3 The differential equation for waves in three dimensions. Having properly expressed the physics of the situation, we may now operate upon Eqs. (3–2) and (3–5) in a purely mathematical manner. In outline (the reader should check the actual steps), this is what is done. First, Eq. (3–2) is differentiated partially with respect to time, holding x, y, and z constant. Then the first of equations (3–5) is differentiated partially with respect to x, the second with respect to y, and the third with respect to z. This process results in right-hand terms of the form $\frac{\partial(\rho u)}{\partial t\,\partial x}$, etc. By adding the three new equations so obtained and combining the result with the equation which is the result of differentiating Eq. (3–2), we obtain the relation

$$\frac{\partial^2 \rho}{\partial t^2} = \frac{\partial^2 p}{\partial x^2} + \frac{\partial^2 p}{\partial y^2} + \frac{\partial^2 p}{\partial z^2}. \tag{3–6}$$

It will be noted that the dependent variable on tne left is ρ, while on the right it is p. The quantities p and ρ are related, however, through the equations

$$\left.\begin{aligned} p &= \mathcal{B}s \\ \text{and} \qquad \rho &= \rho_0(1 + s) \end{aligned}\right\} \tag{3–7}$$

As a result of Eqs. (3–7), it can be shown that

$$\frac{\partial^2 \rho}{\partial t^2} = \frac{1}{c^2}\frac{\partial^2 p}{\partial t^2}.$$

When this expression for $\partial^2\rho/\partial t^2$ is introduced into Eq. (3–6), the latter becomes

$$\frac{\partial^2 p}{\partial t^2} = c^2\left(\frac{\partial^2 p}{\partial x^2} + \frac{\partial^2 p}{\partial y^2} + \frac{\partial^2 p}{\partial z^2}\right) = c^2\,\nabla^2 p, \tag{3–8}$$

where ∇^2 is an operator symbol, representing the process within the parentheses.

Equation (3–8) is the general differential equation for space waves, expressed in terms of cartesian coordinates. It should be admitted freely at this point that precise, useful solutions of this general equation are very difficult to obtain. One of the great difficulties in all solutions of differential equations, and especially in the field of acoustics, is to introduce correctly the physical boundary conditions. The complex contour of almost any practical musical instrument makes impossible any accurate mathematical statement as to source behavior. No one, for instance, would attempt to write the equation for the surface of such a source as the body of a violin!

3–4 The differential equation for spherical waves. A special, simpler form of Eq. (3–8), which has greater practical usefulness, may be obtained, but there are definite restrictions upon its validity. This simpler equation expresses the situation when there is complete spherical symmetry and it is called the equation for "spherical waves." To obtain the spherical wave equation from Eq. (3–8) is a laborious process. Equation (3–8) may be transformed by standard mathematical procedures to spherical coordinates, where the variables are the radius vector r, the polar angle θ, and the azimuthal angle ϕ:

$$\frac{\partial^2 p}{\partial t^2} = c^2 \left\{ \frac{\partial^2 p}{\partial t^2} + \frac{2}{r}\frac{\partial p}{\partial r} + \frac{1}{r^2 \sin^2\theta}\frac{\partial^2 p}{\partial \phi^2} + \frac{1}{r^2 \sin\theta}\left[\frac{\partial}{\partial \theta}\left(\sin\theta \frac{\partial p}{\partial \theta}\right)\right]\right\}. \tag{3–9}$$

Since for spherical symmetry all derivatives with respect to θ and ϕ are zero, only derivatives with respect to r are retained. It is not difficult to show that Eq. (3–9) then reduces to

$$\frac{\partial^2 (rp)}{\partial t^2} = c^2 \frac{\partial^2 (rp)}{\partial r^2}. \tag{3–10}$$

(It should be pointed out that this roundabout mathematical path may be greatly shortened by initially setting up the equation of continuity in spherical rather than in cartesian coordinates.)

It is Eq. (3–10) that forms the basis for our discussion of space waves. Important restrictions on its use will be pointed out as we go along.

3–5 The solution of the differential equation. Except for the different argument in the numerator, it will be noted that Eq. (3–10) is identical in form with the differential equation for plane waves. Therefore we may write down immediately its general solution:

$$rp = f(r \pm ct)$$

or

$$p = \frac{1}{r} f(r \pm ct). \tag{3–11}$$

Only the minus sign is physically significant for practical purposes; we are concerned only with the *diverging* type of wave which travels in the $+r$ direction.

Both the differential equation and its solution may be written in terms of the density and the condensation, using exactly the form of Eq. (3–10). This should be apparent without formal proof, since these quantities are all linearly related to pressure. The differential equation when written in terms of the particle displacement ξ_r and the particle velocity $\dot{\xi}_r$ takes a somewhat different form. We shall not discuss these equations, since ξ_r and $\dot{\xi}_r$ are both derivable from the velocity potential Φ, discussed below.

The most obvious difference between the integral equation, (3–11), and the solution of the plane wave equation is the presence of the important coefficient $1/r$. This is to be expected in view of the physical nature of spherical waves, where the wave energy is spreading into a larger and larger volume. If the reader is puzzled because the coefficient involves the inverse *first* power of r, he should remember that sound intensities are, in general, proportional to the pressure *squared*, therefore the pressure *should* fall off as $1/r$.

Before applying the results just obtained to a practical problem in acoustics, one more basic quantity in the field of sound will be defined, because of its general usefulness. This is the *velocity potential*, Φ, first used by Lagrange. It is interesting to compare this function with the well-known electric and magnetic potentials of other branches of physics; the beautiful symmetry of the mathematics will become apparent in the next section. From a purely manipulative point of view it is possible to dispense with the function Φ in many practical problems. The other field parameters are all related to this function and it is therefore possible to select some other parameter (the acoustic excess pressure p is often chosen) and refer all other variables to this other parameter. This is exactly what is done in some of the current writing on acoustics, and in many cases this is a time-saving approach. One should, however, become familiar with the use of the velocity potential, partly because of the unifying part it plays in correlating the various sound field parameters and partly because of its importance in the general subject of hydrodynamics, a complex subject closely related to acoustics. Whenever viscosity enters into a problem in fluid motion, and most problems in hydrodynamics are of this type, the velocity potential is found to be an indispensable tool.

3–6 The velocity potential, Φ. Both the differential equations for space waves and their solutions may be written in terms of the new function Φ, defined as follows:

$$u = -\frac{\partial \Phi}{\partial x}, \qquad v = -\frac{\partial \Phi}{\partial y}, \qquad w = -\frac{\partial \Phi}{\partial z}, \qquad (3\text{–}12)$$

where u, v, and w are the fluid velocity components. The student of electricity will recognize the similarity between the velocity potential Φ and the electrical potential ϕ. Just as the space derivatives of ϕ represent the electric field components along mutually orthogonal directions, the appropriate derivatives of Φ represent velocity components. In both cases the derivatives represent the magnitudes of *vector* quantities, whereas Φ and ϕ are *scalars*. The usefulness of Φ is exactly that of the electrical potential function. It is often easier to evaluate Φ in the field of sound than to evaluate the vector quantity, i.e., the particle velocity. Once the equation for Φ as a function of (x, y, z) is formulated, however, u, v, and w (or the quantity $\dot{\xi}$, in the case of plane waves) may be readily obtained. The exact way in which the velocity potential is so used will appear presently when a specific problem is considered.

When the function Φ is introduced into the fundamental dynamical equations for a deformable elastic medium, equations of exactly the same differential and integral form as those just developed are obtained:

$$\begin{aligned}
\frac{\partial^2 \Phi}{\partial t^2} &= c^2\, \nabla^2 \Phi && \text{(cartesian coordinates)},\\
\frac{\partial^2 (r\Phi)}{\partial t^2} &= c^2\, \frac{\partial^2 (r\Phi)}{\partial r^2} && \text{(spherical coordinates)}, \qquad (3\text{–}13)\\
\Phi &= \frac{1}{r} f(r \pm ct) && \text{(spherical coordinates)}.
\end{aligned}$$

To the reader who has carefully followed the physical and mathematical arguments of this chapter, these equations will appear quite credible. Those more skeptically inclined will find the details in Appendix I.

Since for the spherical waves now under consideration the particle velocity, as a matter of definition, is always along the direction of the radius vector r, the cartesian equations (3–12) reduce to the single equation

$$\dot{\xi}_r = -\frac{\partial \Phi}{\partial r}. \qquad (3\text{–}14)$$

Once the equation for the particle velocity is known, the expression for ξ may be obtained by integrating Eq. (3–14) with respect to time.

The relations between the condensation s, the dilatation δ, and the excess pressure p, used in connection with plane waves, may be carried over bodily into the discussion of spherical waves, since these relations were in no way restricted to the geometry of plane waves. (The relation between δ and ξ is more complicated than in the case of plane waves, since there is a transverse dimensional change as well as a radial one.) By a straightforward mathematical transformation, however (see Appendix II), a most useful relation between p and Φ may be obtained, good for any system of space coordinates, including the case of spherical symmetry here being considered:

$$\frac{\partial \Phi}{\partial t} = \frac{1}{\rho_0} p = c^2 s. \tag{3-15}$$

3-7 Application of the function Φ. The "pulsing sphere." We shall now apply some of the above relations to the problem of radiation by a "pulsing sphere." The nearest practical realization of this curious but mathematically convenient type of source would be a spherical balloon. Imagine such a balloon to be connected to a reservoir whose pressure suffers a small, regularly periodic variation. The balloon would then expand and contract with simple harmonic radial motion. (Such a model would, of course, only follow the pressure variations at very low frequencies.)

Let us now consider a problem where enough information is furnished so that the complete and exact equation for the function Φ may be written down. We shall first assume a periodic form for Φ, since the derived quantities ξ, p, etc., are known to be periodic in actual sound waves:

$$\Phi = \frac{B}{r} \cos\left[\frac{2\pi}{\lambda}(ct - r) + \alpha\right]. \tag{3-16}$$

[Note that we are using here the alternate expression $(ct - r)$ instead of $(r - ct)$.] The constants B and α are essentially constants of integration, to be determined from the known behavior of the sound source, the pulsing sphere. The amplitude and phase of the somewhat abstract function Φ at the surface of the source, however, are not directly measurable, so we must express our equation in terms of the measurable quantities such as ξ or $\dot{\xi}$.

$$\dot{\xi} = -\frac{\partial \Phi}{\partial r} = -\frac{B}{r}\frac{2\pi}{\lambda} \sin\left[\frac{2\pi}{\lambda}(ct - r) + \alpha\right] + \frac{B}{r^2} \cos\left[\frac{2\pi}{\lambda}(ct - r) + \alpha\right]. \tag{3-17}$$

Example. Let us suppose that the pulsing sphere has a radius of 10 cm and that the maximum radial velocity of its surface is 10^{-1} cm-sec^{-1} while "pulsing" at the rate of 500 cycles-sec^{-1}. (The wavelength in air will then be 66.2 cm.) The two periodic terms on the right-hand side of Eq. (3-17) are just 90° apart in phase and are therefore equivalent to a single periodic term whose amplitude is

$$\dot{\xi}_m = \sqrt{\left(\frac{B}{r}\frac{2\pi}{\lambda}\right)^2 + \left(\frac{B}{r^2}\right)^2} = B\sqrt{\left(\frac{2\pi}{r\lambda}\right)^2 + \left(\frac{1}{r^2}\right)^2}.$$

(This follows from the discussion in Chapter 1.) The constant B may be found from this equation for the velocity amplitude, evaluated at the surface of the sphere:

$$\dot{\xi}_m \Big|_{r=10} = 10^{-1} = B\sqrt{\left[\frac{2\pi}{(10)(66.2)}\right]^2 + \left[\frac{1}{(10)^2}\right]^2},$$

$$\therefore B = 7.25 \text{ cm}^3\text{-sec}^{-1}.$$

The phase angle α may also be found, if information regarding the initial time conditions is available. Since we are interested primarily in the intensity distribution, in this problem we shall ignore the matter of phase.

We are now in a position to compute the value of the important "field" parameters for any point outside the sphere. At the point $r = 100$ cm, for instance, we have

$$\dot{\xi}_m\bigg|_{r=100} = 7.25\sqrt{\left[\frac{2\pi}{(100)(66.2)}\right]^2 + \left[\frac{1}{(100)^2}\right]^2}$$
$$= 6.87 \times 10^{-3} \text{ cm-sec}^{-1},$$

$$\xi_m\bigg|_{r=100} = \frac{\dot{\xi}_m\big|_{r=100}}{2\pi f} = \frac{6.87 \times 10^{-3}}{(2\pi)(500)} = 2.18 \times 10^{-6} \text{ cm}.$$

If $\rho_0 = 1.29 \times 10^{-3}$ gm-cm^{-3},

$$p_m\bigg|_{r=100} = \rho_0\left(\frac{\partial\Phi}{\partial t}\right)_m\bigg|_{r=100}$$
$$= \rho_0\left[\frac{B(2\pi f)}{r}\right]_{r=100}$$
$$= 1.29 \times 10^{-3}\,\frac{(7.25)(2\pi)(500)}{100}$$
$$= 0.291 \text{ dyne-cm}^{-2}.$$

$$s_m\bigg|_{r=100} (= \delta_m) = \frac{1}{c^2}\left(\frac{\partial\Phi}{\partial t}\right)_m\bigg|_{r=100}$$
$$= \frac{1}{c^2}\left[\frac{B(2\pi f)}{r}\right]_{r=100}$$
$$= \frac{1}{(3.31 \times 10^4)^2}\left[\frac{(7.25)(2\pi)(500)}{100}\right]$$
$$= 2.08 \times 10^{-7}.$$

3–8 Intensity for spherical waves. A special comment must be made regarding intensity in the case of spherical waves. If a point is located at a considerable distance from the spherical source, the second term in Eq. (3–17) is small compared with the first, since its coefficient falls off as $1/r^2$, whereas the first term diminishes as $1/r$. Therefore the phase relationship between ξ and Φ is nearly 90°. From Eq. (3–15) there is also the same relationship between s and Φ, so that p ($= \mathcal{B}s$) is then in phase with $\dot{\xi}$. Since this was precisely the situation in the case of plane waves, the plane wave expressions for intensity may also be used for spherical waves, but *only for distant points.* Nearer the source, the second term in Eq. (3–17) can *not* be neglected and no such 90° relationship between ξ and Φ exists. Therefore p and $\dot{\xi}$ are no longer in phase with each other.

This completely alters the intensity picture. In fact, much of the *instantaneous* power delivered by the source to its immediate surroundings *returns* to the source and is not radiated as real wave energy at all. Only at more distant points is the energy flow all radially outward at all instants of time. In the next section we shall refer to this rather peculiar state of affairs again, in connection with the computation of the total radiated power of any small source. The additional techniques introduced in Chapter 5 on acoustic radiation impedance will also be helpful in understanding the above phase relationships and in showing how to compute the intensity for near points.

The reader by now is beginning to wonder why a discussion of the field of sound around a "pulsing sphere" is of any practical importance, since actual sound sources — musical instruments, loudspeakers, etc. — have no such simple geometry. The answer to this question will be given in the next section, where will be explained a simple method whereby the principles just discussed may be applied to the radiation of sound by many ordinary sources.

3-9 The "strength" of a source. This concept is primarily reserved for those sources whose physical dimensions are *small* compared with the wavelength. We shall first apply the idea to a small pulsing sphere whose radius $r_0 \ll \lambda$. The expression for the particle velocity, Eq. (3-17), evaluated at the surface of this sphere then reduces to the second term only:

$$\dot{\xi} = \frac{B}{r_0^2} \cos\left[\frac{2\pi}{\lambda}(ct - r) + \alpha\right]. \tag{3-18}$$

We can now define the "strength" of *any* small source on the surface of which, at any instant of time, all points are moving with the same velocity, as *the product of the surface area times the instantaneous velocity of the surface.* The maximum value of this strength we shall call B' = (surface area of the source)$(\dot{\xi}_m)$. In the case of the small pulsing sphere under discussion, B' becomes

$$B' = 4\pi(r_0)^2 \frac{B}{r_0^2} = 4\pi B. \tag{3-19}$$

Equation (3-16) may then be rewritten in terms of the maximum strength of the source, B':

$$\Phi = \frac{B'}{4\pi r} \cos\left[\frac{2\pi}{\lambda}(ct - r) + \alpha\right]. \tag{3-20}$$

Note again that this expression is correct only when the spherical source is such that r_0 is *small* compared with λ.

For our idealized pulsing sphere, the replacement of B by B' has no particular advantage. It is for the small *irregular* source, of *nonspherical* contour, that the concept of the strength of the source is intended. In Chapter 4 we shall discuss in some detail the matter of wave diffraction, of so much importance in the radiation and reception of sound. At this point we shall simply make use of general ideas, usually presented in even an elementary discussion of light — ideas which are equally applicable to sound propagation. It will be recalled that a parabolic light reflector, whose dimensions are certainly very large compared with the wavelength of light, and having a small source at its focal point, emits a beam of light whose wave front is plane and whose cross section remains quite constant (except for the scattering due to dust particles, etc.). However, when we reduce the initial cross section of the beam (for instance, by placing in its path an opaque plate with a very small hole), we find light diverging from the hole in directions far removed from the normal. In fact, when the dimensions of the hole approach the wavelength of light and become even smaller, there is an almost *hemispherical* distribution of energy on the far side of the obstructing plate. In sound, the analogy to the hole in the plate is completely realized by a rigid acoustic piston set into a rigid wall whose plane coincides with that of the piston face. (The piston is assumed to radiate from the front face only.) The energy leaving the piston face will either have "beam" properties or will have a hemispherical distribution, depending on the ratio of piston diameter to wavelength (see Chapter 4). Moreover, if the piston is small compared with the wavelength and is radiating *without* the presence of the wall (or *baffle*, as it is usually called), the energy will have practically *spherical* distribution properties within a short distance of the source. In Fig. 3–2 there is shown qualitatively the manner in which the wave front, originating at the source contour, gradually transforms, through the spreading effects associated with diffraction, into the spherical shape characteristic of a true pulsing sphere.

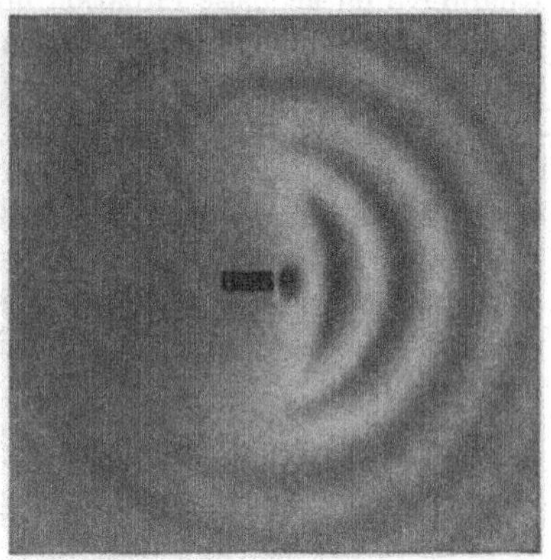

FIG. 3–2. Transformation to a spherical wave of a plane wave originating from an acoustic piston whose diameter is small compared with the wavelength.

3–10. Sources equivalent to a pulsing sphere. It is because of the effects discussed above that the concept of the "strength" of the source is a useful one. The various sound sources illustrated in Fig. 3–3 are all equivalent to a true pulsing sphere, if one considers the observable effects in the region where the wave front has become, effectively, spherical in

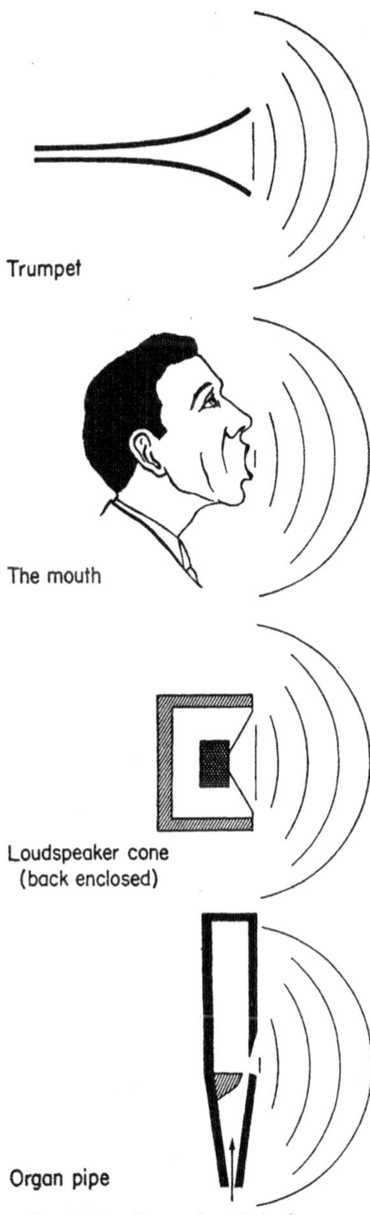

FIG. 3-3. Examples of single sources which, at low frequencies, approximate pulsing spheres. (The curved lines represent the shape of the wave front and are not necessarily one wavelength apart.)

shape. Even though the actual source is *not* a sphere, its "strength" can be computed just as readily in at least a number of cases of practical importance. The familiar loudspeaker cone, when vibrating at a low frequency, moves nearly as a unit and is a close approximation to an idealized acoustic piston whose face area corresponds to the area of the base of the speaker cone. The maximum "strength," B', of such a radiating flat disk is $\pi r_d^2 \dot{\xi}_m$, where r_d is the radius of the disk and $\dot{\xi}_m$ is the maximum velocity of the disk. A hemispherical sound source would be considered to have a strength $B' = 2\pi r_h^2 \dot{\xi}_m$, r_h being the radius of the hemisphere. A rectangular plate (radiating from one face only) is a source of possible "strength," $ab\dot{\xi}_m$, a and b being its two dimensions, etc. In every case the particular value of B' is simply inserted directly into Eq. (3-20), the expression for the velocity potential.

Working with this equation for Φ, instead of the form given by Eq. (3-16), all the field parameters derivable from Φ may be evaluated by means of the relations demonstrated earlier in the chapter. Problems illustrating this technique will be found at the end of this chapter.

3-11 Limitations on the use of the "strength of source" concept. It is imperative to keep clearly in mind the restrictions which *must* be placed upon the above method of analysis if any kind of agreement with experiment is to be expected. These restrictions are summarized as follows:

1. All source dimensions must be small compared with the wavelength of the radiated sound.

2. The vibrating source surface must be moving as a unit if B' is to be evaluated simply as area $\times\ \xi_m$. Approximate rigidity of the vibrating surface is often sufficient if approximate results are acceptable.

3. Equation (3–20) should not be used for points which are too close to the source. "Too close" means within a distance of several wavelengths.

3–12 Extension of the "strength of source" concept. The restrictions listed above are of major importance, and all three are difficult to realize in practice. If the wavelength is necessarily long to satisfy condition 1, one is also restricted to a consideration of the situation at correspondingly large distances from the source in order to satisfy condition 3. For a considerable fraction of the audible band of frequencies, in the middle and in the higher registers, the surface dimensions of musical instruments, far from being *small* compared with the corresponding wavelength, have a magnitude several times that of the wavelength. In addition, the surfaces of very few actual sound sources move as a unit, which is necessary in order to satisfy condition 2 above. In some of these practical cases, where the surface geometry is not too complicated, it is still possible to use the "strength of source" concept on a differential scale.

For such cases the surface area of the source is broken up into an infinite number of infinitesimal sound-generating surfaces and the differential maximum strength of each of the infinitesimal sources may be expressed, using the known velocity at the source. The evaluation of the total function Φ at some point in the medium around the source is then accomplished by summing up the various contributions, $d\Phi$, arising from the various differential sources of typical maximum strength, dB', distributed over the finite radiating surface. In Chapter 5 we shall outline the mathematics of this process in more detail for the case of an acoustic piston surrounded by an infinite plane baffle. Few actual sound sources, however, are susceptible to an analysis of this type, because of the mathematical difficulties.

3–13 The double source. So far in our discussion of the production of sound waves we have assumed that all contributions to the disturbance originate at the surface of the vibrating source with the *same phase*. This assumption has been made for the production of plane waves from an acoustic piston and also for the case of spherical waves, whether originating from an idealized pulsing sphere or from sources of nonspherical contour which may be treated as equivalent to a pulsing sphere. All such sources are called "single" sources.

There is a large and important group of actual sound sources which may be called "double" sources, in the sense that while one portion of the vibrating surface is giving rise to sound waves of one particular phase, another part of the surface is originating waves which are exactly 180° out of phase with the first set. Far from being an uncommon state of affairs, most actual sources of sound are of this double type. A flat plate, surrounded by air on all sides and vibrating in a direction transverse to the plane of its surface, is obviously such a double source. As seen in Fig. 3-4, as the plate moves to the right, the right-hand surface will be starting a compression, a phase where p is positive, at the same time that the left-hand surface is producing a rarefaction, a phase where p is negative. This 180° phase relationship will also hold for the other parameters in the wave, such as s, δ, etc. It will be remembered that in discussing radiation from an acoustic piston it was assumed that sound waves were produced from the front face only. In practice this could only be realized by completely enclosing the back face of the piston, so that no radiation could occur. Without such an enclosure, an acoustic piston is inevitably a double source.

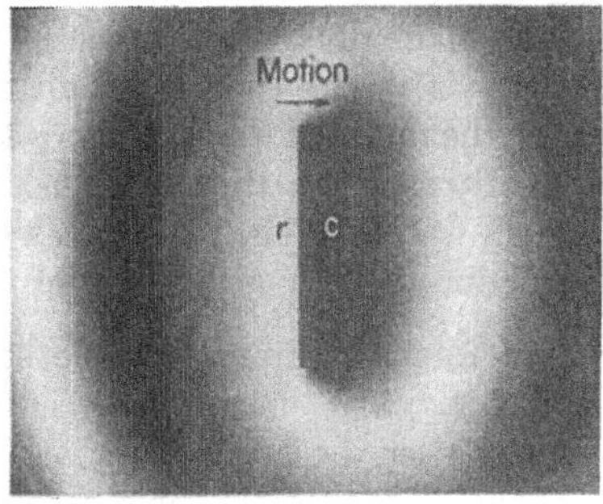

Fig. 3-4. Action of a "double" source.

3-14 Examples of the double source. There are numerous examples of the double source. For all the stringed instruments, the transverse motion of the vibrating string sets up two out-of-phase disturbances originating on opposite sides of the strings, the two contributions to the radiated energy originating in this case at two almost identical positions in space. A tuning fork is really a *double* double source, since each prong sets up its own pair of wave disturbances. The cone of a loudspeaker, vibrating freely in air, without enclosure or baffle of any kind, is another example of the double source.

It is quite reasonable to expect any such double source to be a poor radiator of sound, since two wave trains of opposing nature are set up. This is easily demonstrated with a tuning fork, a good example of a double source. When the fork is held in the hand after being struck, the sound produced will be nearly inaudible, even though the fork is vibrating vigorously. This will be particularly true if the frequency is low, say in the neighborhood of several hundred cycles -sec^{-1}. This phenomenon of low radiation efficiency is a consequence of the highly localized nature of the disturbance of the surrounding fluid, particularly when the two out-of-phase waves are generated at points in space whose distance apart is small compared with the wavelength.

The flow lines around a vibrating plate or bar are shown in Fig. 3–5. If the motion back and forth is slow, corresponding to a low frequency and therefore to a long wavelength, the medium tends to circulate from the side of the plate or bar where the pressure is higher than normal to the other side, where the pressure is below normal. As a consequence of this local compensation for the pressure differences, very little variation in pressure is observed a short distance away. In other words, the radiated wave energy is small. The effect may be clearly seen if a stick is moved slowly back and forth through water. Very few ripples (the evidence of wave production) will be visible around the stick.

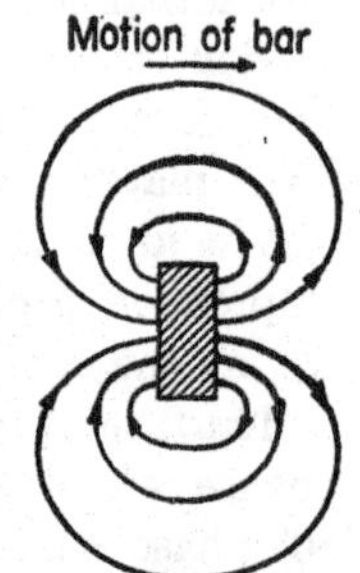

FIG. 3–5. Flow lines around a double source.

3–15 Radiation from a double source as a function of frequency. The effects described above are most pronounced when the frequency is low. For any such vibrating body acting as a double source, the energy radiated in the form of waves increases rapidly with increase of frequency. It is primarily the time factor which accounts for this frequency dependence. As the vibration frequency is raised, it becomes increasingly difficult, because of the finite velocity of propagation for any pressure disturbance, for the local flow to equalize the pressures on the two sides of the plate in so short a time interval. As a result, larger variations in pressure occur near each component of the double source and two wave trains of increasing amplitude and intensity are set up as the frequency is increased.

3–16 Quantitative analysis of the double source. In discussing the radiation properties of "single" sources of sound waves it was necessary to introduce the concept of the strength of the source, an artifice whose use minimizes to some extent the mathematical difficulties associated with the source contour. Even so, there are serious restrictions placed upon the equations obtained with this approach. The difficulties are even greater when a careful analysis of the radiation from a double source is attempted. Much of the analytical information in acoustics may be obtained by consideration of the behavior of a highly idealized model, even though actual sources are but crude approximations to this model. Since the results of the present analysis indicate only the general behavior of double sources, we shall present only an outline of the mathematics.

Two small pulsing spheres of identical maximum strength B' but of opposite "polarity" are imagined close to each other at points O_1 and O_2 in Fig. 3–6. For any point a in the surrounding air, the total velocity potential Φ_a may be written in terms of the contributions due to the two components of the double source, that is,

$$\Phi_a = \frac{B'}{4\pi}\left\{\frac{\cos\left[\frac{2\pi}{\lambda}(ct - r_1) + \alpha\right]}{r_1} - \frac{\cos\left[\frac{2\pi}{\lambda}(ct - r_2) + \alpha\right]}{r_2}\right\}. \quad (3\text{–}21)$$

If the distance apart, d, of the two sources is small compared with the distances r_1 and r_2 to the point a, it is easy to show that the total value of Φ_a is, to a close approximation,

$$\Phi_a = \frac{B'd}{2r\lambda}\cos\theta\sin\left[\frac{2\pi}{\lambda}(ct - r) + \alpha\right], \quad (3\text{–}22)$$

where r is the mean distance from the point a to the mid-point of the line O_1O_2. It is interesting to compare this expression for the function Φ for a *double* source with Eq. (3–20), which is for a *single* source. In both cases the maximum value of Φ ($= \Phi_m$) is a function of B'/r, as would be expected. But in the case of Eq. (3–22) there are the additional factors $\cos\theta$ and d/λ. At a value of $\theta = \pi/2$, Φ_a is zero for all values of r and t; at $\theta = 0$, the coefficient of the sine expression, i.e., $\Phi_m = \dfrac{B'd}{2r\lambda}\cos\theta$, is a maximum.

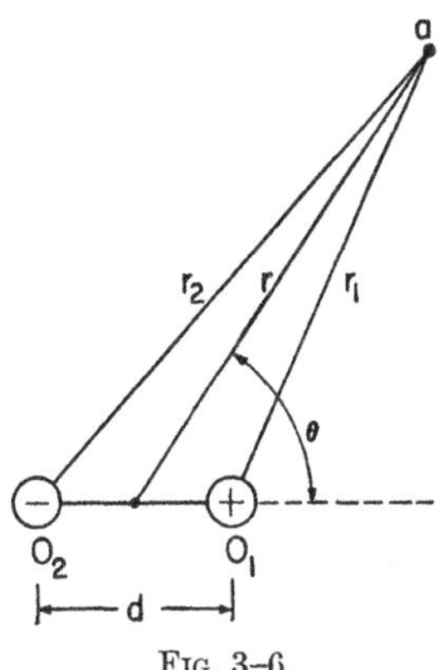

FIG. 3–6.

It will be remembered that this is the exact polar distribution of electric potential around a small *electric* dipole. (Indeed, the product $B'd$ is quite similar to the electric moment of such a dipole.) Along a line perpendicular to O_1O_2 ($\theta = \pi/2$) and passing through its mid-point, one would expect the acoustic potential to be zero, since it is the locus of points equidistant from O_1 and O_2.

The appearance of the factor d/λ is interesting and most significant from a practical point of view. As the wavelength is *increased* (or the frequency decreased), all other quantities in Eq. (3–22) remaining fixed, the maximum value of Φ *decreases*. A decrease in Φ_m means a decreased sound intensity everywhere around the double source. It is easy to see why Φ_m varies in this way. From the point of view of interference, the larger λ is in comparison to d, the distance between O_1 and O_2, the less important becomes the fact that point a in Fig. 3–6 is farther from O_2 than from O_1. As λ is increased, the relative phase of the two contributions arriving at a approaches that of the dipole components themselves, i.e., 180°, and Φ_m, therefore, approaches zero. For this reason all double sources, of whatever shape and complexity of contour, are very poor radiators of sound at *low* frequencies.

3–17 Comparison of total power radiated by different types of sources. (1) *Source of plane waves.* Consider a large, flat circular plate, acting as a single source (i.e., radiating from one side only), and vibrating so as to give rise to a steady stream of plane waves. The total average power radiated by the plate will be the product of the power per unit area and the total area, S. Making use of Eq. (2–29b), this total power U may be written in the form

$$U = 2\pi^2 \rho_0 c \xi_m^2 f^2 S = \frac{\rho_0 c}{2S} (B')^2, \tag{3–23}$$

where B' is the maximum strength of the source.

(2) *Single source of spherical waves.* This classification includes an ideal pulsing sphere and also the more practical sources whose radiation properties approximate those of a true pulsing sphere (as discussed in Section 3–10). For a pulsing sphere of radius r_0, the total instantaneous power associated with the fluctuating component of the pressure p is $p\dot{\xi}S$, where S is the area of the sphere and $\dot{\xi}$ is the instantaneous velocity of its surface. Making use of the relationships for spherical waves, we may write for the instantaneous power

$$U_i = p\dot{\xi}S = \rho_0 \frac{\partial \Phi}{\partial t}\left(-\frac{\partial \Phi}{\partial r}\right)4\pi r_0^2,$$

where Φ is taken to be of the form specified by Eq. (3–20), for a spherical source. To obtain the expression for the *average* power radiated from the surface of the sphere, we multiply U_i by dt, integrate over the time, $t = T$, and divide the result by T. By this mathematical procedure, we obtain

$$U = 2\pi^3 \frac{\rho_0}{c} \xi_m^2 f^4 S^2 = \frac{\pi}{2} \frac{\rho_0}{c} f^2 (B')^2. \tag{3–24}$$

(3) *Ideal double source, or acoustic dipole.* The computation in this case is based on the rather special assumptions which lead to Eq. (3–22). This equation, it will be remembered, is valid only at distances large compared with the distance apart, d, of the two components of the dipole. Since the same total energy must flow per second across the surface of *any* sphere surrounding the dipole, we might just as well take a sphere of *large* radius, where Eq. (3–22) is valid, since this expression is somewhat simpler than the equation for Φ at nearer points. An integration process only slightly more complicated than that used in the case of a single source yields the following expression for the average radiated power:

$$U = \frac{8}{3} \pi^5 \frac{\rho_0}{c^3} d^2 \xi_m^2 f^6 S^2 = \frac{2}{3} \pi^3 \frac{\rho_0}{c^3} d^2 f^4 (B')^2, \tag{3–25}$$

where ξ_m, S, and B' refer to either component of the dipole.

An inspection of Eqs. (3–23), (3–24), and (3–25) will show the radiated power U, in the three cases, to be a function of different powers of S and of f. For plane waves, $U \propto S$, while for the single or double source of spherical waves, $U \propto S^2$. This shows the special importance, in the case of spherical waves, of using sources of large area if one is to achieve a reasonable radiation efficiency. The frequency factor f is raised to the second power for plane waves. In the case of the single source of spherical waves, however, $U \propto f^4$, while for a double source, $U \propto f^6$. The factor f^2 is to be expected because of the fundamental energy considerations associated with simple harmonic motion. The dependence of U upon a higher power of the frequency, when the energy flow possesses *spherical* divergence, will appear more reasonable when we have developed the notion of "acoustic radiation impedance"; for the present we shall simply note the fact. For the *double* source, however, we can say here that this extreme sensitivity of U to frequency changes is a direct consequence of the "local flow" that tends to accompany the vibration of any double source, a process which always reduces the radiated wave energy, especially at low frequencies.

3–18 Practical double sources. The principle of the baffle. The model of a double source that we have set up, i.e., a pair of small spheres "pulsing" periodically with a phase difference of 180°, is, of course, never encountered in practice. A vibrating rigid circular plate, whose radius r_0 is small compared with the wavelength in the emitted sound, will radiate approximately as would a pair of spheres, where the "strength" of each of the equivalent spheres is equal to $\pi r_0^2 \xi$. Since the distance apart of the dipole components is in this case related to the thickness of the plate, the radiated energy will be quite small (since d is small in Eq. (3–25)). This is essentially the situation when a cone speaker is vibrating at a low frequency, with no surrounding box or "baffle" plate of any kind. Under these conditions, where λ is much greater than r_0, very little audible sound is detected a short distance away.

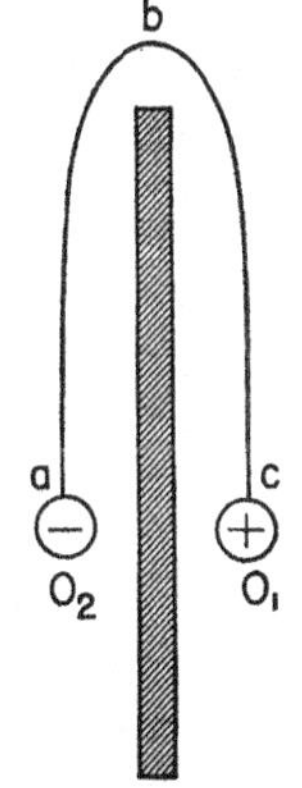

FIG. 3–7. Action of the baffle.

Most actual double sources involve, to a greater or less degree, the principle of the baffle. In Fig. 3–7 a rigid plate has been inserted between the two components of a double source, as shown. It is quite obvious that from a qualitative point of view such a baffle will seriously interfere with the local flow back and forth in the immediate vicinity of the dipole components. The length of the path over which such a flow might take place is shown in the figure to be approximately that of the line *abc*. If each of the source components is vibrating at a rapid rate, the flow will only partially equalize the pressure differences which tend to be set up by O_1 and O_2. As a result, wave generation and propagation set in and the general efficiency of energy radiation is increased.

The geometry of many musical instruments supplies a sort of self-baffling action. In Chapter 7 we shall have more to say about the physics of the violin, but to illustrate the present point it can be said that most of the actual sound energy coming from a violin is not radiated directly from the strings, but from the larger areas of the wooden body of the instrument, which are set into sympathetic vibration. If one side of the instrument is considered to be a thin flat plate (it is not truly flat, of course), the violin can be pictured as a collection of small acoustic dipoles, somewhat as in Fig. 3–8, where the + and − signs are used to represent the 180° phase relationship. As the plate vibrates transversely to the plane of its surfaces, the plate as a whole acts a good deal like the baffle plate of Fig. 3–7 for each one of the small acoustic dipoles. Since the radiation of the whole plate is the sum effect of the radiation of its separate dipoles, any of the larger areas of the instrument then becomes a reasonably efficient radiator.

In a later chapter we shall return to the subject of the baffle in a discussion of the general design features of wide-range loudspeaker systems. In addition, some mention will be made of a speaker enclosure, sometimes known as a "phase inverter." For the lower frequencies, such an enclosure effectively transforms the speaker cone, intrinsically a double source, into what is in effect a single source. This is brought about by coupling the rear side of the speaker cone to a resonator, the radiation from which is virtually in phase with the disturbance originating from the front surface of the cone. In this way effective radiation may be obtained down to the very lowest audible frequencies.

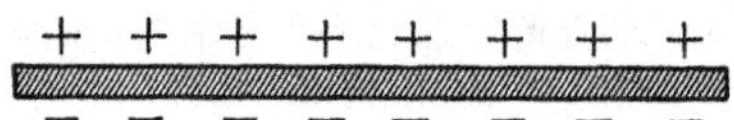

FIG. 3–8. A vibrating plate considered as an array of acoustic dipoles.

PROBLEMS

1. A pulsing sphere has a radius of 1.0 mm and is pulsing in air at a frequency of 100 cycles-sec^{-1}. The maximum radial velocity of its surface is 0.1 cm-sec^{-1}. (a) Find the value of Φ as a function of r. At the position $r = 30$ meters, find (b) the maximum value of Φ, (c) the maximum value of ξ, and (d) the maximum value of p.

2. Assuming that the wave front has become virtually plane at the position $r = 30$ meters in problem 1, compute the

energy density at that point and also the intensity in cgs units.

3. A large pulsing sphere has a radius of 1 meter and is pulsing in air at the ultrasonic rate of 35,000 cycles-sec^{-1}. The maximum radial velocity of its surface is 0.1 cm-sec^{-1}. (a) Find the expression for the maximum value of Φ as a function of r. At the position $r = 2.0$ meters (measured from the center of the pulsing sphere), find (b) the maximum value of Φ, (c) the maximum value of ξ, and (d) the maximum value of p.

4. A pulsing sphere has a radius of 1.0 mm and is pulsing in air at a frequency of 100 cycles-sec^{-1}. The amplitude of motion of the surface of the sphere is 0.01 cm. (a) Find the maximum value of Φ as a function of r. At the position $r = 30$ meters, find (b) the maximum value of Φ, and (c) the maximum value of p.

5. In which of the following cases might it be correct to apply the concept of the strength of the source in order to make use of the spherical wave relations? (a) A soprano singing a note of high frequency. (b) A bass singing a note of very low frequency. (c) A loudspeaker cone of diameter 6 inches, radiating from one side only without a baffle and vibrating at a frequency of 50 cycles-sec^{-1}. (d) The radiation from the mouth of a large parabolic reflector at the focus of which is a small high frequency whistle. (Consider the mouth of the reflector as the source.)

6. Name as many musical instruments as you can, classifying each as predominantly a single or a double source.

7. An acoustic piston is radiating into air from the front face only. It has a radius of 1.0 cm and is oscillating at a frequency of 200 cycles-sec^{-1}. No baffle plate is used. The amplitude of motion of the piston face is 0.01 cm. (a) Calculate the strength of the source. (b) Find the maximum value of Φ at a distance 20 m from the center of the piston face. (c) Compute the maximum value of p at the same point.

8. An acoustic piston of radius R, moving with a velocity $\dot{x}$ and radiating from the front face only, is surrounded by an infinite plane baffle lying in the plane of the piston face. The piston may be considered to have a maximum strength $2B'$, where B' is $\pi R^2(\dot{x})_m$. Explain why this is correct.

9. Making use of the statement in problem 8, compute the answers to parts (a), (b), and (c) of problem 7, if the piston face is surrounded by an infinite plane baffle.

10. A small pair of pulsing spheres constitutes an acoustic dipole. Assume a fixed value for the polar distance r such that Eq. (3–22) is valid. What is the effect upon the rms pressure at such a point (the polar angle θ also being fixed) of (a) doubling the frequency, (b) doubling the amplitude of motion of each sphere, (c) doubling the distance separating the dipole components?

11. (a) Derive an expression for the dilatation δ for spherical waves in terms of ξ and r. (b) Show that for large values of r the expression reduces to $\partial\xi/\partial r$. (Compare with the plane wave value, $\partial\xi/\partial x$.)

CHAPTER 4

INTERFERENCE PATTERNS. DIFFRACTION

4–1 Definition of interference for wave motion. Broadly speaking, *interference* may be said to occur for waves whenever two or more motions are simultaneously impressed upon a particle or a single set of particles of the medium. In this sense, any complex wave consisting of two or more simple harmonic components, of the same or of different frequencies, might be described as an interference phenomenon. As the term is usually applied, however, interference refers primarily to combination effects associated with waves of the *same* frequency originating from different sources or from different areas of the same source. We have already considered one such effect in the phenomenon of beats, Chapter 1. The subject of stationary waves, which will be discussed in Chapter 7, is a most important example of wave interference. Throughout the remainder of this book we shall have frequent occasion to refer to interference in connection with almost every aspect of acoustics.

4–2 Diffraction. In this chapter we shall be concerned primarily with the interference effects classified under the subject of *diffraction.* The particular interference patterns set up in the medium whenever waves originate from sources whose dimensions are of the *same order of magnitude* as the wavelength, or whenever waves stream past obstructions of any sort, are, in general, called *diffraction patterns.*

4–3 Diffraction in acoustics and in light. The theory of wave interference, and of diffraction in particular, may be discussed with perfect generality for *any* type of wave — longitudinal or transverse, mechanical or electromagnetic. A comparison of the interference effects in acoustics with those occurring in light is therefore logical. The subject of diffraction is of much greater practical importance in acoustics than it is in light. The student of elementary physics may remember that particular care was needed to demonstrate even the *existence* of diffraction in light. Pinholes, slits, diffraction gratings — these are part of the paraphernalia of the scientist and the laboratory, not of everyday living and experience. In acoustics no such special equipment is required.

When a listener is directly in front of a radio loudspeaker, the music sounds "brilliant," and speech is crisp and easy to understand. At positions off to one side, both music and speech are somewhat muffled and are lacking in the important higher frequencies. This is an effect associated largely with the distribution of the sound diffraction pattern (although reflection by the walls of the room also plays a part). Similarly, the phethe phase of what arrives from the lowest area element with what is contributed by elements higher up, there is plainly a progressive phase lag, from the bottom to the top. This phase relationship will always exist, regardless of changes in the wave disturbance occurring at the slit.

4-7 Vector method of determining the acoustic pressure at point *a*. In Fig. 4-2a, a series of small vectors of equal length are laid off end to end, each one representing the maximum pressure contribution, Δp_m, at point a due to each of the area elements of the slit opening. The first vector, at the bottom, represents the effect due to the element next to the lower slit edge. Each successive vector above the first is rotated by the same small angle, to represent the progressive phase shift at point a due to the greater and greater distance of travel. The *total* pressure effect at a may be found (using the methods discussed in Chapter 1) by drawing in the vector p_m, which represents the vector sum of the separate contributions. The angle γ is the phase lag between the contribution from the topmost area element as compared with that from the first and lowest one, and is an angle which will enter in an important way into the analysis of this problem. The angle θ is the phase lag between the resultant vector p_m and the first small contribution. (Note that this use of vectors to indicate magnitudes and phases does not imply any particular *spatial* direction on the part of p. Acoustic pressure is a scalar.)

It should be quite obvious that the correct way to sum up the various contributions from different parts of the slit area will be to let the area elements shrink in width until they become true differentials. The vector polygon of Fig. 4-2a will then become an arc of a circle (Fig. 4-3), where the resultant pressure p_m constitutes the subtending chord. The length of the arc, labeled $(p_m)_0$ in the diagram, is the graphical equivalent of adding the separate small contributions, *assuming no phase differences among them* (Fig. 4-2b).

At the start of this analysis one of the simplifying assumptions was that the typical point, a, lies upon a circle whose center coincides with the center

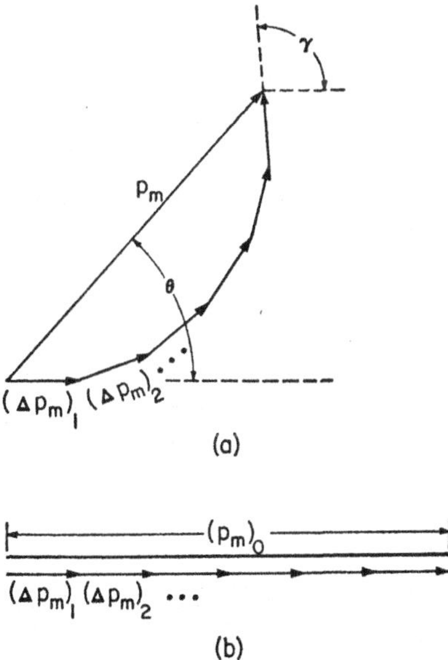

FIG. 4-2.

of the slit. The very interference effects which we are discussing depend, of course, on slight variations in the distances between point a and the different area elements of the slit, thus producing phase effects. If, however, as was assumed, the small dimension of the slit is much less than the radius of the circle, these small variations in distance will constitute a negligible *fractional* change in the total distance from point a to any point on the slit area. As a result, we may assume that for any point on the circular arc the *magnitude* of the *maximum acoustic pressure* Δp_m remains essentially constant, even though the *phase* of the *instantaneous pressure* may vary. Due to the cylindrical divergence of the wave, one would expect, for any locus of points other than a circular arc, some kind of dependence of Δp_m upon the distance from the slit (*not* the same law as for spherical divergence, however).

By holding the distance virtually constant, we will avoid this additional complication. Our primary investigation will be into the effect of varying the angle β, in Fig. 4–1.

4–8 Essential geometry and equations. Consider the geometry of Fig. 4–3. The following two equations should be self-evident:

$$(p_m)_0 = 2r\theta,$$

and (4–1)

$$p_m = 2r \sin \theta,$$

where $(p_m)_0$ is the total length of the arc, which in turn signifies the sum of all the small contributions, assuming no phase differences. This sum, $(p_m)_0$, is in actuality the total value of p_m to be observed at the point a_0, in the direction normal to the plate, where the angle β is zero (Fig. 4–1). For a point along this direction, all distances to the various area elements are virtually the same, and therefore the arc of the vector polygon becomes a straight line.

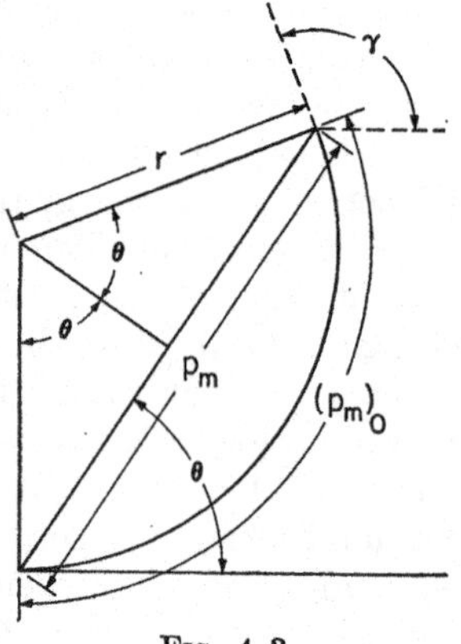

FIG. 4–3.

Eliminating r from the two equations (4–1), we may write

$$p_m = (p_m)_0 \frac{\sin \theta}{\theta}. \tag{4–2}$$

This is the fundamental equation which describes the diffraction pattern to the right of the plate. It relates the total observed pressure p_m for any point a to the value which obtains at the central point a_0, through the parameter θ, which appears in the vector diagram. The angle θ is not directly measurable, but it is simply related to the polar angle β in Fig. 4-1. From Fig. 4-3,

$$\gamma = 2\theta, \tag{4-3}$$

and from Fig. 4-1 the angle γ is

$$\gamma = \frac{d}{\lambda} 2\pi, \tag{4-4}$$

where d is the amount by which the path am' exceeds the path am and λ is the wavelength. In addition, from Fig. 4-1,

$$\sin \beta = \frac{d}{b}. \tag{4-5}$$

Therefore, combining Eqs. (4-3), (4-4), and (4-5), we have

$$\sin \beta = \frac{1}{\pi} \frac{\lambda}{b} \theta. \tag{4-6}$$

Equations (4-2) and (4-6) are the important ones for the determination of the complete nature of the slit pattern. Instead of introducing (4-6) into (4-2) to eliminate the parameter θ, it is more convenient to keep both equations, using (4-2) as a sort of "universal" relation, independent of λ or d, and making use of (4-6) whenever we wish to describe the actual pattern when confronted with a particular value of the wavelength and the slit width.

4-9 The variation of p_m with θ. The graph of p_m vs θ (the broken line of Fig. 4-4) is a curve having alternate algebraic maxima and minima above and below the line $p_m = 0$. When $\theta = 0$, $p_m = (p_m)_0$. Between

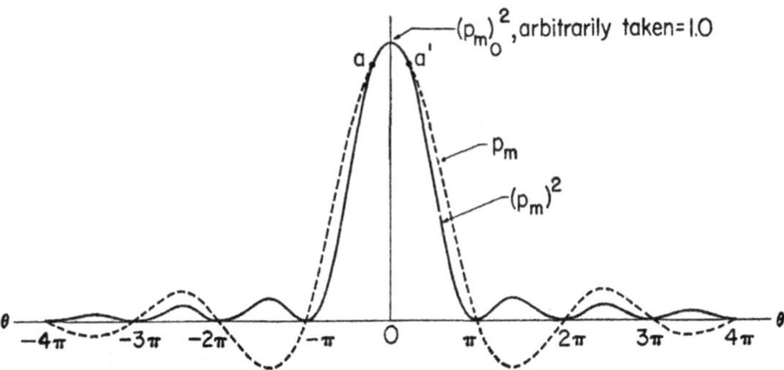

FIG. 4-4. The quantities p_m and $(p_m)^2$ plotted as a function of θ, from the equation $p_m = (p_m)_0 \sin \theta/\theta$.

each successive maximum and its adjacent algebraic minimum lies a point where p_m is zero. These latter points occur when $\theta = n\pi$, where n is any integer. To find the values of θ for the maximum and minimum values of p_m, it is necessary to find $dp_m/d\theta$ and set this equal to zero. The necessary condition is found to be that $\theta = \tan\theta$. The first few roots of this equation are $\theta = 0, 1.43\pi, 2.46\pi, 3.47\pi, \ldots$, which approximate, after the first root, to $\frac{3}{2}\pi$, $\frac{5}{2}\pi$, $\frac{7}{2}\pi$, etc. Whether a root gives a maximum or a minimum for p_m is of no importance as far as sound *intensity* is concerned since, in general, the intensity goes up as the square of the pressure and is therefore always positive whether p_m is positive or negative. The *intensity maxima* are therefore approximately halfway between the zero or null points.

We may assume that the intensity I is proportional to $(p_m)^2$. The graph of $(p_m)^2$ vs θ is given in Fig. 4–4 for the first few intensity maxima symmetrically located on either side of the central maximum, which is by far the greatest. The heights of the first few secondary maxima, expressed as fractions of the central intensity maximum, are given in Table 4–1.

TABLE 4–1

	Intensity
Central maximum	1.0
1st subsidiary maximum	0.047
2nd subsidiary maximum	0.016
3rd subsidiary maximum	0.0083
4th subsidiary maximum	0.0050

4–10 The variation of $(p_m)^2$ with the polar angle β. To translate the graph of Fig. 4–4 into its equivalent in terms of the angle β, it is necessary only to make use of Eq. (4–6), relating β to θ. For a given value of λ and b the coefficient on the right of (4–6) is a constant, so that $\sin\beta \propto \theta$. The plot of $(p_m)^2$ vs β will look a good deal like Fig. 4–4, since β will increase with θ, although not linearly. In general, however, the *spread* or *compression* of the pattern along the β axis will depend markedly on the *ratio* of λ to b. Any given feature of the curve of Fig. 4–4 will occur at large values of β when λ/b is a large number and, conversely, at small values of β when λ/b is a small number.

4–11 Representation of intensity distribution on a polar graph. A convenient way to picture the spread of wave energy from a slit or from an opening of any other shape is to make use of a polar graph. Figure 4–5 is such a polar plot for a slit in the case where λ is one-half of b, the slit width. In this diagram the radius vector, drawn from the mean position

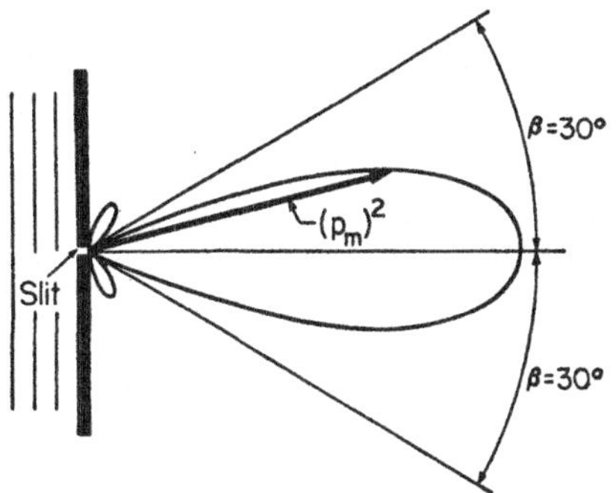

FIG. 4-5. Polar distribution pattern for a slit width b equal to twice the wavelength.

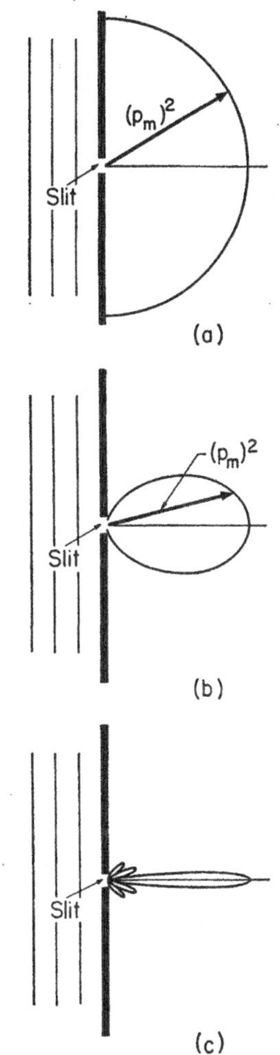

FIG. 4-6. Polar distribution of sound intensity for a slit width b (a) much less than λ, (b) equal to λ, (c) much greater than λ.

of the slit to some point on the curve, has a length proportional to the observed intensity in the direction β. For the vector drawn, the intensity is about $\frac{3}{4}$ that observed at the peak of the central maximum. A little study will show that this type of plot may be used to indicate all the features of a cartesian plot of intensity vs β, and at the same time it gives a direct spatial picture of the distribution. In general, there will be a series of "lobes" of decreasing prominence at increasing distances from the central lobe (or maximum). Between lobes there is a direction along which the radius vector has zero length, corresponding to the null points in Fig. 4-4.

In Fig. 4-6 three polar distribution graphs for different values of the important ratio λ/b have been drawn to approximate shape. In Fig. 4-6a, λ is very large compared with b. An inspection of Eq. (4-6), relating β to θ, will show that the right-hand side can be no greater than 1, since the largest possible value for β is obviously $\pm\pi/2$. If the ratio λ/b is large, $\sin\beta$ will become equal to unity with a very small value of θ. This means that for any value of β between $\pm\pi/2$

we shall be dealing with practically constant ordinates in Fig. 4–4, ranging from some such point as a to point a'. The polar intensity graph will therefore be almost a circle, as shown. For the distribution in the plane of the paper, this behavior is essentially that of a point source.

If, on the other hand, λ is very small compared with b, so that the ratio λ/b is a very small number, there will be quite a different pattern. In this case β will increase very slowly with an increase in θ and the pattern will be highly compressed around the normal direction. In this kind of distribution (see Fig. 4–6c) there is virtually a "beam" of waves leaving the slit opening, with practically no energy spread away from the forward direction. This is the usual situation for light, where the ratio λ/b is almost always a very small number.

Between the two extremes just discussed, when λ is of the same order of magnitude as b, the polar diffraction pattern will be of the type shown in Fig. 4–5. If $\lambda = b$, there will be just the one central lobe, since the first null point will occur at exactly $\beta = \pi/2$, as shown in Fig. 4–6b.

4–12 General significance of the diffraction pattern for a single slit. The detailed analysis we have just made is admittedly an approximate one. The assumptions are essentially those characteristic of Fraunhofer diffraction in optics, where the rays leaving different parts of the slit are taken to be parallel. In optics, the use of a collecting lens to the right of the plate makes it possible to bring such a bundle of parallel rays to a focus at a position quite near the plate. Lenses are rarely used in acoustics, so we must consider the above results to be valid only for distant points. The general conclusions reached are nevertheless useful in predicting the type of radiation to be expected from a slit and from openings of other shapes as well. The same ratio of wavelength to dimension is the significant factor for any type of opening.

4–13 Openings of other shapes. The short rectangle or the square. If the long dimension of the slit is reduced so that it becomes comparable to the wavelength in the disturbance, one finds, as would be expected, a diffraction pattern in the *horizontal* plane (the plane including the long dimension of the slit) as well as in the vertical plane. The spread of the maximum and minimum positions, however, depends upon fundamentally the same relations as discussed above. Patterns occur in both planes, with considerable modification of each because of their interdependence. With a rectangle having equal sides (a square), the patterns along the vertical and along the horizontal are identical, as one would expect.

In this connection, a further comment may be made on the assumptions in the single slit analysis. As will be remembered, the slit was taken to be very long and from the preceding paragraph it should be clear that in the plane including a rectangular dimension long compared with the wavelength the diffraction pattern will be beamlike, with practically uniform intensities over the cross section of the beam. Hence the assumption that there is no intensity variation along a line parallel to the long dimension of the slit was justified.

The exact description of the intensity distribution in all possible directions relative to the plane of a plate with a rectangular aperture is hard to give, because of geometrical difficulties in the analysis. Fortunately, in acoustics the diffraction from a *circular* aperture is of more practical importance than from apertures of any other shape, and while the mathematics is too involved to give here, the results are simple enough to summarize and discuss.

4-14 Diffraction pattern for a circular aperture. The intensity distribution, in this case, will have circular symmetry around a line perpendicular to the plane of the aperture at its center. Airy, Verdet, and others have computed the intensity for a sinusoidal wave as a function of the angle of deviation from the normal to the plane of the aperture. Due to the symmetry in the problem, coaxial regions of maximum intensity alternate with regions of minimum intensity. The distribution may be visualized by imagining Fig. 4-5 to be rotated around its axis of symmetry. The *space surface* so produced can then be considered as an intensity plot for different directions. Figure 4-7 represents such a surface.

The directions of zero sound intensity (corresponding to the zero intensity directions for the slit) are found to be given by an equation somewhat similar to Eq. (4-2) for the slit. It will be recalled that in the case of the slit the intensity is zero when θ is equal to $n\pi$, where n is any integer. The corresponding values of the angle β are given by

$$\sin \beta = n \frac{\lambda}{b}. \tag{4-7}$$

For the circular hole, Eq. (4-7) may be written

$$\sin \beta = n' \frac{\lambda}{D}, \tag{4-8}$$

where D is the diameter of the hole and where n', instead of being an integer, as for the slit, is now a number which must be computed for each successive point of zero intensity. The values of n' for the first three null points are 1.22, 2.23, and 3.24. The secondary intensity maxima at the angles between those for the null points are smaller in relation to the central maxi-

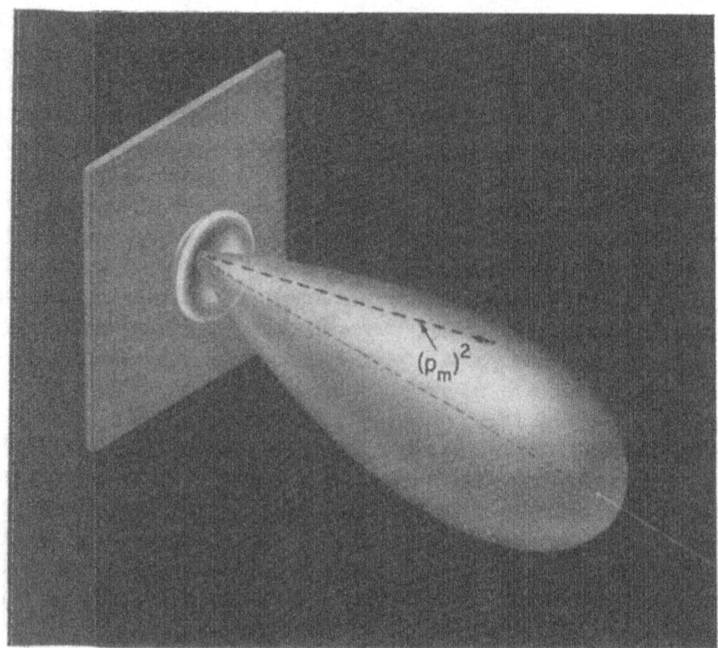

FIG. 4–7. Space diagram of intensity distribution in front of a plate with a circular hole. A vector drawn from the center of the hole to any point on the surface has a length proportional to the sound intensity in that direction, as observed at a large fixed distance from the hole.

mum than in the case of the slit pattern. Table 4–2 lists the relative values for the first few such maxima.

TABLE 4–2

	Relative intensity	*Intensity in db referred to central maximum*
Central maximum	1.0	0.0
1st subsidiary maximum	0.017	−17.7
2nd subsidiary maximum	0.0041	−23.9
3rd subsidiary maximum	0.0016	−28.0
4th subsidiary maximum	0.00078	−31.1

Since most of the energy is radiated within the boundaries of the central lobe, we may say, making use of Eq. (4–8), that most of the energy lies within a cone, half of whose plane vertex angle is given by $\sin \beta = 1.22\lambda/D$.

Conclusions regarding the behavior of the diffraction pattern with variations in wavelength and in hole size are quite similar to those for the slit pattern. When the hole is very small compared with λ, the intensity distribution to the right of the plate will be very nearly spherical. On the other hand, when the diameter of the aperture is very large compared with λ, the energy will proceed as a narrow beam whose cross-sectional area approaches that of the hole. When the hole diameter is approximately twice the wavelength, there will be a distribution such as is represented by Fig. 4-7.

4-15 Practical examples of the diffraction pattern for a circular aperture. In acoustics the usual aim is to spread or *diffuse* the sound over as large a solid angle as possible. A speaker standing in front of a large audience desires to be heard clearly by everyone, those sitting at the sides as well as persons directly in front of the speaker. The various observed relative intensities of the instruments of an orchestra should, ideally, be independent of the particular position of the listener with respect to each of the instruments. In short, the diffraction pattern of the speaker, the singer, or the musical instrument should be as broad as possible. The shapes of most sound-emitting sources are approximately circular; some, like the instruments of the horn family, the drum, and the radio loudspeaker cone, are exactly circular. A glance at Table 4-2 will show that for the circular hole diffraction pattern practically all of the energy is concentrated in the central maximum. It is therefore necessary, for wide angle radiation, that the first null point ($\theta = \pi$) correspond to a value of β close to $\pi/2$. This implies, from Eq. (4-8), that the effective diameter of such a circular source can be no greater than 1.22λ. Considering an average source to have a radiating area whose diameter is of the order of 6 inches, this means that there will be good polar spread for frequencies of say 2000 cycles and lower, but for higher frequencies the radiation pattern will take on more and more of a directional nature.

Examples of such diffraction effects are numerous and commonplace. From a position behind or well to one side of a speaker, speech may be difficult to understand because the higher frequencies, so essential to speech comprehension, are propagated forward in a beamlike manner. The lower frequencies, which incidentally carry a good part of the energy in speech, diffract readily to the side and in the case of the very lowest frequencies, where the mouth and head dimensions are small compared with the wavelength, low frequencies will even diffract to the rear. Similarly, the higher frequencies are always more apparent in the sound from a radio loudspeaker when one is sitting directly in front of the cone — in the "beam." The low frequencies, on the other hand, are radiated with practically spherical divergence and are heard equally well from any position.

For reasons of efficiency, it is desirable in some applications to confine the sound radiation to a relatively small solid angle. It is highly wasteful of acoustical power to spread it in all directions when the listeners are concentrated within one special area, as for instance at public meetings, outdoor musical performances, and the like. The spread due to diffraction may be minimized by several means. If no electrical equipment is to be used, large curved reflectors may be erected behind the platform, to return the backward radiation to the audience. Factors associated with diffraction are also involved and since such a device is effectively a sound source of large area, it will confine the reflected energy to a beam in the desired direction. When electrical reproducing equipment is employed it is customary to use horns, which are coupled to the loudspeaker units. Such horns should have mouths of large size.

4–16 Multiple slits and openings. Books on geometrical optics commonly consider the two-slit diffraction pattern, the multiple-slit problem (i.e., the diffraction grating), and also the case of two adjacent circular apertures. This latter consideration is essential to the understanding of optical resolving power, so important in the design of optical instruments. For certain special aspects of applied acoustics these cases are of some importance and full discussion of them will be found in any good book on optics. In Chapter 11 we shall have occasion to refer to the acoustic analog of the problem of two adjacent apertures in connection with the use of dual speakers in wide-range reproducing systems. Such speakers are often so placed that their diffraction patterns overlap, a matter of some interest in the total radiation distribution.

We might mention at this point that an acoustic diffraction grating may be constructed so as to operate on exactly the same principle as the optical grating. In the acoustic grating, the opaque regions may be wooden slats separated by air spaces of an inch or so. A plane sound wave, "sprayed" against such an array, may be analyzed for its frequency components just as light is analyzed with an optical grating. The resolution will be rather poor, however, since it is difficult to use many slits, due to considerations of size (resolution improves with the number of slits).

4–17 Diffraction effects around the edges of obstructions. In optics we are accustomed to the sharp geometrical shadows cast by obstructions opaque to light. The diffraction patterns that occur in the neighborhood of the shadow boundary are confined to a very small angular region and are ordinarily unnoticed by the eye. In acoustics the same general phenomenon will occur, but because of the much longer wavelength the angular spread of the diffraction pattern may be quite large and the effects are therefore of much greater importance. As will be seen, sound waves of some frequencies cast acoustical "shadows" nearly as sharp as the optical ones. For other frequencies, however, such shadows are almost nonexistent because of energy which flows around the edge of the obstruction, well into the so-called shadow region.

An analysis (due to Fresnel) will follow in the next section. Like the Fraunhofer diffraction analysis for the single slit, it is somewhat of an approximation, but it will yield useful information nevertheless. The results may be used for points somewhat nearer the obstruction than was possible for the slit, since in Fresnel's construction lines drawn from the exposed wave front to the point of observation are not assumed to be parallel.

4-18 Fresnel laminar zones. In Fig. 4-8 the line bb' lies parallel to the wave front of a plane wave, advancing from left to right. Consider the excess pressure p_F existing in a layer of air located along this line, given by the equation

$$p_F = (p_m)_F \cos 2\pi \frac{c}{\lambda} t, \qquad (4\text{-}9)$$

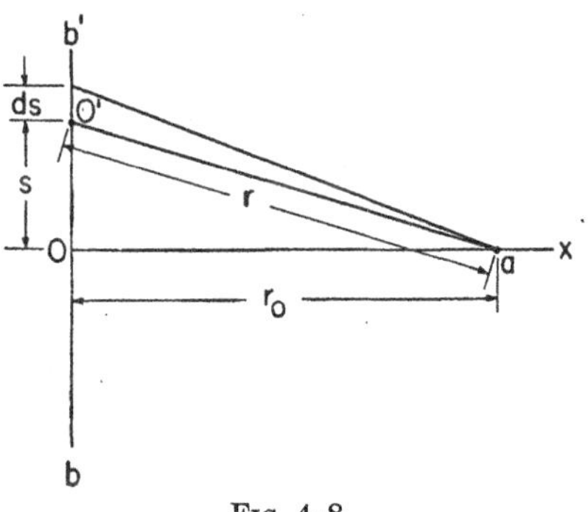

FIG. 4-8.

where $(p_m)_F$ is the maximum value of p_F. The problem will be to compute the total effect at a more distant point a, due to all portions of the wave front at bb', the latter to be considered as made up of an infinite number of infinitesimal areas in the nature of thin laminar strips running perpendicular to the plane of the paper. The effect at a, due to the motion of a strip located at a height s above point O, and of width ds, will be proportional to the maximum value of the pressure $(p_m)_F$ at bb' and also to the width ds of the strip being considered. The contribution to the instantaneous acoustic pressure at a, in terms of the wave equation, may then be written

$$dp_a \propto (p_m)_F \cos \frac{2\pi}{\lambda} (ct - r)\, ds, \qquad (4\text{-}10)$$

r being the distance between the infinitesimal strip and the point. (It should be noted, from the form of this expression, that although variations in the distance r are sufficient to affect the *phase* of arrival at point a, they are not so great as to seriously affect the maximum value of the pressure due to any one strip, and consequently may be neglected. Otherwise there would have to be a factor in the nature of an inverse function of r in the proportionality. We are also taking no account of the so-called "obliquity" factor.*)

If r_0 is the perpendicular distance from the point to the plane of the wave front, we see from the figure that

$$r = (r_0^2 + s^2)^{\frac{1}{2}}.$$

Expanding the right-hand side by the binomial theorem and retaining only the first two terms (an approximation, therefore), we obtain

$$r = r_0 + \frac{s^2}{2r_0}. \tag{4-11}$$

If this relation is introduced into Eq. (4–10), we have

$$dp_a \propto (p_m)_F \cos \frac{2\pi}{\lambda}\left[ct - \left(r_0 + \frac{s^2}{2r_0}\right)\right] ds,$$

which may be rewritten

$$dp_a \propto (p_m)_F \cos \left[\frac{2\pi}{\lambda}(ct - r_0) - \frac{\pi s^2}{\lambda r_0}\right] ds. \tag{4-12}$$

If we now expand Eq. (4–12) as the cosine of the difference between two angles, we have

$$dp_a \propto (p_m)_F \left[\cos \frac{2\pi}{\lambda}(ct - r_0) \cos \frac{\pi s^2}{\lambda r_0} + \sin \frac{2\pi}{\lambda}(ct - r_0) \sin \frac{\pi s^2}{\lambda r_0}\right] ds.$$

It is this expression for dp_a that must be integrated with respect to s over the exposed portion of the wave front, if we are to compute the instantaneous total pressure effect at point a:

$$p_a = \int dp_a \propto (p_m)_F \left[\cos \frac{2\pi}{\lambda}(ct - r_0) \int \cos \frac{\pi s^2}{\lambda r_0}\, ds + \sin \frac{2\pi}{\lambda}(ct - r_0) \int \sin \frac{\pi s^2}{\lambda r_0}\, ds\right]. \tag{4-13}$$

At this point in the analysis a substitution is convenient. Let

$$\left.\begin{aligned} \int \cos \frac{\pi s^2}{\lambda r_0}\, ds &= N \cos \alpha \\ \text{and} \qquad \int \sin \frac{\pi s^2}{\lambda r_0}\, ds &= N \sin \alpha. \end{aligned}\right\} \tag{4-14}$$

If the above relations are introduced into Eq. (4–13), the latter may be rewritten in the form

* See Jenkins and White, *Fundamentals of Optics*, 1950, p. 348.

$$p_a \propto N(p_m)_F \cos\left[\frac{2\pi}{\lambda}(ct - r_0) - \alpha\right]. \tag{4–15}$$

In our examination of diffraction patterns of the "edge" type, we shall be primarily concerned with the effect of cutting out certain portions of the wave front at bb' in Fig. 4–8. This means that we shall be interested in the *integrals* appearing in Eqs. (4–14), since they are directly related to the quantity N in Eq. (4–15). The maximum value of p_a is clearly proportional to N. Squaring both sides of Eqs. (4–14) and adding, we find that

$$N^2 = \left(\int \cos\frac{\pi s^2}{\lambda r_0}\,ds\right)^2 + \left(\int \sin\frac{\pi s^2}{\lambda r_0}\,ds\right)^2. \tag{4–16}$$

Again assuming the intensity to be proportional to the square of the maximum pressure p_m, the intensity at point a will be proportional to N^2. The integrals on the right-hand side of Eq. (4–16) enable us to compute N^2, once the proper limits for s are specified, for any particular exposure of the wave front.

4–19 The Fresnel integrals. The spiral of Cornu. In order to put the integrals in Eq. (4–16) in a form having more general usefulness, it is usual to change the variable from s to v, where

$$s = \sqrt{\frac{\lambda r_0}{2}}\,v. \tag{4–17}$$

Equation (4–16) then becomes

$$N^2 = \frac{\lambda r_0}{2}\left[\left(\int \cos\frac{\pi v^2}{2}\,dv\right)^2 + \left(\int \sin\frac{\pi v^2}{2}\,dv\right)^2\right]. \tag{4–18}$$

The integrals in the form given by Eq. (4–18) are called *Fresnel's integrals.* (Note that the coefficient $\lambda r_0/2$ is not a function of the degree of exposure of the surface of the wave front, so that in the new form N^2 will still be proportional to the sum of the squares of the two integrals.) The particular value of each of the Fresnel integrals, for the limits 0 to v, have been computed for various numerical values of the parameter v (see Appendix III). A graphical plot of the results is known as *Cornu's spiral* and is given in Fig. 4–9. In this graph

$$x = \int_0^v \cos\frac{\pi v^2}{2}\,dv$$

and

$$y = \int_0^v \sin\frac{\pi v^2}{2}\,dv. \tag{4–19}$$

Points on the curve correspond to specific values of x and y, as given by Eqs. (4–19), using in each case a particular numerical value of the limit, v.

There are two symmetrical parts to the spiral, corresponding to plus and minus signs for v (and hence for the coordinate s on the wave front). For very large values of $\pm v$ (or $\pm s$) the curve approaches the points e_1 and e_2 in the diagram, as limiting positions of points x, y. The numerical values of the abscissas and ordinates for these points are 0.5 in each case. The last parts of the spiral are not shown, since v must be infinite for the curve to reach these points.

The Cornu spiral may be used to predict the intensity at a further point, due to the exposure of any given portion of the wave front. As shown in the preceding section, the intensity $I \propto N^2 = (x^2 + y^2)$, where x and y represent the two Fresnel integrals. Suppose that a portion of the wave front is exposed, extending from the coordinate s_1 to the coordinate s_2. The corresponding values of v will be v_1 and v_2, the exact relationship between the v's and the s's being stated in Eq. (4–17). We may then write

$$x = \int_{v_1}^{v_2} \cos \frac{\pi v^2}{2}\, dv$$

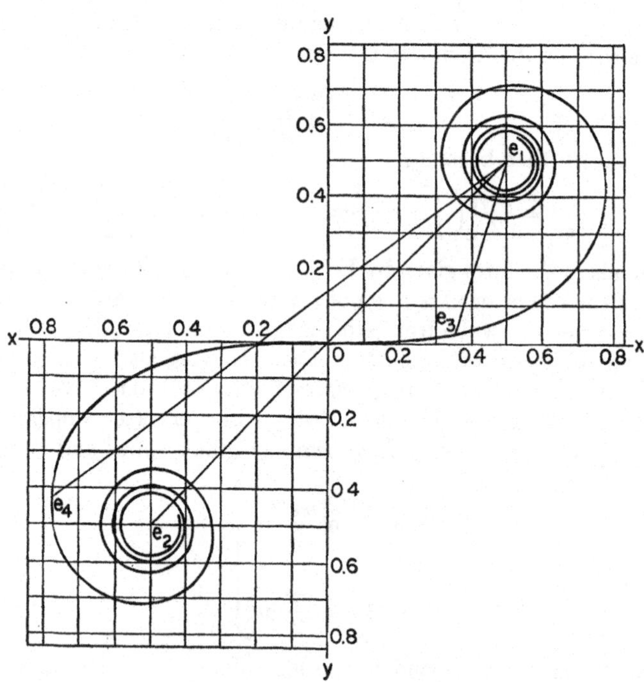

FIG. 4–9. The spiral of Cornu.

and

$$y = \int_{v_1}^{v_2} \sin \frac{\pi v^2}{2}\, dv.$$

These integrals may be rewritten

$$\left.\begin{aligned} x &= \int_0^{v_2} \cos \frac{\pi v^2}{2}\, dv - \int_0^{v_1} \cos \frac{\pi v^2}{2}\, dv \\ \text{and} \qquad y &= \int_0^{v_2} \sin \frac{\pi v^2}{2}\, dv - \int_0^{v_1} \sin \frac{\pi v^2}{2}\, dv. \end{aligned}\right\} \tag{4-20}$$

From Eqs. (4-20) it should be clear, since $N^2 = (x^2 + y^2)$, that the correct magnitude of N, on the graph of the Cornu spiral, will be given by the length of a line drawn between the points x_1y_1 and x_2y_2.

4-20 Use of the Cornu spiral to determine the diffraction pattern for a straight edge. A single application will serve to show the general usefulness of the Cornu spiral. It may be used in a number of ways and several of the problems at the end of the chapter are illustrative.

Let us suppose, as in Fig. 4-10a, that a portion of a very large wave front is completely blocked off by a plate opaque to sound, extending from point O to a point very far down in the negative y-direction. The integrals in Eq. (4-16), leading to the intensity at point a, will then involve for s the limits of virtually 0 to ∞, and the limits for the Fresnel integrals, in terms of v, will be the same. The plot of the point x, y on the Cornu spiral will therefore be point e_1. Since $N^2 = (x^2 + y^2)$, the intensity at point a in Fig. 4-10a will be proportional to the *square* of the distance from the origin to point e_1 on the Cornu spiral, i.e., Oe_1.

For points above and below point a in Fig. 4-10a, the geometrical equivalents

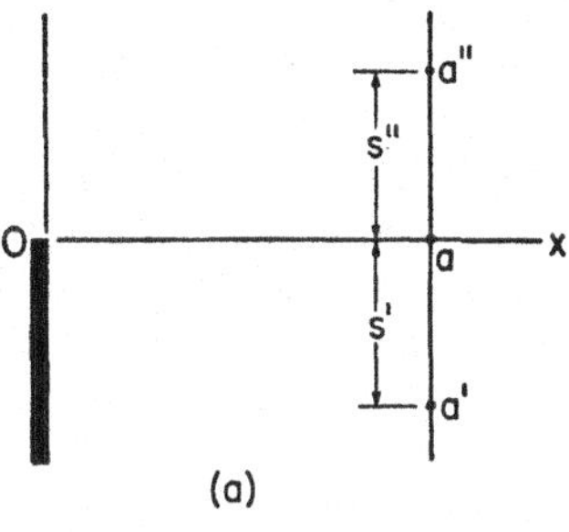

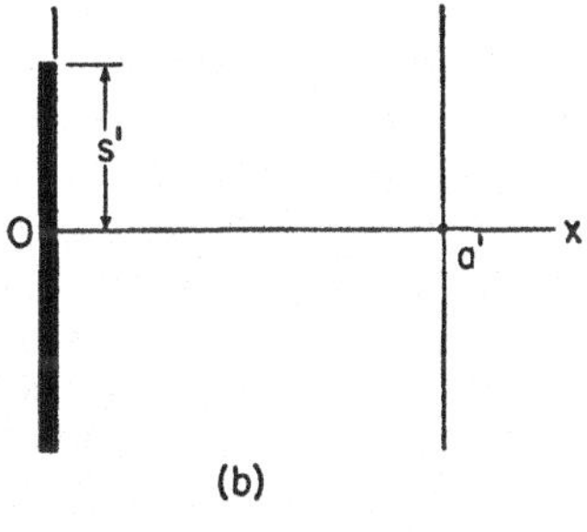

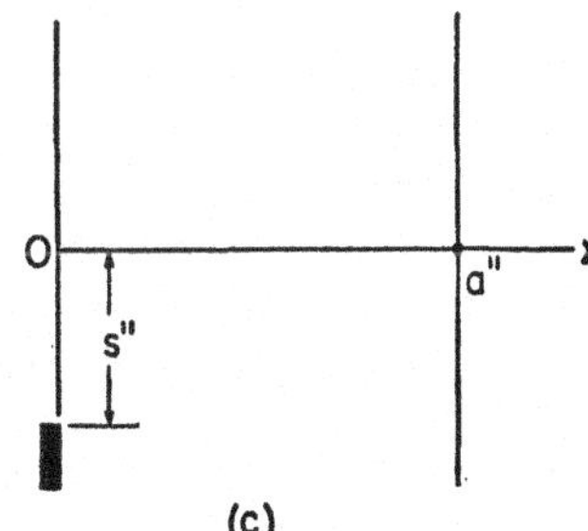

FIG. 4-10.

given in Fig. 4–10b and 4–10c may be used. As may be seen, moving the point *down*, as far as exposure of wave front is concerned, is equivalent to moving the obstruction *up*, and vice versa. A comparison of Fig. 4–10a and Fig. 4–10b will show that as the observation point is moved down to a' (and therefore into the geometrical shadow), the length of wave front exposed will decrease by the amount s', as compared with position a, directly opposite the edge of the obstruction. On the Cornu spiral the relative intensity will be indicated as the square of the length of the line drawn from point e_1 to a point such as e_3 on the curve, corresponding to the particular value of $s = s'$ indicated in Fig. 4–10b. As the point a' is moved farther down into the geometrical shadow, the line e_1e_3 will rotate around e_1 as a pivot, the point e_3 moving along the curve in a direction away from the origin. The intensity will therefore approach zero asymptotically, since the radius of the spiral turns gradually decreases.

For positions on the plane of observation *above a* in Fig. 4–10a the equivalent geometry is that of Fig. 4–10c. It is easy to show that the intensity due to contributions from portions of the wave front *both above and below* the line Oa in Fig. 4–10a is correctly indicated by the length of a straight line drawn between points on either half of the spiral, the values for v that are used corresponding to the two values of s on the wave front. The line e_1e_4 is such a line, corresponding to the situation at point a'' in Fig. 4–10a. The wave front exposure for this point includes *all* of the upper half and part of the lower. As the observation point is moved away from the geometrical shadow the line on the spiral will revolve about e_1 as a pivot, the point e_4 moving along the spiral towards e_2. The length of the line e_1e_4 will undergo a series of oscillations. At a considerable distance from the edge of the obstruction, the length will approach that of the line e_1e_2, corresponding to total exposure of the wave front.

It should be pointed out that the innermost turns of the spiral, in the neighborhood of e_1 and e_2, correspond only very crudely to the geometry of Fig. 4–8. This is because in this analysis we have neglected the so-called "obliquity factor," referred to earlier, which is associated with the inclination of the line aO' to the plane of the wave front, and also because of other approximations such as the effect of an increase in the distance r on the intensity, mentioned in Section 4–18. However, the shapes of the last turns of the spiral do not affect the total intensity as markedly as do those of the first few turns, and the error is thus not too serious.

4–21 Direct graph of intensity. To obtain numerical results for the problem discussed in the previous section it is necessary to know the wavelength and the distance r_0, since these quantities enter into the relationship

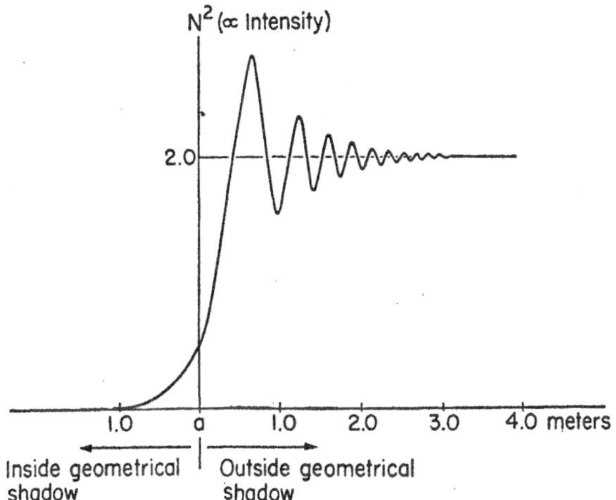

FIG. 4-11. Approximate variation in sound intensity inside and outside the geometrical shadow of an obstructing plate for a wavelength of $\frac{1}{8}$ meter, as observed along a plane at a distance of 4 meters from the plate ($r_0 = 4$ m).

between s and v. To make this clear, let us suppose that $\lambda = \frac{1}{8}$ meter and that $r_0 = 4$ meters. For the point a in Fig. 4-10a the limits for s will then be virtually 0 to ∞, and likewise for v. The intensity, from the Cornu spiral, will be proportional to $(Oe_1)^2$ or 0.5. For a point a', 0.5 meter below a in Fig. 4-10a, the limits for s will be 0.5 to ∞; therefore for v they are from 1.0 to ∞ (Eq. 4-17). (Note that s will have the units of length chosen for λ and r_0; v is dimensionless.) For these limits the length of the chord on the spiral is 0.29 and the intensity is therefore proportional to $(0.29)^2 = 0.084$. (The length of the chord may be conveniently found by making use of the x- and y-coordinates of its end points, as read from the table for the Fresnel integrals, Appendix III.) In this manner the complete diffraction pattern may be drawn for all points along the plane of observation.

Figure 4-11 is a graph showing relative intensities inside and outside the shadow for $\lambda = \frac{1}{8}$ meter and $r_0 = 4$ meters. Within the shadow there will be appreciable sound intensities for a distance of about one meter. Moving away from the geometrical shadow, one finds a series of maximum and minimum points, finally leveling off to a steady intensity at a distance of approximately 3 meters from the edge.

4-22 The shape of the diffraction pattern, as a function of λ. One of the most important features of a diffraction pattern of the edge type, as in the case of the slit, is its degree of "spread" or "compression." From Eq. (4-17) the limits for s, for a given value of r_0, are proportional to $\sqrt{\lambda}$.

Since moving downward into the geometrical shadow is equivalent to a corresponding rise in the position of the limit, s, on the wave front, a certain value of the intensity, as read from the Cornu spiral, will appear *close* to the edge of the obstruction when λ is small and far away from the edge when λ is large. In other words, the pattern will be highly compressed for the short wavelengths and spread out for the long wavelengths.

The conclusions reached above show that the bending of sound waves around the edges of obstructions is a prominent feature of the wave propagation. The bending is highly selective as regards frequency, because of the manner in which the wavelength enters into the problem. Let us consider the intensity along a vertical plane at a distance $r_0 = 4$ meters from the obstructing plate. For a frequency of 650 cycles-sec^{-1} the intensity will drop by approximately 50% if one moves a distance s' of 60 cm from point a (Fig. 4–10a) into the geometrical shadow. On the other hand, with a frequency of 2600 cycles-sec^{-1} this same drop in intensity will take place within a distance of only 30 cm. With a complex sound wave, such as that associated with music, there will be a kind of dispersion behind an obstruction. For a listener sitting well within the "shadow," the bass portion of the music will predominate over the treble. This effect is often observed when music is being played in an adjacent room, with a partially intervening partition.

4–23 Diffraction of waves around obstacles of various contours lying in a field of sound. The discussions above on the interference patterns associated with apertures and edges will suffice to give a good deal of qualitative insight into other diffraction problems of more complicated geometry. In general, the introduction of some obstacle into the path of plane or spherical sound waves will disturb the conditions existent before such introduction. For instance, it is shown in books on optics that there is a bright spot directly behind a disk held in the path of a beam of parallel light. The simple construction used to demonstrate the truth of this prediction may be applied with equal validity to either sound or light waves. At other points behind the disk, not lying on its axis, there may be shown to exist regions of maximum and minimum acoustic pressure, very similar in distribution to the patterns for a circular *aperture*. On the other side of the disk there will appear a somewhat similar pattern due to energy *reflected* from the disk. All of these patterns are to be expected from the previous analyses in this chapter.

Theoretical computations have been made by Rayleigh, Schwarz, Stenzel, and others on the extent of the disturbance of the sound field in the neighborhood of obstructions having a few simple shapes, such as disks, spheres, cylinders, etc. As one would expect, the mathematics is complicated. Experimental measurements by Wiener and others have confirmed many of the theoretical results. Microphones and probes of various shapes may be assumed to approximate some such simple geometry and it is important to understand their effect upon the sound field. The sphere, it might be mentioned, is a diffracting obstacle of especial interest, since the human head is roughly spherical in shape and the diffracting properties of the head are of interest whenever listening is involved. The pressure distribution on the surface of such an obstacle is of concern, as well as the pressure distribution in the surrounding medium.

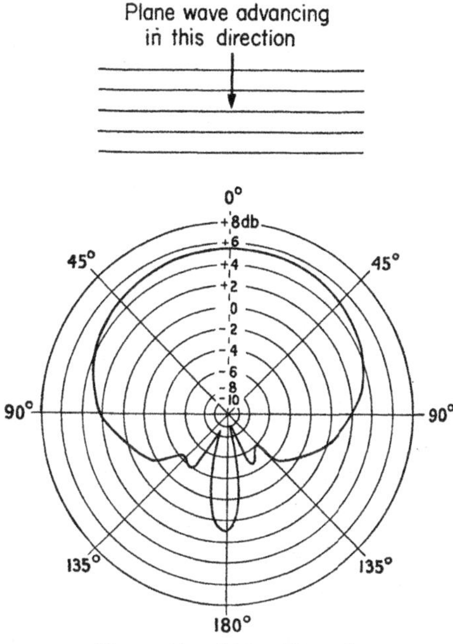

FIG. 4–12. (*After* Wiener)

Figure 4–12 is a polar plot, adapted from Wiener, * which illustrates how the excess pressure in a plane sound wave is affected by the presence of a sphere whose radius is of the order of the wavelength. For the head, this corresponds to a frequency of about 2000 cycles-sec^{-1}. The radial distance from the pole to the curve is proportional to the ratio of the sound pressure at various points on the surface of the sphere to that existing in the wave before the introduction of the sphere (the ratio in this case is plotted in decibels). The angles are measured with respect to the direction from which the wave is coming. The circle labeled 0 decibels crosses the curve not far from the 90° position. Therefore at this position the sound pressure will be about that in the undisturbed wave and if one is facing the oncoming wave, the ears will lie, effectively, in the undisturbed field. For other points on the head this is not so.

A problem of considerable theoretical and also of practical interest is the "scattering" of longitudinal waves by a large array of small particles (small compared with the wavelength). Rayleigh has shown that the

* Jour. Acous. Soc. of Amer., **19**, 446 (1947).

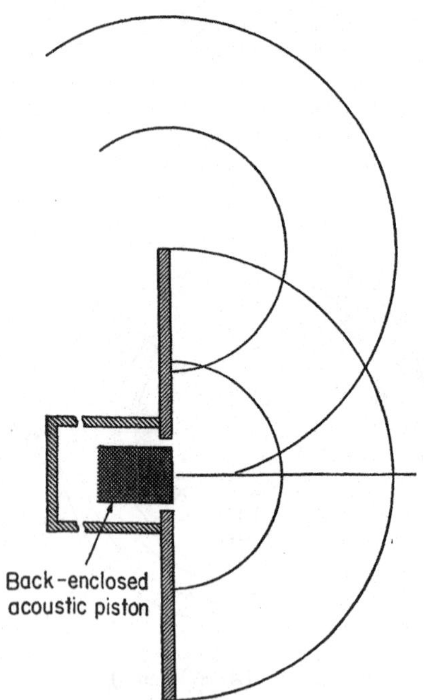

FIG. 4-13. Diffraction pattern in front of an acoustic piston surrounded by a baffle of limited dimensions.

intensity of the scattered wave (essentially a diffraction process) varies, for any one such particle, as the fourth power of the frequency. This effect is of particular interest in the propagation of waves through sea water, where there may be air bubbles and suspended solid particles. The dissipation of useful energy in this manner is of much importance in undersea signaling.

4-24 Diffraction effects for an acoustic piston set in a circular plate of finite size. As a final example of a diffraction pattern of practical importance in sound reproducing equipment, let us consider the following situation. Suppose that a plane wave strikes a plate having a circular hole, but that the size of the plate is insufficient to entirely cut off the advancing wave at the outer edges. One would then expect *two* overlapping diffraction patterns, one from the aperture, the other from the energy diffracting around the edges of the plate. In the practical arrangement to be described, somewhat the same sort of thing will occur.

A loudspeaker unit (effectively an acoustic piston) is mounted at the center of a plate whose diameter is several times that of the speaker cone. Let us assume the diameter of the cone to be much less than λ, but the diameter of the plate to be at least as great as λ. The back of the speaker unit is completely enclosed, to eliminate double source complications. Instead of the simple spherical divergence to be expected from what amounts to an aperture small in size compared with λ, an interference pattern will be observed in front of the plate. This pattern results when the wave disturbance reaches the boundary of the plate, since the edge becomes, by diffraction principles, a new source of waves. As indicated in Fig. 4-13, the combination of this new wave train with the primary waves set up at the cone gives rise to interference effects. The ensuing irregularities in intensity distribution are often of practical interest in sound equipment and in laboratory measurements.

If the back of the cone is *not* enclosed, the radiation from the rear becomes important because additional interference effects occur due to the double source action. This problem involves the general behavior of the baffle, which was discussed qualitatively in the preceding chapter. When we consider the general problems of practical loudspeaker design in Chapter 11, a quantitative discussion will be included.

4–25 General conclusions on diffraction. It should now be quite clear that diffraction is a complicated matter. Only a few of the cases, of particular interest in acoustics, have been presented. For the treatment of other problems of diffraction, reference should be made to a book on optics. For many practical problems the analytical work is too difficult for even an approximate solution and in these cases one must explore the field of sound with experimental probes. This is a laborious process and is subject to much error, due among other things to the diffracting properties of the probe itself. More will be said on this point in Chapter 10.

PROBLEMS

1. Consider the single slit pattern. Draw vector diagrams to show the total pressure p_m for a value of d (see Fig. 4–1) equal to (a) $\lambda/4$, (b) $\lambda/2$, (c) λ, and (d) $3\lambda/2$.

2. Consider a point a (Fig. 4–1) such that $d = \lambda/4$. (a) Suppose now that the slit width is doubled, the wavelength and the position of point a remaining fixed. Compare the new pressure p_m at a with the original value; also the new intensity with the original value. (b) Repeat part (a), assuming the slit width to be increased by a factor of 10 (the width is still small compared with the distance from the slit to the point a).

3. Repeat problem 2, assuming, however, the slit width to remain fixed at the original value and the *wavelength* to be (a) halved, and (b) reduced to 0.1 of its original value.

4. Plot a curve on rectangular axes to show the relation between the maximum pressure p_m and the angle β for the case where (a) $\lambda = b$, (b) $\lambda = b/2$.

5. Draw a polar graph of the relation between the intensity I ($\propto (p_m)^2$) and the angle β, at a fixed distance from a slit whose width b is equal to the wavelength.

6. For the circular hole pattern, find the value of β for the appearance of the second minimum, if the wavelength is 2.0 cm and if the area of the hole is 75 cm^2.

7. Consider the distribution of pressures over a plane of observation 2 meters from a plate containing a circular hole through which sound waves are passing. The ratio of wavelength to hole diameter is 0.1. Find the radius of the circles representing the first three loci of zero pressure.

8. A loudspeaker cone of diameter 12 inches is surrounded by an infinite plane baffle. Plot the intensity I as a function of the angle β (for a fixed radial distance) when the frequency of vibration is (a) 50 cycles-sec^{-1}, (b) 5000 cycles-sec^{-1}. Consider the cone to move as a unit in both cases.

9. A plane sound wave in air, of frequency 10,000 cycles-sec^{-1}, strikes a paling fence with normal incidence. The width of the boards may be considered small compared with the wavelength. The

spacing between the boards is 5 inches. For the reflected energy, find the angle to the normal for the appearance of the first order and second order spectrums, assuming the fence to act like an optical diffraction grating of the reflecting type.

10. Show that a differential distance measured along the Cornu spiral is equal to dv. (Since this is so, any point on the spiral for the limits zero to v may be located immediately by measuring off the value of v along the spiral curve, beginning at the origin.)

11. (a) Show that for a given value of r_0 and of λ, the slope at any point on the Cornu spiral is equal to the tangent of the phase angle between the contribution at a (Fig. 4–8) from the point s on the wave front and the contribution from the point directly opposite a. (b) What is the significance of the slope of a straight line connecting two points on the Cornu spiral?

12. A point a lies directly opposite the center of a slit through which are passing plane waves. The distance from the point a to the upper or lower edge of the slit is greater than the distance to the mid-point of the slit by just $\lambda/2$. (a) On the graph of a Cornu spiral, draw a line whose length is proportional to p_m at point a. (b) Read off the values of the Fresnel integrals for the end points of this line. (c) Making use of the table of Appendix III, find the corresponding values of v. (d) If $\lambda = 1.0$ cm and $r_0 = 10$ m, compute the width of the slit from v. Check the result from the direct geometry of the figure.

13. (a) A point a_1 is directly opposite the lower edge of a slit 20 cm wide and is 100 cm distant from it. Plane waves of a frequency of 10,000 cycles-sec^{-1} are passing through the slit. Using the proper limits for the variable v, find the values of the appropriate Fresnel integrals (Appendix III). (b) In a similar way, find the values of the Fresnel integrals for a point a_2 directly opposite the mid-point of the slit. Then determine the ratio of the maximum pressure at the first point to that at the second point.

14. Redraw the abscissa scale for Fig. 4–11, assuming a wavelength of $\frac{1}{16}$ m.

CHAPTER 5

ACOUSTIC IMPEDANCE. BEHAVIOR OF HORNS

5–1 The principle of analogy. In Chapter 1 some use was made of the similarity between the differential equation for an electrical circuit and for the motion of a particle. Since the differential equations were the same in both cases, the form of the two solutions could also be assumed identical. This is only one problem among many where the solutions to circuit problems can be carried over into a different field, such as mechanics. In probably no other branch of engineering or physics has the mathematical technique for handling problems been developed to such a high point as in the field of electrical circuits. Therefore in showing that a problem in another field is mathematically equivalent to some well-known circuit problem, one is well on the road to a solution. Great care must be taken, however, to establish clearly all the details of the analogy, so as to correctly interpret the new solutions.

5–2 Types of analogies. It is often possible to set up an analogy in more than one way, even between the same two sets of phenomena.* In this book two general types of analogy will be found helpful. In the first, already made use of in connection with particle motion, a mechanical system is broken up into inertial, elastic, and dissipative components. Sometimes the inertial components are the solid masses and the elastic components are the springs of ordinary mechanics; sometimes the system comprises enclosed volumes of air, with their associated mass and elasticity. Frequently a system made up of both solid and air elements must be considered. Such systems may be said to possess mechanical impedance whose behavior under the action of impressed forces of various kinds may be compared to the part played by electrical impedances in electrical circuits. In the electrical case, such mechanical impedances may be "lumped" (ordinary inductors, capacitors, and resistors used at low electrical frequencies have impedances of this type). On the other hand, as in the case of air conduits, the impedance may be "distributed," in the sense in which the term is used for electrical transmission lines.

The usual procedure with the above type of analogy is to set up the differential equations for the dynamics of the system or for some part of it. These equations are then compared with those for well-known electrical circuits whose solutions have already been studied. In this comparison, two points of view may be taken. The most obvious one is to draw the

* See Morse, *Vibration and Sound*, McGraw-Hill, 1948, for a discussion of various types of analogies useful in acoustics.

analogy between *velocity* and *current* and between *force* and *potential difference*. This was the approach in Chapter 1, in connection with particle vibrations. It is also possible, by rewriting the differential equations in a somewhat altered mathematical form, to show a different analogy, where *velocity* is compared with *potential difference* and *force* with *current*. This latter cross-comparison has distinct advantages wherever the mechanical system is a complicated one. The chief of these advantages is that the equations for a mechanical system arranged in a physical *series* (or "tandem" manner) have then the same form as those for a *series* electrical circuit. Similarly, a *parallel* mechanical arrangement corresponds to a *parallel* electrical circuit. However, from the first point of view, which we shall use in this book, a *series* arrangement for a mechanical system must be compared to a *parallel* electrical circuit and vice versa, and this can be an annoying source of confusion. Despite this disadvantage, we shall retain this point of view, since the mathematics is simpler to set up and since, for the most part, we shall be dealing with simple systems, having few components.

We shall presently consider certain aspects of the behavior of the Helmholtz resonator as an example of an analogy of the above type. Later we shall refer to "acoustic filters" of the type studied by Stewart and others. These filters are made up of air conduits of various shapes and sizes, with acoustic frequency characteristics closely similar to those of corresponding electrical filters. The various parts of these conduits possess mechanical impedance, in the sense used above, and the electrical analog may be fully represented as an equivalent circuit containing inductors, capacitors, and resistors.

The second type of analogy, of a somewhat more restricted nature and not to be confused with that discussed above, is quite useful in problems connected with the radiation and transmission of sound. In the development of this analogy we shall not be concerned with the impedance of some mechanical system or its parts, but with what is called the *specific acoustic impedance* * at a point in the medium through which sound waves are passing. This quantity is also referred to as *specific radiation impedance*. It will presently be defined and its general usefulness will appear when it is applied in some specific problems.

Problems often arise in which it is convenient to make use of *both* types of analogy, in which case it is important to keep the definitions and procedures for each clearly in mind. The analysis of the Helmholtz resonator

---

* See Morse, pp. 236–7, for definitions of three different kinds of acoustic impedance. In this chapter the term *acoustic impedance* will, unless otherwise qualified, refer to *specific* acoustic impedance.

and certain features of the radiation of sound by loudspeakers and horns are cases in point.

5-3 Sound radiation and acoustic impedance. The radiation of energy from sources of sound has been discussed in the previous chapter from the standpoint of the field or wave equations. There is no other way to obtain a point-to-point description of the field of sound. As has been seen, it may be very difficult to apply the equations to practical problems, largely because of the complexities of the boundary conditions. Many approximations must be made to obtain any solutions at all.

If one is interested in the *over-all radiation* of real sound energy *from the source* there is another approach, through the concept of *acoustic impedance*. The results of this analysis are illuminating in connection with the behavior of certain widely used sources of sound and will be applied to some of these sources later in the chapter. Other applications of the idea of acoustic impedance will appear in later chapters.

As a preliminary to a definition and discussion of acoustic impedance, it is necessary to summarize the essential features of the complex notation used in a.c. circuit analysis.

5-4 Elements of complex notation as applied to electrical circuits. A complex number, i.e., a real plus an imaginary quantity, may be represented in several ways:

$$
\begin{aligned}
&a + jb, \\
&M(\cos\theta + j\sin\theta), \qquad (5\text{-}1) \\
&M\epsilon^{j\theta}.
\end{aligned}
$$

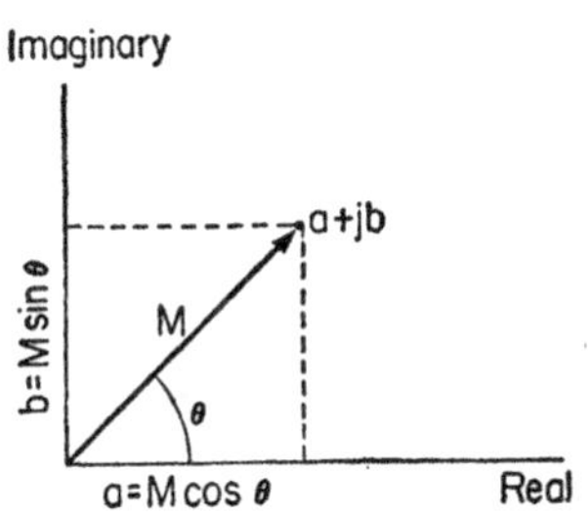

FIG. 5-1. Polar representation of a complex quantity.

The first of these expressions is the definition of a complex number, j being equal to $\sqrt{-1}$. The second form is based upon a plot of a complex number as a point in the so-called complex plane (Fig. 5-1). The modulus, M, has a magnitude which is the distance from the origin to the point representing the complex quantity, and θ is the angle between this radius vector and the horizontal real axis. The whole expression can be seen to be equivalent, geometrically, to $a + jb$. The modulus M can also be written $\sqrt{a^2 + b^2}$. The third (polar) form can be shown to be equivalent to the second form by expanding $\cos\theta$, $\sin\theta$, and $\epsilon^{j\theta}$ into their respective equivalent power series.

To demonstrate the advantages of the use of complexes in a.c. circuit theory, we shall return to the differential equation for an L-R-C circuit (Eq. (1-29)). Instead of writing the impressed potential as $E_m \cos(\omega t)$, let us use the complex expression $E_m\epsilon^{j\omega t}$, so that

$$L\frac{di}{dt} + Ri + \frac{1}{C}\int idt = E_m\epsilon^{j\omega t} = E_m[\cos\omega t + j\sin\omega t]. \tag{5-2}$$

Since the right-hand side of this equation is complex, the expression on the left must be complex; and since all coefficients are real, the solution for the current i must be a complex expression. If a solution of the form $A\epsilon^{j\omega t}$ is assumed and substituted into the differential equation, the latter will be satisfied, provided that the coefficient A is equal to the complex expression

$$A = \frac{E_m}{R + j\left(\omega L - \frac{1}{\omega C}\right)}. \tag{5-3}$$

Therefore the current (the steady state current, assuming that the transient part of the solution has died away) is

$$i = \frac{E_m}{R + j\left(\omega L - \frac{1}{\omega C}\right)}\epsilon^{j\omega t}. \tag{5-4}$$

This may be rewritten, using the polar equivalent for a complex number,

$$i = \frac{E_m}{M\epsilon^{j\theta}}\epsilon^{j\omega t} = \frac{E_m}{\sqrt{R^2 + \left(\omega L - \frac{1}{\omega C}\right)^2}}\epsilon^{j(\omega t - \theta)}, \tag{5-5}$$

where

$$\tan\theta = \frac{\omega L - \frac{1}{\omega C}}{R}.$$

If the usual completely real expression for the current in such an a.c. circuit is desired, it may be extracted from Eq. (5–5) by using only the real part of $\epsilon^{j(\omega t - \theta)}$, that is,

$$i_{\text{real}} = \frac{E_m}{\sqrt{R^2 + \left(\omega L - \frac{1}{\omega C}\right)^2}}\cos(\omega t - \theta). \tag{5-6}$$

The real part of Eq. (5–5) represents the real current in the circuit, just as the real part of $E_m\epsilon^{j\omega t}$ represents the actual applied potential. It is to be noted also that the electrical impedance of the circuit (that quantity which gives the current when divided into the potential difference) may be written either as a complex or as a real number. From Eq. (5–4),

$$z = R + j\left(\omega L - \frac{1}{\omega C}\right) \tag{5-7}$$

or, from Eq. (5-6),

$$Z = \sqrt{R^2 + \left(\omega L - \frac{1}{\omega C}\right)^2}. \tag{5-8}$$

The real form for the impedance (Eq. (5-8)), is simply the modulus of the complex form, and the magnitudes of the real and imaginary parts of z in Eq. (5-7) are respectively the resistance R and the reactance X of the circuit. The phase angle θ between current and applied potential is arctan $b/a = \arctan X/R$.

For the solution of circuit problems, there are many advantages in retaining the complex form of e, i, and z. For the acoustical problems at hand the main advantage will be in the more ready determination of the nature and extent of the power radiated by a source of sound. In an electrical a.c. circuit it is well known that the only real power delivered is due to the presence of resistance. (The average value of the real power may be written either as $E_{\text{rms}} I_{\text{rms}} \cos \theta$, or $I_{\text{rms}}^2 R$.) For acoustical problems a quantity in the field of sound which is analogous in many ways to complex electrical impedance, z, will be defined. The nature of this acoustic impedance, whether it is real, imaginary, or a combination of both, will be closely tied up with the nature of the acoustic energy flow in the region where this impedance is computed. Moreover, by the use of the complex form for the instantaneous excess pressure p, the particle velocity ξ, and other field parameters, useful phase relationships may be quickly obtained.

5-5 Specific acoustic impedance. Specific acoustic impedance, z_s, at a point in a field of sound is defined as

$$z_s = \frac{p}{\xi}, \tag{5-9}$$

where p and ξ are the instantaneous excess pressure and particle velocity, respectively. For a single sound frequency, p and ξ are sinusoidal functions of the time t but they are not necessarily in phase for all types of waves. Therefore specific acoustic impedance, pressure, and particle velocity are related exactly as are electrical impedance, potential difference, and current in an a.c. circuit. The dimensions of the acoustical quantities, it should be noted, are different from those of the electrical ones, although there are some similarities. The most important distinction is that electrical impedance, defined as e/i, exists *between two points* in the current-carrying circuit, whereas the specific acoustic impedance is a *point* property. The acoustic impedance, as here discussed, is completely defined as soon as the coordinates of a point and the corresponding values of p and ξ are specified.

The analogy carries further than Eq. (5–9). Just as the product ei represents instantaneous electrical power, so the product $p\xi$ represents the instantaneous acoustical power at the point in question. However, the power ei is *total power delivered to the circuit* across which the p.d. exists, whereas the acoustical power $p\xi$, since it involves the pressure, is the *energy flow per unit area and per unit time.* Nevertheless, the analogy is close enough to be quite useful. As mentioned earlier, the exact limitations of any analogy must be recognized.

5–6 Specific acoustic impedance for plane waves. If the specific acoustic impedance at any point in the path of plane waves is computed, the result is particularly simple. We shall make use of the velocity potential Φ, which was first introduced in Chapter 3 in connection with three-dimensional waves. It is perfectly correct to use this function also for plane waves, since they are simply a special case of three-dimensional waves in cartesian coordinates. The first of the equations (3–13) in Chapter 3 becomes, with Φ a function of x only,

$$\frac{\partial^2 \Phi}{\partial t^2} = c^2 \frac{\partial^2 \Phi}{\partial x^2}. \tag{5–10}$$

The solution, which represents Φ as a function of x and t for a wave traveling in the $+x$ direction, may then be written

$$\Phi = A \cos \frac{2\pi}{\lambda}(ct - x). \tag{5–11}$$

As is to be expected, the coefficient A of the cosine expression is a constant with no inverse function of the space coordinate, since in the case of plane waves the maximum values of p, s, etc., all derivable from the function Φ, do not decrease with the distance from the source. For our present purposes we shall write the periodic function on the right of Eq. (5–11) in the complex form,

$$\Phi = A\epsilon^{jk(ct-x)}. \tag{5–12}$$

The quantity k here replaces the ratio $2\pi/\lambda$. (The real part of Eq. (5–12) is identical with (5–11).)

It is now a simple matter, by means of the derivative relationship, to obtain the appropriate equation for the instantaneous excess pressure and the particle velocity. Making use of $p = \rho_0 \dfrac{\partial \Phi}{\partial t}$, we find that

$$p = j\rho_0 kcA\epsilon^{jk(ct-x)}. \tag{5–13}$$

Also, for the particle velocity,

$$\dot{\xi} = -\frac{\partial \Phi}{\partial x} = jkA\epsilon^{jk(ct-x)}. \tag{5-14}$$

Therefore the specific acoustic impedance $(z_s)_0$ for plane waves in free space, at any point along the direction of propagation, becomes

$$(z_s)_0 = \frac{p}{\dot{\xi}} = \frac{j\rho_0 kcA\epsilon^{jk(ct-x)}}{jkA\epsilon^{jk(ct-x)}} = \rho_0 c. \tag{5-15}$$

(This result could have been obtained directly from parts (b) and (e) of Eq. (2–19), without use of the function Φ and without using the complex form. The present method of determining z_s, however, is fundamental to the following discussion of the more complicated cases, and is used here for the sake of generality.)

The simple product $\rho_0 c$ just obtained has very nearly the numerical value of 42 gm-cm^{-2}-sec^{-1} (in cgs units) under standard conditions of temperature and pressure in free air (0°C, 760 mm Hg). The equivalent number in mks units is 420 kgm-m^{-2}-sec^{-1} (these values are slightly lower at temperatures in the neighborhood of 20°C). The specific acoustic impedance for plane waves in free space is thus a constant, independent of x and t and also of the parameters characteristic of any particular disturbance, such as frequency, particle amplitude, maximum excess pressure, and all the other related quantities. Thinking of $(z_s)_0$ in terms of the electrical analogy, it is *real* and therefore corresponds to an electrical resistance, with all that this implies. In an a.c. circuit the average power delivered may be computed as $I^2_{\text{rms}}R$, where R is the real part only in the complex quantity representing the total circuit impedance. In the case of plane waves, z_s is *all* real, therefore the average acoustical power (per unit area) is $(\dot{\xi}_{\text{rms}})^2\rho_0 c$. (Compare Eq. (2–29c).)

All acoustical energy per unit area that arrives in the plane of the wave front passes on in the direction of wave propagation, *away* from the source, with no fraction "reflected" back *towards* the source. This would not be true if there were a reactive (i.e., imaginary) component of the specific acoustic impedance. In the latter case, some energy would periodically flow in the $-x$ direction, just as in an electrical circuit instantaneous energy fed into an inductor or into a capacitor would presently return to the generator.

It should be pointed out that Eqs. (5–13) and (5–14) give a ready indication as to phase. Due to differentiation, both expressions have $+j$ in the coefficient. The exponential $\epsilon^{jk(ct-x)}$ is a complex number which may be represented in the complex plane by a vector and an angle. The presence of the $+j$ before the exponential has the effect of rotating this vector 90° in a counterclockwise direction. (This may be readily verified by writing $\epsilon^{jk(ct-x)}$ in the form $a + jb$, multiplying by $+j$, and then interpreting the result graphically.) From Eqs. (5–13) and (5–14) it is seen that p and ξ are in phase with each other, but are 90° ahead of the variation in Φ. (It should be noted that a coefficient of $-j$ has the effect of rotating the vector in a clockwise direction and therefore signifies a lag relative to Φ.)

5–7 Analogous acoustic impedance. Besides the concept of specific acoustic impedance, there is another useful quantity, the acoustic impedance z_a *for an area,* S. Following the terminology of Morse, we shall call z_a the *analogous acoustic impedance.** This is defined as the ratio of the instantaneous excess pressure to the so-called "volume current," $S\xi = \dot{X}$. Note that since $z_a = z_s/S$, the analogous impedance is *less* than the specific impedance if S is greater than unity. The definition of z_a is designed to be in line with power considerations. When the impedance of an electrical circuit is decreased, keeping the applied potential difference constant, the current will increase and therefore so will the power. In the acoustical case, more energy will flow per second through a large area than through a small area. Hence the analogous impedance is an inverse function of the area. Since for plane waves the instantaneous power per unit area is $(\xi)^2(z_s)_0$, the power U for a total area S may be written in terms of the volume current $\dot{X}$ and the analogous impedance $(z_a)_0$, as

$$U = S(\xi)^2(z_s)_0 = (\dot{X})^2(z_a)_0. \tag{5–16}$$

5–8 Specific acoustic impedance for spherical waves. For spherical waves, the velocity potential Φ is an inverse function of the distance r from the pole. Rewriting Eq. (3–16), for a sinusoidal wave in the complex form,

$$\Phi = \frac{B}{r}\epsilon^{j[k(ct-r)+\alpha]}, \tag{5–17}$$

we find that in this case

$$p = \rho_0\frac{\partial\Phi}{\partial t} = \frac{jkc\rho_0 B}{r}\epsilon^{j[k(ct-r)+\alpha]} \tag{5–18}$$

and

$$\xi = -\frac{\partial\Phi}{\partial r} = \left(\frac{1}{r} + jk\right)\frac{B}{r}\epsilon^{j[k(ct-r)+\alpha]}. \tag{5–19}$$

Therefore the specific acoustic impedance z_s becomes

$$z_s = \frac{p}{\xi} = \rho_0 c\left(\frac{k^2r^2}{1 + k^2r^2} + j\frac{kr}{1 + k^2r^2}\right), \tag{5–20}$$

---

* Called *analogous* because the "volume current" $S\xi$ is more similar to *current* in an electrical circuit than is ξ. The quantity ξ is comparable to *current density.*

where the result has been written in the standard form for a complex number, $a + jb$. (Eq. (5–20) should be verified, making use of the usual technique for handling complex quantities.) An inspection of Eq. (5–20) will show that, unlike plane waves, there is an imaginary or reactive component to z_s for spherical waves. The resistive component, the only part involved in the radiation of real sound energy, is

$$(z_s)_R = \rho_0 c \frac{k^2 r^2}{1 + k^2 r^2}. \tag{5–21}$$

At the surface of a *very small* sphere, whose radius $r_0 \ll \lambda$, this may be written in the simpler form

$$(z_s)_R \Big|_{r_0} \cong \rho_0 c k^2 r_0^2. \tag{5–22}$$

This expression will be useful in the next section.

Before applying the results of the previous section, Eq. (5–20) merits some discussion. An examination of the two terms in parentheses will show that as the distance from the source becomes greater and greater, the term containing j approaches zero and the first term approaches unity, so that z_s becomes $\rho_0 c$ for very large distances. This is to be expected, since for distant points the wave front is effectively plane and the expression for z_s should approach that for plane waves. For nearer points, where the wave front has greater curvature, the resistive component is smaller in comparison with the reactive component, and the latter becomes of considerable importance. Near the source the power flow consists of two kinds. The first is real power. The second, in the language of electrical circuits, is "wattless" power, involving energy which surges out from the source and then back towards the source, without ever being radiated as sound waves. Since the reactive part of z_s is always positive, it is reasonable to say that it is the mass or inertial property of the air that is involved, just as in Chapter 1 a particle was shown to have positive reactance due to its mass property.

One other comparison may be made. There are two kinds of electric and magnetic fields around a circuit carrying alternating current, the "induction" or "coulomb" fields, which are not involved in the radiation of energy, and the "radiation" fields, which are responsible for any electromagnetic wave which may be set up and which carry all the energy associated with the wave. While there are no precise counterparts to these fields in the acoustic case, there is a division of the energy into two parts, that which is radiated and that which remains local.

5–9 The Helmholtz resonator. This well-known resonator furnishes a good illustration for the application of the electrical analog. It is essentially a rigid container, with an opening to the surrounding air. Figure 5–2 shows, in cross section, two different types of Helmholtz resonators. The resonator in Fig. 5–2a has a simple circular opening, while that in 5–2b has a short attached cylindrical "neck." If a tuning fork of a frequency to which the cavity resonates is held near the opening, the sound intensity in the neighborhood will be greatly enhanced.

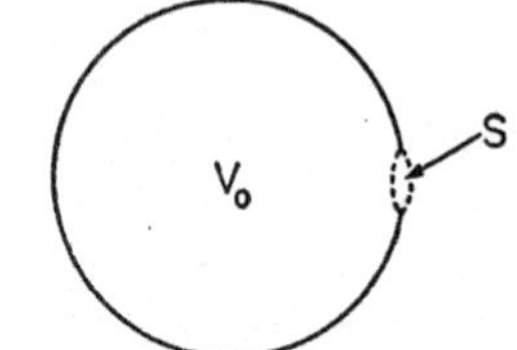

Helmholtz resonator without "neck"
(a)

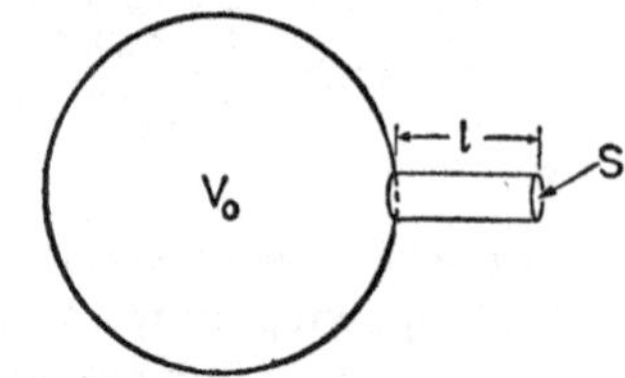

Helmholtz resonator with "neck"
(b)

FIG. 5–2.

As a result of the following analysis, it is possible to predict with fair accuracy the natural frequency of the resonator, subject to important restrictions placed upon the dimensions of the resonator, as will be seen presently.

We shall make use of both types of analogy discussed at the beginning of this chapter. First we shall consider the dynamics of a mechanical system consisting entirely of enclosed air and having the properties of mechanical reactance and resistance. The resonator with the attached neck is simpler to analyze and what follows will apply to this type. As shown in Fig. 5–2b, V_0 represents the volume of the main cavity, while S and l are the cross-sectional area and length, respectively, of the neck. All dimensions are assumed small compared with the wavelength in air. When a source of energy such as a vibrating tuning fork is held near the opening, some of the energy radiated towards the resonator will set into vibration a cylindrical "plug" of air within the neck, of volume lS. This plug of air may be assumed to move as a unit (since $l \ll \lambda$) under the action of the driving force due to the tuning fork, the elastic force on the inner end of the plug (due to the compressibility of the air enclosed in the volume V_0), and a force of dissipation. This last force is due mainly to the radiation of sound energy and may be expressed in terms of acoustic radiation impedance, as we shall soon see.

The equation for the motion of the air in the neck, treating the air plug as a particle, is

$$(\text{mass})\ (\text{acceleration}) = F_{\text{driving}} + F_{\text{dissipative}} + F_{\text{elastic}}. \qquad (5\text{–}23)$$

In this equation, the mass is that of the air plug, $\rho_0 lS$, where ρ_0 is the average undisturbed density of the air. Therefore we may write (mass) (acceleration) as $\rho_0 lS\ddot{\xi}$, using for acceleration the particle acceleration of the air at the right-hand end of the plug.

To evaluate the dissipative term, use is made of the previous discussion of the acoustic impedance. Neglecting friction in the neck, the main force of dissipation is that associated with the radiated sound energy and therefore with the real part of the acoustic impedance at the right-hand end of the plug of air. As the plug oscillates, its right-hand face acts as a single source, giving rise to sound waves. The generating surface may be considered a circular plane area. However, the diameter of this circle, like the other dimensions of the resonator, is small compared with λ. As was seen in the previous chapter, a surface of *any* shape whose dimensions are much less than λ will give rise to a wave shape which is spherical at a distance not far from the source, due to diffraction effects. Assuming this to be the case, we may replace the plane generating area S by a *hemisphere* whose radius r_0 is that of the neck itself (Fig. 5-3).*

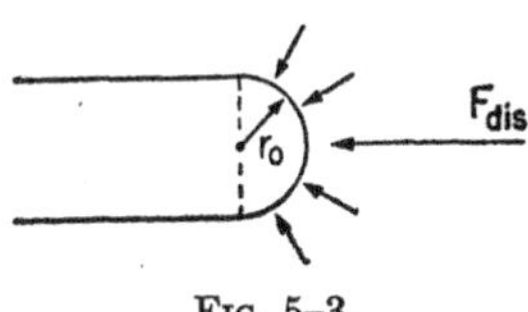

FIG. 5-3.

Consider now the energy dissipated at the surface of this hemisphere. Since the area S' of the hemisphere is greater than that of the cross section of the plug of air (actually twice as great), the particle velocity $\dot{\xi}'$ at its surface must be less than the velocity $\dot{\xi}$ next to the plane end of the plug. (This follows from the fact that both surfaces must be of equal "strength" for complete physical equivalence.) The dissipative part of the acoustic pressure at the surface of the hemisphere may be written

$$p_{\text{dis}} = (z_s)_R\dot{\xi}' = (z_a)_R S'\dot{\xi}', \tag{5-24}$$

where $(z_a)_R$ is the real part of the analogous acoustic impedance for the whole surface of the hemisphere. The value of $(z_a)_R$ is

$$(z_a)_R = \frac{(z_s)_R}{\text{area}} = \frac{\rho_0 ck^2 r_0^2}{2\pi r_0^2} = \frac{\rho_0 ck^2}{2\pi},$$

making use of the simpler expression for $(z_s)_R$, Eq. (5-22), since r_0 is here much less than λ. Therefore

$$p_{\text{dis}} = \frac{\rho_0 ck^2}{2\pi} S'\dot{\xi}' = \frac{\rho_0 ck^2}{2\pi} S\dot{\xi}. \tag{5-25}$$

(The product $S'\dot{\xi}'$ may be replaced by $S\dot{\xi}$ because of the source equivalence.) Having determined p_{dis}, we may now compute the total axial force on the

* Strictly speaking, if a hemispherical wave shape is assumed, a flange should be mounted at the outer end of the neck to prevent back radiation.

plug, F_{dis}, due to the dissipation, as the product of p_{dis} and the cross-sectional area S of the plug of air:

$$F_{\text{dis}} = -\,p_{\text{dis}}S = -\,\frac{\rho_0 c k^2}{2\pi}\,S^2\dot{\xi}. \tag{5-26}$$

The velocity $\dot{\xi}$ is that in the air next to the surface of the plug, or that of the plug itself, since the two are identical.

The elastic force in Eq. (5–23) is a result of the springlike effect of the air enclosed in the cavity upon the left-hand end of the plug of air. It is important to note that since the diameter of the cavity is small compared with λ, no phase differences such as are associated with a wave exist within the enclosure. Therefore, in effect, pressures are increased and decreased instantaneously throughout the volume V_0 with the motion of the air in the neck. The magnitude of the elastic force per unit area is

$$p_{\text{elastic}} = -\,\mathcal{B}\,\frac{v}{V_0} = -\,\rho_0 c^2 \frac{v}{V_0}, \tag{5-27}$$

where $\mathcal{B}$ and c have the usual meaning. But due to the motion of the plug of air, $v = S\xi$. Therefore the total elastic force on the air in the neck is

$$F_{\text{elastic}} = Sp_{\text{elastic}} = -\,S\rho_0 c^2 \frac{v}{V_0} = -\,\frac{\rho_0 c^2 S^2}{V_0}\,\xi. \tag{5-28}$$

5–10 The resonance frequency. The complete differential equation for the motion of the plug of air can now be written

$$\rho_0 l S\ddot{\xi} + \frac{\rho_0 c k^2 S^2}{2\pi}\,\dot{\xi} + \frac{\rho_0 c^2 S^2}{V_0}\,\xi = F_m \epsilon^{j\omega t}, \tag{5-29}$$

where the driving force on the right, due to the arrival of wave energy from the tuning fork, has been expressed as a complex quantity. This equation is identical in mathematical form with Eq. (1–27). By comparing the coefficients in Eq. (5–29) with those in the earlier particle equation and making use of Eq. (1–35), it may be verified that the frequency for displacement resonance is given by

$$f_{\text{res}} = \frac{1}{2\pi}\,c\,\sqrt{\frac{S}{V_0 l}}\left(\sqrt{1 - \frac{k^4 S V_0}{8\pi^2 l}}\right). \tag{5-30}$$

If the damping factor is neglected, a more approximate but simpler expression for f_{res} is obtained:

$$f_{\text{res}} = \frac{1}{2\pi}\,c\,\sqrt{\frac{S}{V_0 l}}. \tag{5-31}$$

Experimentally, Eq. (5–30) agrees quite well with the lowest frequency to which the Helmholtz resonator, with neck, will respond. Cavities of this type will also resonate to higher frequencies, not given by Eq. (5–30) or Eq. (5–31). These resonances are associated with the existence of standing waves within the cavity, occurring when the wavelength of the driving frequency becomes *less* than the cavity diameter. In these cases the assumptions made here are not valid and the results of the above analysis do not apply.

The physical behavior of the Helmholtz resonator without a neck is fundamentally no different from that of the one with the neck, with the exception that the volume occupied by the plug of air is much less well-defined. Instead of being cylindrical, the volume of air, moving more or less as a unit, is somewhat lens-shaped. Because of the difficulty of determining the exact shape and size of this volume, the entire analysis is necessarily only approximate. The loudspeaker enclosure, known as the bass reflex or phase inverter type, is essentially a Helmholtz resonator. It will be referred to again in Chapter 11. The natural resonance frequency of such an enclosure may be determined most accurately by experiment.

5–11 The behavior of horns. Horn analysis, by any means, is a complicated procedure and involves a great deal of mathematics, in the course of which many approximations must be made in order to reach useful conclusions. It would hardly be worthwhile, in a book of this kind, to give all the details of a complete analysis. Horns, however, are very familiar objects. The simple megaphone or conical horn is used at football games, and the use of horns of other shapes in connection with the indoor and outdoor amplification of music, etc., has become very common. What will be attempted here, in order to explain the essential features of horn behavior, is a brief outline of the subject from the standpoint of acoustic impedance, pointing out the chief physical ideas involved, the fundamental mathematical process, the necessary assumptions, and the important results that may be obtained. Much of the material is based on an analysis by P. M. Morse in his book *Vibration and Sound*, to which the reader is referred for further details.

From one standpoint, a horn may be considered to be the outgrowth of a cylindrical tube, in which the cross section of constant area has been replaced by one whose area gradually increases. It so happens that computations which are difficult for open horns become very simple when applied to a closed cylindrical tube. While this latter case is trivial so far as any practical application as a source of sound is concerned (for, being closed, no sound can get out!), its analysis, as a preliminary problem, will help to make clear the more complicated computations necessary for actual horns.

5-12 Radiation into a cylindrical tube closed at one end. The physical arrangement of such a tube is shown in Fig. 5-4. At the left is an ideal "acoustic piston" whose forward face is plane and perfectly rigid. It fits closely into one end of the air-filled cylindrical tube of length l, the other end of which is closed with a rigid flat plate. The piston is assumed to be driven sinusoidally at some fixed frequency f and with amplitude Q_m. The dimensions of the piston face in relation to λ are of no consequence in this problem. Plane waves will be set up at the piston face and will be *maintained* down the tube, friction at the tube walls being neglected.

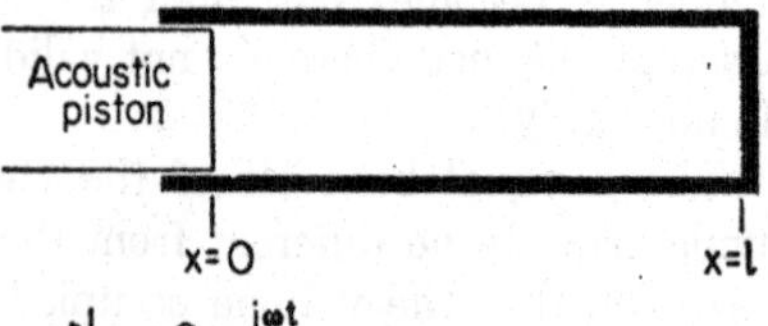

FIG. 5-4. Acoustic piston radiating into a closed cylinder.

The air displacements produced within the tube must satisfy the general equation for plane waves,

$$\xi = f(ct \pm x), \tag{5-32}$$

and, in addition, the boundary conditions at the piston face and at the closed end must be met. Making use of the complex form, these boundary conditions may be stated:

$$(\xi)_{x=0} = Q_m\epsilon^{j\omega t}, \ (\xi)_{x=l} = 0. \tag{5-33}$$

A particular solution can easily be built up to satisfy both Eqs. (5-32) and (5-33). The following is consistent with Eq. (5-32):

$$\xi = A \sin \frac{2\pi}{\lambda} (ct + x) - A \sin \frac{2\pi}{\lambda} (ct - x). \tag{5-34}$$

By expanding Eq. (5-34) as the sines of the sums and differences of angles, we obtain

$$\xi = 2A \sin \frac{2\pi x}{\lambda} \cos \frac{2\pi ct}{\lambda} \tag{5-35}$$

or, expressing the cosine term in the complex form,

$$\xi = 2A \sin \frac{2\pi x}{\lambda} \epsilon^{j\omega t}. \tag{5-36}$$

Equation (5-36) may be fitted to the boundary conditions by letting $A = -\dfrac{Q_m}{2 \sin\left(\dfrac{\omega l}{c}\right)}$ and by adding a phase angle, $\omega l/c$, to the angle $\dfrac{2\pi x}{\lambda}$.

(The solution will still be a function of $(ct \pm x)$, as required to satisfy the differential wave equation.) The final form becomes

$$\xi = -\frac{Q_m}{\sin\left(\frac{\omega l}{c}\right)} \sin\left[\frac{\omega}{c}(x - l)\right] \epsilon^{j\omega t}. \tag{5-37}$$

It may be verified that this satisfies the two essential boundary conditions (1) that at $x = 0$ the air motion be identical with that of the piston and (2) that at $x = l$, ξ and $\dot{\xi}$ be zero, since the rigid plate at that end is always stationary.

In order to see what type of force and energy transfer is involved at the piston face, we may now compute the specific acoustic impedance at the plane specified by $x = 0$. Using Eq. (2-19e) for plane waves, the acoustic excess pressure at the piston face becomes

$$p\Big|_{x=0} = \rho_0 c\omega Q_m \cot\left(\frac{\omega l}{c}\right) \epsilon^{j\omega t}. \tag{5-38}$$

The particle velocity may be obtained by differentiating Eq. (5-37) with respect to time:

$$\dot{\xi}\Big|_{x=0} = j\omega Q_m \epsilon^{j\omega t}. \tag{5-39}$$

Therefore the specific acoustic impedance at the piston face is found to be

$$z_s\Big|_{x=0} = \frac{p}{\dot{\xi}} = -j\rho_0 c \cot\left(\frac{\omega l}{c}\right). \tag{5-40}$$

From Eq. (5-40) it is seen that the air in the tube offers to the driving piston, per unit area, an effective impedance which is always imaginary and therefore in the nature of an electrical reactance. For low frequencies, where the cotangent function is positive, z_s is negative, so that the air acts like a simple spring, or like an electrical capacitance, as might be expected. At higher frequencies, the value of z_s goes through a series of oscillations, being alternately positive and negative as the frequency is raised indefinitely, and hence showing the acoustic impedance to be alternately inductive and capacitive. The magnitude of the reactance alternates between zero and infinity. These extreme limits will not obtain in any actual tube, where there will always be some dissipative force of friction, but assuming any such factor to be small, there will be certain frequencies for which very large motions will be imparted to the air and others for which such motions will be very small. This periodic feature of the behavior of the air in the tube is closely connected with the various standing wave patterns to be expected with a column of air whose length is comparable to or longer than the wavelength. (Standing wave patterns will be studied further in Chapter 7.) In general, whenever z_s turns out to be a periodic function of the frequency, as will be true in certain cases for horns, standing wave phenomena will be implied.

5–13 Force on the piston. Total radiation impedance. Before leaving this problem, it should be pointed out that the *total force* F on the piston due to the air load is Sp, where S is the area of the piston and p is the acoustic pressure given by Eq. (5–38). Also, the total effective *mechanical* impedance $(z_m)_{\text{air}}$, added to whatever mechanical impedance the piston structure itself may have, is

$$(z_m)_{\text{air}} = \frac{F}{(\dot{\xi})_{x=0}} = \frac{Sp}{p/z_s} = Sz_s = -jS\rho_0 c \cot\left(\frac{\omega l}{c}\right). \tag{5–41}$$

When we come to discuss actual sound radiators whose generating surfaces approximate that of an ideal acoustic piston, it will be the magnitude and nature of this mechanical air-load impedance which will determine the speaker's effectiveness as a source of sound. This air-load impedance is often called the *total radaition impedance* offered to the sound source. Note that $(z_m)_{\text{air}}$ equals z_s *multiplied* by the area, whereas the analogous impedance z_a equals z_s *divided* by area. The analogous impedance z_a is useful in connection with energy flow *in the medium*. The mechanical impedance $(z_m)_{\text{air}}$ is a measure of the *reaction on the source* due to the radiation.

5–14 Tube open at one end. Once the cylindrical tube of our problem is opened at the end where $x = l$, the analysis becomes more complex and we shall give only an outline of the derivations. Large gaps in the mathematical procedure are to be expected and many statements and equations must be taken for granted. If, however, the arguments of this chapter have been carefully followed, the results should seem plausible. Even though many of the details are missing, the following summary should shed considerable light upon some rather complicated acoustical phenomena.

Two limiting cases will be considered first. If an acoustic piston has a face large in diameter compared with λ, it will radiate into free air a beam of plane sound waves whose cross-sectional area is the same as that of the piston itself. A closely fitting cylindrical tube placed in front of such a piston will obviously have no effect upon the radiation. As a result, we may conclude that the value of z_s near the face of a piston placed at the end of a cylindrical tube whose diameter is large compared with λ will be $\rho_0 c$, just as for plane waves in free space. The quantity z_s is all real. There are no reflections or standing waves and there is no "wattless" power periodically returned to the source.

At the other extreme, when the tube diameter is small compared with λ, practically all of the wave energy reaching the open end is reflected back towards the piston. It can be shown (see Morse) that the specific acoustic impedance at the piston is in the nature of an electrical *reactance*, just as was the case with the closed tube. A mass of air near the open end of the tube moves as a unit, with a reacting force proportional to its acceleration. When the frequency is low enough so that the length of the tube is considerably less than λ, the specific acoustic impedance at the piston face is approximately

$$z_s\bigg|_{x=0} \cong j\omega\rho_0(l + 0.6r), \tag{5-42}$$

where l is the tube length and r its radius. This equation, which neglects the small amount of power radiated from the small open end of the tube, shows z_s to be in the nature of an *inductive* reactance, since it is positive. With longer tubes, stationary waves may occur, and z_s may be either plus or minus, depending on the relationship of l to λ, and therefore may be either inductive or capacitive.

The dimensions encountered in practical sound problems are rarely the limiting ones that are very large or very small compared with λ. Rather they are *comparable* to λ. Equation (5-43) is an approximate expression for the value of z_s at the driving end of a cylindrical tube, open at the far end, and for which the diameter is neither very large nor very small compared with λ.

$$z_s\bigg|_{x=0} \cong \rho_0 c \tanh \pi\left[\alpha + j\left(\beta + \frac{2l}{\lambda}\right)\right], \tag{5-43}$$

where α and β are functions of λ and of the tube dimensions. The extent to which reflection occurs at the open end determines the value of α, while β may be plus or minus, depending on the phase associated with the reflection. As might be expected, z_s is neither all resistive nor all reactive, but in general is a combination of both, with the proportions determined by the ratio of tube dimensions to wavelength. Interpreted in simple terms, the form of z_s indicates that some fraction of the energy leaving the piston at any instant is reflected upon reaching the open end and the remainder is radiated into space as real sound energy.

A few of the wind instruments, like the piccolo and the pipe organ, make use of tubes of constant cross section. The mechanism by which waves are set up within such tubes is not easy to visualize, since no vibrating driver is used but instead a steady stream of air is blown in near one end. This will be discussed later in Chapter 7.

5–15 Horns. When the factor of flare is added to a cylindrical tube, it becomes a horn. Horns are usually of circular cross section, but often are square or rectangular, for structural convenience. What is important in the following analysis is not the shape, but rather the *rate* at which the cross sectional area increases as one moves down the axis of the horn.

To obtain equations of reasonable simplicity it is advantageous to assume that plane waves set up at the small end of the horn by an ideal acoustic piston are maintained as plane all the way down the horn. This seemed a reasonable assumption in the case of the cylindrical tube, but with the introduction of flare, a gradual transformation from a plane to a spherical wave shape is to be expected. A more precise analysis of horn behavior that takes this process into account, however, will lead to the same general conclusions as those summarized in this chapter.

We shall also assume that friction at the walls dissipates negligible energy and, to encourage simpler mathematics, that the mouth of the horn is large compared with the wavelength of any disturbance traveling along the axis. (Specifically, that the *perimeter* of the mouth is larger than λ.) This latter assumption is rather violent for horns of practical dimensions but it represents a considerable mathematical simplification because, from considerations of diffraction, all wave energy reaching the mouth will then leave the horn, with no reflections. At the end of our horn discussion we shall briefly consider the effect of mouth size upon horn behavior. The effect is complicated and only qualitative comments will be given.

Even though a plane wave front is assumed throughout the interior of the horn, it is necessary to alter the original differential equation for plane waves in free space to take account of the expanding cross section. The essential change is the introduction of a new expression for the dilatation, $\delta = v/V_0$. This expression takes account of the variation of the cross-sectional area S with the distance along the x-axis and may be written

$$\delta = \frac{1}{S}\frac{\partial(S\xi)}{\partial x}.$$

(If S is constant, as in a beam of plane sound waves, δ becomes $\partial\xi/\partial x$, as for plane waves in free space.) With this change, it can be shown that the differential equation describing pressure variations as a function of time and of distance along the axis may be written

$$\frac{\partial^2 p}{\partial t^2} = c^2 \frac{1}{S}\frac{\partial}{\partial x}\left(S\frac{\partial p}{\partial x}\right). \tag{5–44}$$

This is the general differential equation for any horn. Once the function $S = f(x)$ is specified, the shape of the horn is defined. It remains, then, to integrate Eq. (5–44) in order to completely determine the pressure p

as a function of x and t. Once p is known, all other important wave properties within the horn may be determined.

Of all the possible forms for the function $S = f(x)$, two only will be discussed here, i.e., those defining the conical and the so-called exponential horn. Other shapes have been studied but these two are most widely used in practice.

5-16 The conical horn. The conical horn is the oldest and, at least until recent years, the most widely used of horn shapes. It is defined by the equation

$$S = S_0\left(1 + \frac{x}{x_0}\right)^2. \quad (5\text{-}45)$$

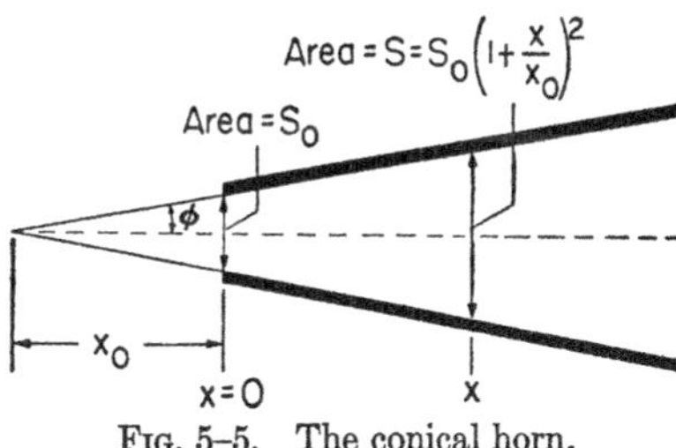

FIG. 5-5. The conical horn.

In Fig. 5-5, S_0 is the cross-sectional area (assumed to be circular in shape) at the small end of the horn ($x = 0$), where the energy is introduced. The distance x_0 is measured from this small end back to the position of the geometrical apex of the cone. The quantities S_0 and x_0 are related to the rate of flare. As is seen in Fig. 5-5, the factor x_0 may also be defined as $y_0/\tan\phi$, where y_0 is the radius of the small end and ϕ is half the plane vertex angle of the cone. The value of x_0 is therefore essentially determined by the ratio of the size of the driver, placed at the small end of the horn, to the rate of flare as controlled by the angle ϕ. The parameter x_0, as we shall see, is an important factor controlling the behavior of a conical horn.

If the function given by Eq. (5-45) is inserted into the differential equation for a horn, (5-44), the result is an expression having a mathematical form identical with that for spherical waves in free space. (This is mathematically true, even with the physical assumption that the wave front remains approximately plane as the disturbance travels down the horn.) The solution for p may then be written down immediately, assuming the mouth of the horn large compared with λ so that no reflection takes place. As in the case of spherical waves in free space, the pressure falls off inversely with the distance along the axis of the horn but the coordinate x for the horn replaces the coordinate r used for spherical waves.

From the expression for the pressure, the equation for the particle velocity $\dot{\xi}$ may be found by means of the usual wave relations. Hence one may evaluate the specific acoustic impedance at the small end, where $x = 0$. The results are

$$\left.\begin{aligned} (z_s)_{x=0} &= R + jX, \\ \text{where} \quad R &= \rho_0 c \frac{1}{1 + \left(\frac{1}{2\pi}\frac{\lambda}{x_0}\right)^2} \\ \text{and} \quad X &= \rho_0 c \frac{\frac{1}{2\pi}\frac{\lambda}{x_0}}{1 + \left(\frac{1}{2\pi}\frac{\lambda}{x_0}\right)^2}. \end{aligned}\right\} \quad (5\text{–}46)$$

It will be noted that both R and X are functions of the ratio of λ to the parameter x_0.

An interesting comparison can be made between the equations just obtained and the expressions for the two components of z_s for a point on a pulsing sphere radiating spherical waves into free space (see Eq. (5–20)). A *small* pulsing sphere radiates poorly since, from Eq. (5–20), when r is small the real part of z_s is small compared with the imaginary part. The quantity x_0 appearing in Eqs. (5–46) affects the ratio of R to X in the same manner. A small vibrating source is easier to construct than a large one (for some of the reasons, see Chapter 11). The use of a conical horn is advantageous because although a small acoustic piston, like a small pulsing sphere, will radiate poorly, if we couple to it a conical horn for which x_0 is large compared with λ, we obtain a radiating system which possesses the efficiency of a pulsing sphere of *much larger radius* (where R is greater in proportion to X). Since $x_0 = y_0/\tan\phi$, we may either make y_0 *large* or, as is more common with the usual driver, where y_0 is necessarily small, we may ensure a large value of x_0 by using a very gradual taper so that ϕ is *small*.

The next section will throw still further light on this matter.

5–17 Transmission coefficient for a horn. To show how the ratio x_0/λ enters into the efficiency of a conical horn, we shall define *transmission coefficient* for a horn of *any* taper. The transmission coefficient is the ratio of the real power radiated into (and therefore out of) a given horn, to the power radiated by the same acoustic piston, having the same velocity, into an infinitely long cylindrical tube of the same cross-sectional area as the small end of the horn. In the case of the cylindrical tube, all instantaneous power will flow away from the piston, since plane waves of constant cross section will be set up and z_s at the piston face will therefore be $\rho_0 c$. The transmission coefficient consequently is a measure of the efficiency of a horn as compared with a simple direct generator of plane waves, the latter process being the most efficient from the standpoint of real radiated sound power.

For the cylindrical tube, the average power fed in by the piston is

$$U_t = (\xi_{\rm rms})^2 \rho_0 c S_0.$$

For the conical horn, making use of the expression for R from Eq. (5-46), the power is

$$U_c = (\xi_{\rm rms})^2 \frac{\rho_0 c S_0}{1 + \left(\dfrac{1}{2\pi}\dfrac{\lambda}{x_0}\right)^2}.$$

Therefore the transmission coefficient τ for a long horn with a large mouth is

$$\tau = \frac{1}{1 + \left(\dfrac{1}{2\pi}\dfrac{\lambda}{x_0}\right)^2}. \tag{5-47}$$

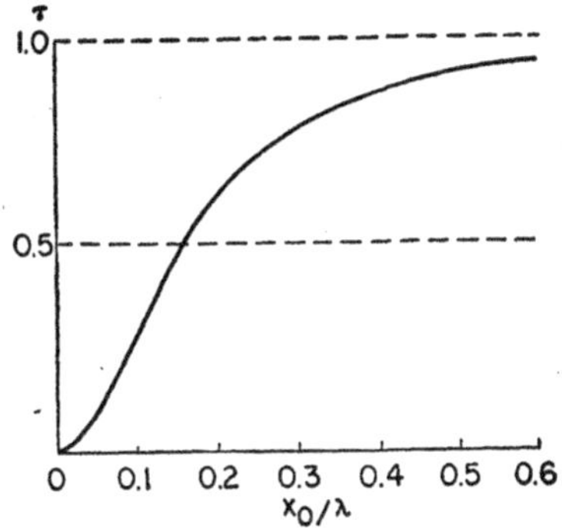

FIG. 5-6. Transmission coefficient τ for a long conical horn with open end large enough to eliminate reflection. (*After* Morse)

Figure 5-6 is a plot of τ vs the ratio x_0/λ, the significant parameter. It is seen that for a given rate of taper, where x_0 is constant, the abscissa is essentially the frequency, since $x_0/\lambda \propto f$. While the horn will radiate at all frequencies, its efficiency obviously falls off rapidly at the lower end of the spectrum. If x_0 is increased, one may go to longer wavelengths (or lower frequencies) before the transmission coefficient falls much below unity. However, for a driver of fixed size, it will be necessary to *decrease* the cone angle ϕ and to increase the length of the horn so that the mouth will be large enough to preclude reflections. A very long and therefore bulky horn is desirable for good efficiency over a wide frequency range. (Such a long horn may, however, actually *decrease* the efficiency at *very high* frequencies, due to frictional forces along the horn walls.) In general, the efficiency of the conical horn, indeed of all horns, falls off at the low frequencies due to practical limitations of size.

5-18 The exponential horn. It is the performance in the low frequency range that makes the exponential horn so superior to the simple conical horn. An analysis similar to that for the conical horn therefore seems worthwhile. An exponential horn is one whose cross-sectional area varies according to the equation

$$S = S_0 \, \epsilon^{2x/h}, \tag{5-48}$$

where S_0, as for the conical horn, is the area of the small end and S is the area of any other cross section a distance x from the small end (Fig. 5-7).

The constant h is the "flare" factor. If the cross-sectional area is circular, Eq. (5–48) may be written

$$y = y_0 \epsilon^{x/h}, \tag{5-49}$$

where y_0 is the radius of the small end and y is the radius at any distance x along the axis. If we let $x = h$ in Eq. (5–49), $y = \epsilon y_0$ and h is thus the distance along the horn axis such that the radius of cross section increases by the factor 2.718. Note that if h is *large*, the rate of flare is *small*, and vice versa.

If the expression for S given by Eq. (5–48) is introduced into the differential equation for a horn, the acoustic pressure may be found as a function of x and t. From the pressure, we can obtain the particle velocity. The real and imaginary parts of $z_s = p/\xi$, evaluated at the piston end of the horn, where $x = 0$, turn out to be, for outgoing waves,

$$R = \rho_0 c \sqrt{1 - \left(\frac{1}{2\pi}\frac{\lambda}{h}\right)^2}$$

and

$$X = \rho_0 c \left(\frac{1}{2\pi}\frac{\lambda}{h}\right). \tag{5-50}$$

When λ approaches $2\pi h$, so that $f(=c/\lambda)$ approaches $c/2\pi h$, the value of the real part of z_s approaches zero. Below this critical or *cutoff frequency*, $f_0 = c/2\pi h$, no real power enters or leaves the horn. This type of behavior is peculiar to an exponential horn. It can be shown that an exponential horn transmits waves in the same way as a dispersive medium; different frequencies travel down the horn with different velocities. At the frequency f_0 the velocity of propagation of any given phase becomes infinite, which is equivalent to saying that no true wave motion exists within the horn, there being no phase differences along the axis.

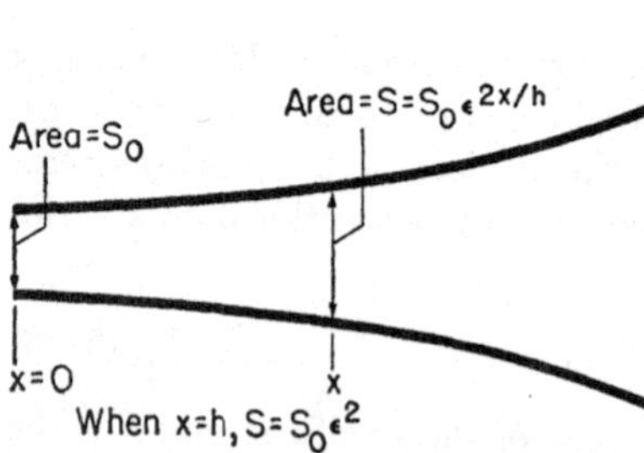

FIG. 5–7. The exponential horn.

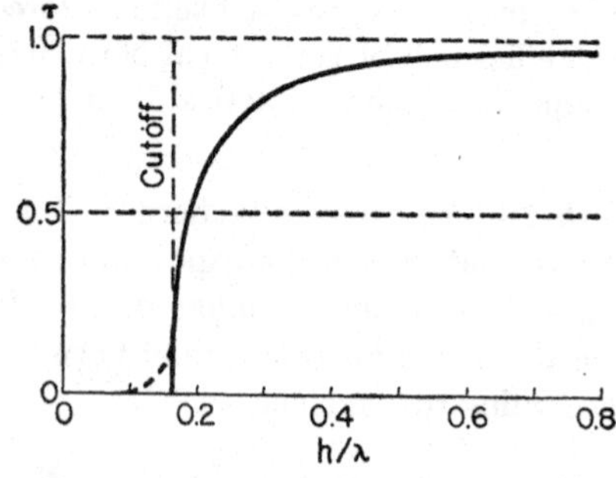

FIG. 5–8. Transmission coefficient τ for a long exponential horn with open end large enough to eliminate reflection. (*After* Morse)

5-19 Transmission coefficient for an exponential horn. Comparison with the conical horn. The characteristics of an exponential horn are shown graphically in Fig. 5-8, which is a plot of the transmission coefficient τ against the ratio h/λ (h is the flare factor defined above). For a horn of fixed flare rate, the abscissa is proportional to $1/\lambda$ or to frequency. In this case $\tau = \sqrt{1 - \left(\frac{1}{2\pi}\frac{\lambda}{h}\right)^2}$. The most interesting feature of the graph is the sharp cutoff and its relation to flare rate and frequency. The cutoff occurs for $h/\lambda \cong 0.16$. From this number the critical wavelength λ_0 and the critical frequency f_0 may be determined, once the value of h has been fixed. A numerical illustration will show how this may be done.

Suppose that we wish to design a horn whose cutoff frequency is 50 cycles-sec^{-1}. This corresponds to $\lambda_0 = c/f_0 \cong 20$ ft, and therefore the required value of h is $0.16(20) \cong 3.2$ ft. To ensure good transmission all the way down to 50 cycles-sec^{-1}, it will be necessary to select an abscissa somewhat larger than the cutoff, say a value of about 0.3. In this case h will be 6 ft. A horn whose radius increases by a factor ϵ (approximately 3) in a distance of 6 ft along the axis will be a long one if, for the reasons given earlier, it is to have a large mouth.

Experimental horns do not exhibit the mathematically sharp cutoff of the solid line in Fig. 5-8, but instead show a gradual tapering off at the lower end, somewhat as shown by the broken curve. This deviation from theory is not surprising, in view of the approximations in the mathematics. The essential shape, however, is as predicted.

A striking comparison can be made between a long, wide-mouthed conical horn and an exponential horn of identical over-all dimensions by plotting the transmission coefficients of each. The abscissas in this case are simply frequencies. These graphs are shown in Fig. 5-9, with the specific data for the flares. The superiority of the exponential shape for uniform frequency transmission is most apparent.

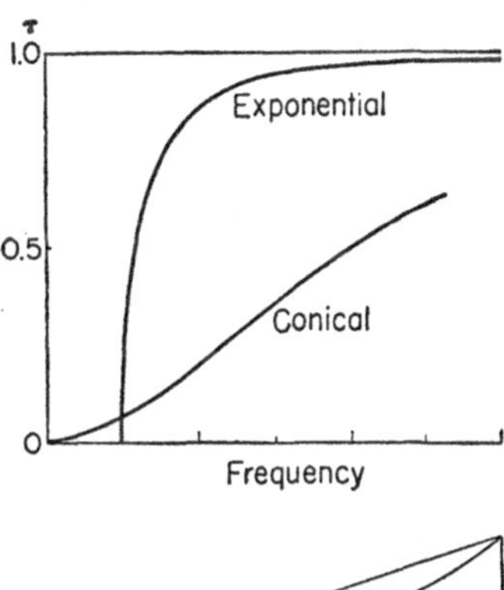

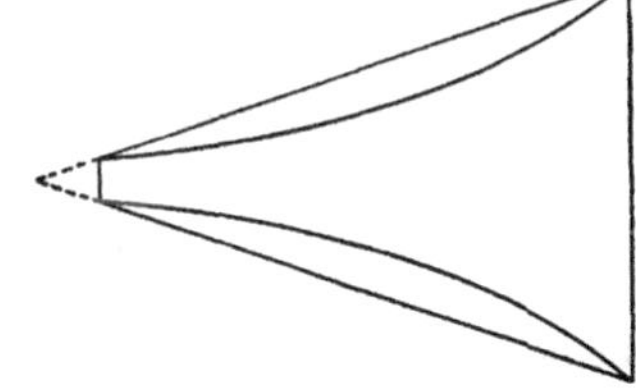

FIG. 5-9. Frequency characteristics for a conical and an exponential horn of the same over-all dimensions. (*After* Morse)

5–20 Effect of reflections upon horn behavior. With horns that are not very long and where the perimeter of the mouth is not large compared with the wavelength, the conclusions reached above are subject to considerable modification. The specific acoustic impedance at the radiating mouth is no longer that for plane waves, $\rho_0 c$, and some of the energy is reflected back to the source, giving rise to standing waves. This is desirable in certain musical instruments where this resonance produces the characteristic pitch (and overtones). In all musical instruments some radiation is essential, in addition to resonance, and both radiation and resonance come about naturally when the horns are of limited dimensions. When horns are used in loudspeaker systems for the wide-range reproduction of music, however, resonance is a definite drawback, giving rise to an unnatural enhancement of certain frequencies and attenuation of others.

When reflection takes place the analysis of horn behavior is more complicated and we shall not attempt it here. When an approximate expression for z_s at the small end of such a horn is obtained, the magnitude of the real part, R, is found to fall off at the lower frequencies, as before, but there are numerous peaks and valleys at frequencies corresponding to the horn resonances. For practical purposes, it is clear that horns which are long and large-mouthed (and so, unfortunately, bulky) are the most satisfactory for a wide range of frequencies.

5–21 The horn as an impedance matching device. From the electrical analog point of view, we may make an important generalization about the behavior of horns. We have seen that a small vibrating surface, spherical in shape or otherwise, is an ineffective radiator of sound waves because of the large reactive component of the acoustic impedance at its surface. On the other hand, an acoustic piston large in diameter in comparison with the wavelength encounters the specific acoustic impedance $\rho_0 c$ characteristic of plane waves and is therefore an effective radiator. With the horn in front of it, a small acoustic impedance may be made to radiate as would a large piston, emitting waves that are nearly plane (since z_s is then all real) over a large frequency range. In this sense the horn may be considered as a kind of impedance matching device, transforming the acoustic impedance characteristic of a small source to that necessary to match the acoustic impedance for plane waves in free space. In this way a great deal more power is radiated than would be the case without the horn. This is the same impedance matching idea that is so important in electrical circuits.

5–22 The "hornless" or direct-radiator loudspeaker. Specific acoustic impedance at the surface. In recent years there has been a tendency, particularly for compact sound reproducers in the home, to dispense with the use of the horn entirely, and so avoid its numerous design complications. The radiator is then simply the equivalent piston itself, set in a large baffle (Fig. 5–10a) as nearly impervious to sound as possible, whose purpose is to ensure a single, rather than a double, source action, as discussed in Chapter 3. Unless the diameter of the piston face is impossibly large, one would expect, from the discussion at the end of the previous section, that the radiation efficiency would be low for low frequencies. This is indeed true, as we shall presently see, even when the baffle area is very large.

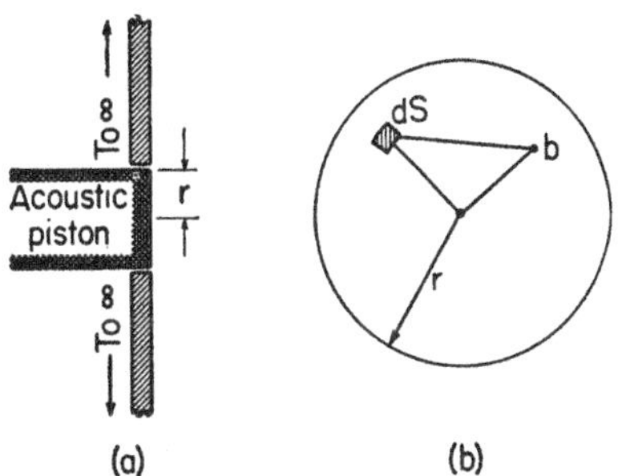

FIG. 5–10. Acoustic piston set in an infinite baffle.

The outline of the mathematical procedure necessary to obtain the acoustic impedance at the piston face is as follows. The baffle area is assumed to be infinite in extent. Referring to Fig. 5–10b, the first expression to be set up is for the instantaneous differential pressure dp occurring at a point b on the piston face, due to the spherical waves originating from a differential area dS at another point, also on the piston face. This is readily done in terms of the spherical wave equations, using suitable coordinates. The total pressure at b due to all such areas dS is obtained by integrating over the piston face. The total instantaneous acoustic force F on the piston is then the second integral of $p\,dS'$ over the area of the piston, dS' being an area element at point b. The integration difficulties in this process are considerable but they may be overcome. The *average* acoustic pressure is obtained by dividing the total force F by the area of the piston. This average pressure placed over the piston velocity (the same as the velocity of the air next to it) yields an average value of z_s. The real and imaginary components of z_s so obtained are

$$R = \rho_0 c\left[1 - \left(\frac{2}{w}\right)J_1(w)\right]$$

and (5–51)

$$X = \rho_0 c M(w).$$

In these expressions, $w = 4\pi r/\lambda$, where r is the radius of the piston face. The symbol $J_1(w)$ is the usual one for a first-order Bessel function. A Bessel function is a certain regular, convergent power series (appearing often in solutions to the differential equations of physics) whose sum may be computed and tabulated for various values of the argument w. Therefore for every value of r/λ there is a specific number representing $J_1(w)$.

The symbol $M(w)$ represents a certain definite integral whose numerical value is a function of w and of the coordinates chosen in the integrals described above.

The plot of $R/\rho_0 c$ in Fig. 5–11 is identical with a plot of the *transmission coefficient* defined earlier in connection with horn radiation. As in the graphs for the horns, the abscissas are proportional to the frequency, for any one piston of a given radius.

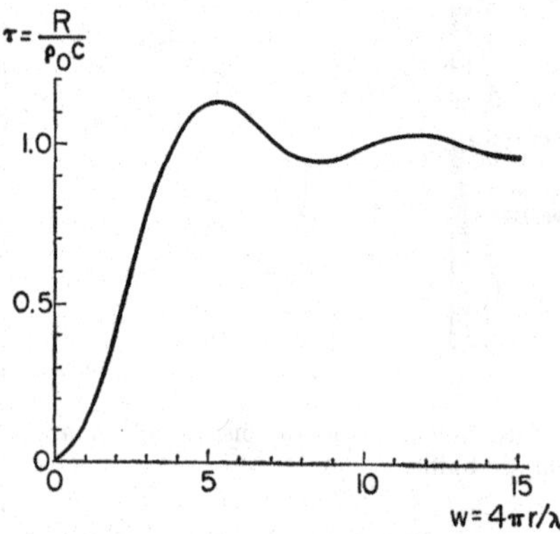

FIG. 5–11. Transmission coefficient τ for an acoustic piston mounted in an infinite baffle. (*After* Morse)

It is often thought, by practical workers in the field of acoustics, that the use of a very large baffle will, of itself, ensure the efficient radiation of the lowest frequencies in the audible spectrum. From Fig. 5–11 it can be seen that even the use of a baffle of *infinite* area does not prevent a serious falling off below a certain frequency. The value of this fairly critical frequency depends on the piston radius r. If numerical values are inserted, the results are discouraging from a practical point of view. Suppose that we desire a source of good radiation efficiency down to a frequency of 50 cycles-sec^{-1}. Taking the abscissa $w = 5.0$ as the lower limit of the horizontal portion of the curve for $R/\rho_0 c$, and using the proper value for λ (in this case about 20 ft), we find that the necessary piston radius is 8 ft! It is quite clear, then, that with the more modest value of, at most, 0.5 ft for the radius of the ordinary radio or phonograph loudspeaker, the acoustic radiation for low frequencies, *assuming constant piston velocity*, will be very poor. Fortunately, as we shall see in Chapter 11, it is possible to compensate for the inherent low frequency limitations of the direct radiator quite satisfactorily by the use of mechanical resonance and the proper design of the mechanical impedance of the piston itself. When we come to this practical discussion, it will be seen that there are serious complications in the *high* frequency range, not apparent in the curves shown in this chapter. This is due primarily to the difficulty of making the piston, which is as large as possible for the reasons just given, respond efficiently to the driving force at the higher frequencies. For this and other reasons, the present trend is to feed the high frequencies, by means of a frequency-selective network, into a small horn speaker whose driving piston may be much smaller and lighter.

5-23 Distribution pattern for energy leaving horns or direct radiators. It is to be noted that no clue has been given in this chapter as to just how the energy leaving the horn or direct radiator is distributed in space. This is primarily a matter of diffraction and certain general features of the distribution should be apparent from the discussions of Chapter 4. If the diameter of the mouth of a horn or of the end of a directly radiating acoustic piston is large compared with wavelength, there is a good acoustical match with free space for the production of a beam of plane sound waves. The cross-sectional area of this beam is the same as the radiating area. For piston or horn areas small compared with λ, one would expect the spherical distribution characteristic of a point source. The interesting and important in-between cases, where the radiating areas have diameters comparable to λ, give rise to diffraction patterns, the features of which were discussed in Chapter 4.

5-24 General significance of acoustic impedance for radiation. The results of the several specific problems analyzed in this chapter have wide application in many aspects of acoustics. The energy emitted from the mouth in speech or in singing is affected by the value of the acoustic impedance at the opening, and the design of musical wind instruments, historically fashioned on a purely empirical basis, unconsciously has taken account of the problem of impedance matching. Instruments like the piccolo and the flute, with their relatively high registers, have short horns or tubes with mouths of comparatively small diameter, whereas the lower register brasses, like the contrabass tuba, require and have large radiating ends.

An interesting and important parallel to these problems in radiation occurs in sound reception. When sound waves strike the ear, that fraction of the energy that eventually passes to the mechanism of the inner ear is a function, among other things, of the impedance match for the particular frequency or frequencies involved at the entrance of the ear. Again, the proper design of microphones, whose operation depends on the transfer of acoustic energy to some mechanical diaphragm, ribbon, etc., takes account of the acoustic impedance in the air and its relation to the mechanical impedance of the instrument.

Other applications of the idea of acoustic impedance will appear later in this book. It is a concept of growing usefulness in problems of acoustics, fully as fruitful in its field as is the similar idea of radiation impedance in the propagation of electromagnetic waves.

PROBLEMS

1. Find the real and the imaginary parts of the following complex quantities: (a) $10\epsilon^{j(30°)}$; (b) $5\epsilon^{-j(10°)}$; (c) $2\epsilon^{2+j(10°)}$; (d) $7\epsilon^{-2-j(30°)}$.

2. Rewrite the following complex quantities in the polar form: (a) $4 + j3$; (b) $10 - j3$; (c) $(a + jb)^2$.

3. Reduce the following expressions to the form $a + jb$: (a) $(2 + j)\epsilon^{j(30°)}$; (b) $\epsilon^{j(10°)} + \epsilon^{-j(30°)}$; (c) $(c - jd)(e + jf)$.

4. A complex expression has the form $\frac{(e + jd)^2}{e - jd}$. Rationalize the denominator by multiplying the expression by $\frac{e + jd}{e + jd}$, collect terms, and reduce to the form $a + jb$.

5. An acoustic pressure is given by $p = p_m\epsilon^{j\omega t}$. The particle velocity at the same point is $\dot{\xi} = (\dot{\xi})_m\epsilon^{j(\omega t+\theta)}$. (a) Find the specific acoustic impedance in the form $a + jb$. (b) What is the average real flow of power at the point?

6. Give a physical argument to show why the reactive component of the specific acoustic impedance at the surface of a small sphere should be inductive (i.e., $+jX$) rather than capacitive.

7. What is the effect upon the resonance frequency of a Helmholtz resonator of doubling (a) the volume V_0, (b) the length l of the neck, (c) the cross-sectional area S of the neck?

8. Compute the approximate resonant frequency in air of a Helmholtz resonator where the boundaries of V_0 are those of a box $1 \times 2 \times 3$ feet and where the cylindrical neck has a length of 6 inches and a diameter of 4 inches.

9. For the resonator described in problem 8, plot a resonance curve to show $\dot{\xi}$ as a function of frequency for a driving force of constant maximum value.

10. Referring to Section 5–12, discuss the physical situation in the closed tube of fixed length l for those frequencies at which $z_s\Big|_{x=0}$ approaches zero or infinity. Make use of Eqs. (5–36), (5–37), (5–38), and (5–39).

11. An acoustic piston sending plane waves into the closed tube of Section 5–12 has itself a mechanical impedance, exclusive of the air load, of $80 + j50$ cgs units (mechanical ohms). If the tube length is 100 cm and the cross-sectional area is 10 cm², find the total mechanical impedance of the piston at a frequency of 50 cycles-sec⁻¹, including that due to the presence of the air in the tube.

12. Compute the total effective mechanical impedance of the piston in problem 11 if the further end of the tube is open. Make use of Eq. (5–42), first checking its validity for this case.

13. A long conical horn has a small end 4 cm in diameter and a semiplane vertical angle of 10°. From the graph of Fig. 5–6, find the transmission coefficient for a frequency of 700 cycles-sec^{-1}.

14. The cutoff frequency of a long exponential horn is 100 cycles-sec^{-1}. (a) What is the flare factor h? (b) Along the horn axis, how far apart must the positions be for the ratio of one radius to the other to be 2:1? (c) If the small end of the horn has a radius of 2 cm, how long must the horn be to have a mouth radius of 50 cm?

15. The horn in problem 14 is being driven by an acoustic piston of constant velocity, independent of frequency. The decibel scale may be used to compare any two powers, as well as any two acoustic intensities. At what low frequency will the power output of this horn drop 1 db below the output at very high frequencies?

16. (a) For a long conical horn, plot the reactive component X at the piston end as a function of x_0/λ. (b) Plot a similar curve for the exponential horn.

using as abscissas the ratio h/λ. Compare these graphs with the plots of the real part of the specific acoustic impedance.

17. Change the abscissa scale of the graph of Fig. 5–11 to frequency, assuming an acoustic piston of radius 6 inches. At what low frequency will the power output drop 5 db below the output at very high frequencies, assuming a constant piston velocity, independent of frequency? (See problem 15 for use of the decibel scale.)

CHAPTER 6

LONGITUDINAL WAVES IN DIFFERENT GASES. WAVES IN LIQUIDS AND SOLIDS

Throughout the preceding chapters we have been dealing entirely with the medium *air*, since longitudinal waves through air are of the most practical interest in acoustics. In this connection we have been using for the velocity of sound in air the value 331 meters-sec^{-1}, which has been stated to be the experimental value under ordinary free-air conditions. This value also agrees with the theoretical equation $c = \sqrt{\mathcal{B}/\rho_0}$, where $\mathcal{B}$ is the adiabatic bulk modulus. To understand more fully the transmission process of longitudinal waves in air and in other gases as well, let us examine carefully the factors determining the wave velocity c.

6–1 Isothermal and adiabatic bulk modulus for an ideal gas. If one assumes (as Isaac Newton did) that the deformation process associated with longitudinal waves in an ideal gas is isothermal, the proper value of the bulk modulus $\mathcal{B}$ may be obtained directly from the equation of state for such a gas. Assuming the temperature to remain constant, we have

$$PV = \text{constant}$$

and therefore

$$P\,dV + V\,dP = 0,$$

so that

$$\frac{dP}{-dV/V} = \mathcal{B}_i = P. \tag{6–1}$$

When this value of $\mathcal{B}_i$ is introduced into the expression for the wave velocity c, one obtains a value which, for air, is smaller than the experimental value by about 20%. Newton had an ingenious but wholly erroneous explanation for this discrepancy.* Laplace showed that the correct elastic constant to use is the *adiabatic* bulk modulus, $\mathcal{B}_a$. For an ideal gas undergoing an adiabatic change, the relationship between pressure and volume is

$$PV^\gamma = \text{constant},$$

where γ is the ratio of the specific heat of the gas at constant pressure to that at constant volume. Taking differentials, we have

$$PV^{\gamma-1}\,dV + V^\gamma\,dP = 0$$

* See Miller, *Anecdotal History of Sound*, Macmillan, 1935.

and upon rearranging, we get

$$\frac{dP}{-dV/V} = \gamma P = \mathcal{B}_a. \tag{6–2}$$

This value of $\mathcal{B}$, when inserted into the expression for c, gives good agreement with the observed value for waves of ordinary amplitudes over the range of audible frequencies. We may therefore write for sound waves

$$c = \sqrt{\frac{\gamma P}{\rho_0}}. \tag{6–3}$$

The value of the dimensionless ratio γ is dependent upon the number of so-called degrees of freedom of the gas molecule, such number in turn being dependent on the molecular complexity. For monatomic molecules γ turns out to be 1.66, for diatomic molecules it is 1.40, for triatomic molecules, 1.29, etc. In Table 6–1 are listed a few common gases, together with the experimentally determined values of γ, all at a pressure of one atmosphere. Notice that for air, whose major constituents are diatomic gases, the value of γ is 1.40, which is characteristic of such gases.

TABLE 6–1

VALUES OF γ FOR DIFFERENT GASES AT 1 ATM PRESSURE

Gas	*Temperature*	γ
Air	17°C	1.403
Carbon dioxide	15°C	1.304
Hydrogen	15°C	1.410
Nitrogen	15°C	1.404
Oxygen	15°C	1.401

6–2 Factors affecting the velocity of longitudinal waves in gases. The quantity γ varies for different types of gases, but for any one gas it is quite constant for wide variations in temperature. For air, γ remains at the value 1.40 over a temperature range of -80°C to at least 150°C. The other two factors determining the velocity of waves of small amplitude are the density ρ_0 and the pressure P. The density varies for different gases, and both density and pressure are related to the absolute temperature T through the gas laws. The temperature is therefore an implicit variable in determining the velocity of a wave. Table 6–2 gives the velocity of sound waves for various gases. The most precise value of c for air, as computed by averaging a number of the best determinations,* is

$$c = 33{,}145 \pm 5 \text{ cm-sec}^{-1}$$

* Hardy, Telfair, and Pielemeier, *Jour. Acous. Soc. Amer.* **13**, 226–233 (1942).

for the conditions 0°C, 760 mm of Hg, 0.03 mole percent content of CO_2, and 0 percent water content.

We are usually interested in the velocity of sound waves in the open air. The variations in pressure under these conditions are not ordinarily great enough to affect the velocity of sound waves significantly. The variations in density due to temperature fluctuations, however, are more important. Since at constant pressure the density of a gas is inversely proportional to the absolute temperature T, we may write for a fixed pressure of one atmosphere,

$$c = c_{273}\sqrt{\frac{T}{273}}. \tag{6–4}$$

The velocity of propagation of ordinary sound waves is quite independent of the frequency, as Eq. (6–3) indicates. At very high frequencies, well into the ultrasonic region, some frequency dependence is to be expected because the deformation process tends towards the isothermal as the frequency is raised to very high values. That this is so does not seem at all obvious. Because of the shorter *time* factor at the higher frequencies, one might think the process to be even more completely adiabatic than at lower frequencies. This reasoning, however, neglects the importance of the *temperature gradient*, as has been pointed out by Herzfeld and Rice.* The total heat flow between adjacent compressions and rarefactions during any one half-cycle is proportional to $1/f$ because of the time factor, but due to the increased steepness of the temperature gradient (dT/dx) at the higher frequencies, the rate of flow will be greater in proportion to f^2. As a result, the total flow *increases* as the frequency is raised and is proportional to f. With such a flow, the process tends to be *less* adiabatic. The frequencies at which this effect becomes significant are very high (of the order of 10^8 or 10^9 cycles-sec^{-1}).

TABLE 6–2

VELOCITY OF SOUND WAVES FOR VARIOUS GASES

Gas	*Velocity, c (calculated, cm-sec^{-1}) 0°C, 1 atmosphere*
Carbon dioxide	2.58×10^4
Helium	9.70
Hydrogen	12.69
Nitrogen	3.37
Oxygen	3.17

* *Phys. Rev.* **31**, 691 (1928).

6–3 Experimental determination of the velocity of sound waves in gases. Careful experimental measurements of the velocity of sound waves in the open air were made as early as 1738, when data based on cannon fire were recorded. The numerous observations taken under military auspices since that date have usually been made with an explosive type of source, giving rise to waves of rather high intensity. Using a source of more moderate intensity, T. C. Hebb (at the suggestion of Michelson) in 1905 and again in 1919* made a very careful determination of the velocity c in open air, through the direct measurement of the *wavelength* in a standing wave pattern set up between two coaxial parabolic reflectors. The source was a whistle of known frequency. Having measured the wavelength λ, the velocity c can be computed from the relation $c = f\lambda$. Most of the more recent experiments are basically of this same general type. Sound sources of standard frequency whose value is known to a high degree of precision are now familiar objects in every well-equipped acoustical laboratory. (Such frequency standards will be discussed in Chapter 10.) With such sources available, the precision of the determination of c is then dependent on the precision with which the frequency is known and on the precision of the measurement of λ, usually rather high.

Because of the ease with which the density, pressure, and temperature may be controlled and varied, many measurements of c have been made with air and other gases in closed tubes. These experiments also make use of a standing wave pattern and are variations on the experiments of Kundt in 1866. Kundt set a column of air into vibration by means of an exciting rod made to vibrate longitudinally by being rubbed with a rosin-coated piece of leather, as shown in Fig. 6–1a. As the plunger A is moved along the tube, adjusting the length of the air column, resonance is found to occur for certain positions. In the associated standing wave pattern there are regions where the air is stationary, called *nodes*, and others where the air is in violent motion, called *antinodes*. As we shall see in our discussion of stationary waves in the next chapter, the distance between adjacent nodes is just $\lambda/2$. The nodal planes are made visible experimentally by introducing into the tube light particles such as cork dust. In the presence of excitation these particles collect in rather sharply defined piles at the nodal regions. (There usually occur subsidiary striations, much more closely spaced than $\lambda/2$. The spacing of these small striations is largely a function of particle size, rather than of the wavelength of the tube, and may be explained in terms of a Bernoulli effect upon the individual particles.) In the modern versions of the Kundt's tube experiment,

---

* *Phys. Rev.* **20,** 91 (1905); also **14,** 74 (1919).

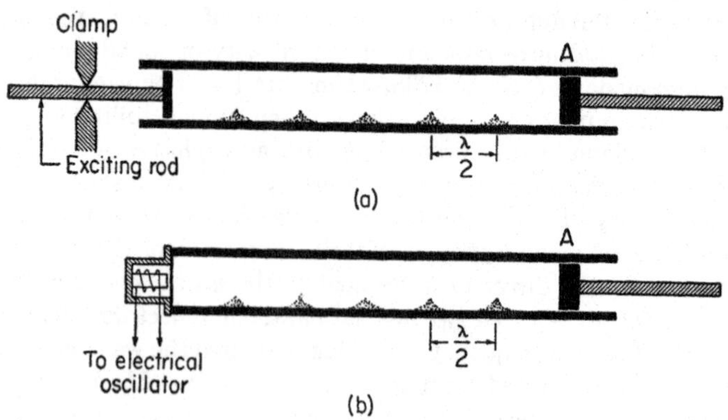

FIG. 6–1. Kundt's tube.

an electrically driven loudspeaker, equivalent to an acoustic piston at one end of the tube (Fig. 6–1b), replaces the exciting rod. The frequency of the wave is then that of the electrical oscillator connected to the driving unit.

The wave velocity c measured for gases confined to tubes is found to be somewhat smaller than that measured under free-air conditions, unless the diameter of the tube is very large. This is due to the retarding effect caused by viscosity at the walls, and to heat conduction at the walls, which make the bulk modulus tend toward the isothermal rather than the adiabatic value. Helmholtz, Kirchhoff, and others derived formulas that give a theoretical value of c as a function of the tube diameter, the frequency, and a dissipation coefficient.

To the chemist and the physicist the velocity c is of interest beyond its own intrinsic value in acoustics. The direct measurement of the specific heats, which for any gas determine the value of γ, is a rather difficult matter. Equation (6–3), on the other hand, furnishes an excellent means of computing γ in terms of c, P, and ρ_0, all quite easy to determine. Many of the best values for γ have been found in this way.

6–4 Transmission of longitudinal waves through gases as related to kinetic theory. In discussing wave propagation through gases, we have so far treated the medium as continuous in the ideal sense. As mentioned earlier, ordinary sound sources and receivers are very large indeed compared with the average spacing of the molecules under the usual atmospheric conditions, so that the assumption of virtual continuity is a valid one. In the microscopic sense, however, the actual transfer of momentum associated with the passage of a wave through the medium must be accomplished through impacts between molecules. It is therefore reasonable to assume that there is some connection between the molecular speeds of kinetic theory and the velocity of sound waves. This is indeed so, as we shall now show.

According to the kinetic theory of gases, the molecules of a gas are moving about in space with assorted velocities ranging in value, for a very large number of particles, from near zero to near infinity. The distribution of the various velocities among the molecules between these limits is a definite one, known as the Maxwellian distribution. From the mathematical expression for this distribution it is possible to arrive at three kinds of "average" velocities, characteristic of a given gas under a given set of conditions. These are respectively u_p, the most probable velocity; u_a, the velocity which is the arithmetic mean over all the molecules; and u_k, which represents the kinetic energy mean and is the square root of the average squared velocity. These three velocities bear the ratios

$$u_k : u_a : u_p = 1 : 0.920 : 0.816. \tag{6-5}$$

It is easy to associate the velocity u_k with the velocity of sound waves, since

$$P_0 = \frac{1}{3}\rho_0(u_k)^2,$$

where ρ_0 is the density of the gas and u_k is the root mean square molecular velocity. Therefore

$$(u_k)^2 = \frac{3P_0}{\rho_0}. \tag{6-6}$$

Since the total kinetic energy must be distributed equally among the energies associated with the x, y, and z directions, the average squared velocity for any one of these directions, such as x, will be

$$(u_k)_x^2 = \frac{P_0}{\rho_0}$$

and the root mean square velocity along x is

$$(u_k)_x = \sqrt{\frac{P_0}{\rho_0}}. \tag{6-7}$$

This will be recognized immediately as identical with the expression for the velocity of a sound wave, assuming the process to be isothermal. Since the above simple equations of kinetic theory assume no temperature variations throughout the body of the gas, a result characteristic of an isothermal process is to be expected. Using the relations between u_k and the other velocities characteristic of a Maxwellian distribution, we may write, for air,

$$c_{\text{adiabatic}} = \sqrt{\gamma}\, c_{\text{isothermal}} = 0.68u_k = 0.74u_a = 0.84u_p. \tag{6-8}$$

The point to emphasize in this result is that the velocity of wave propagation and any one of the "average" molecular velocities of kinetic theory are virtually of the same order of magnitude. This will seem quite reasonable if one considers the order of magnitude of the particle velocities encountered in sound waves of ordinary intensities. It is rare for particle velocities greater than 10 cm-sec^{-1} to exist in the waves associated with speech and music, and usually they are much smaller. On the other hand, the "average" velocities characteristic of the random molecular motion of kinetic theory are of the order of 10^4 to 10^5 cm-sec^{-1}, some one or ten thousand times as great. The process of wave propagation can be pictured as follows. In darting about near the sound source, some of the gas molecules come into contact with the vibrating surface and so receive additional momentum, small compared with the average momentum they already possess. This additional increment of momentum is handed on to other molecules as the first set chances to strike them in the course of their random motion. Since the original momentum was a directed quantity, the disturbance propagates in a definite direction in space at a speed primarily determined by the kinetic theory velocities, not by the particle velocity imparted to the molecules at the source. Note that this picture is entirely consistent with the observation that the wave velocity remains constant even over distances great enough so that the particle velocity associated with the wave has become extremely small.

If the increment of particle velocity imparted by the vibrating source becomes an appreciable fraction of the random molecular velocity, the wave velocity becomes higher than the normal value for waves of small amplitude, as would be expected. Air disturbances originating with explosions often travel outward, in regions near the source, with speeds several times that for normal sound waves. Other interesting facts connected with high intensity waves appear in the next section.

6-5 Waves of large amplitude. In our derivation of the differential equation for waves of small amplitude in Chapter 2, we assumed that $\mathcal{B}$ $\left(= \frac{p}{-v/V_0}\right)$ was a constant. As has been seen in Section 6-1, the adiabatic bulk modulus $\mathcal{B}_a = \gamma P$, for an ideal gas. Since the total pressure P varies slightly in the presence of the wave, $\mathcal{B}$ cannot be said to be strictly a constant. For waves of ordinary amplitudes, the slight fluctuations in $\mathcal{B}$ are of no importance but for waves of abnormally high amplitudes the variations in $\mathcal{B}$ cannot be ignored. To see the general effect upon wave propagation when the intensities are very large, we shall derive the differential equation for plane waves without assuming $\mathcal{B}$ to be necessarily constant.

It will be remembered that Newton's second law for the thin layer of air, as given by Eq. (2–10), is

$$\frac{\partial^2 \xi}{\partial t^2} = -\frac{1}{\rho_0}\frac{\partial p}{\partial x}. \tag{6–9}$$

An exact relation for a given mass of gas undergoing an adiabatic change is

$$\frac{P}{P_0} = \left(\frac{V_0}{V}\right)^{\gamma},$$

where P_0 and V_0 are the pressure and volume at the beginning of the change and P and V are the corresponding values after the change. Solving for P and introducing the dilatation δ, we have

$$P = P_0\left[\frac{V_0}{V_0(1+\delta)}\right]^{\gamma} = P_0(1+\delta)^{-\gamma}.$$

The partial derivative of P with respect to x becomes

$$\frac{\partial P}{\partial x} = \frac{\partial p}{\partial x} = -\gamma P_0(1+\delta)^{-(\gamma+1)}\frac{\partial^2 \xi}{\partial x^2} \tag{6–10}$$

(since $\delta = \partial\xi/\partial x$). This result may be written in terms of the condensation s rather than in terms of δ by making use of the exact relation between s and δ given by Eq. (2–3),

$$\frac{\partial p}{\partial x} = -\gamma P_0(1+s)^{(\gamma+1)}\frac{\partial^2 \xi}{\partial x^2}.$$

Inserting this value for $\partial p/\partial x$ into Eq. (6–9), the wave equation becomes

$$\frac{\partial^2 \xi}{\partial t^2} = c^2(1+s)^{(\gamma+1)}\frac{\partial^2 \xi}{\partial x^2}. \tag{6–11}$$

Comparison of this differential equation with the one for waves of small amplitude will show that, in effect, c^2 has been replaced by $c^2(1+s)^{(\gamma+1)}$. Since the condensation s is itself a function of x and t, Eq. (6–11) is essentially a more complicated mathematical equation than the usual plane wave equation, and the solution to the latter no longer applies. From an approximate point of view, however, we may conclude that the solution implies some kind of a traveling disturbance of the general form $f(x \pm c't)$, where c' is not constant but is equal to $c(1+s)^{(\gamma+1)/2}$. The condensation is itself a quantity that varies from point to point in the path of the wave. Therefore *different parts of the disturbance must travel with different speeds.* Wherever the condensation is large, the propagation speed will be greater than where the condensation is small. Since a wave in which the maximum condensation is large will also be one in which the amplitude, maximum

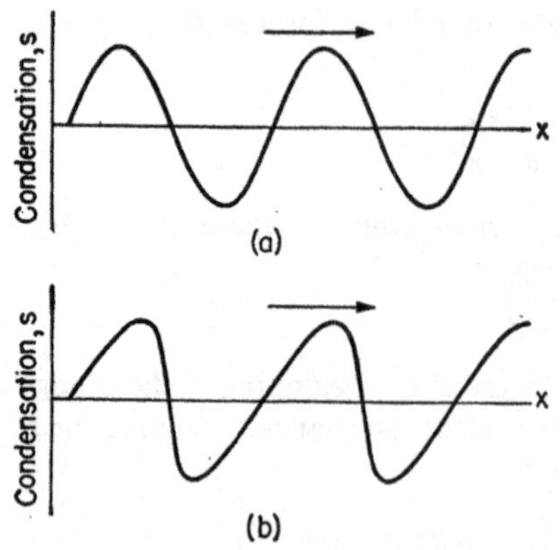

FIG. 6–2. (a) There is no change in wave form for waves of small amplitude. (b) Distortion of wave form for a wave of large amplitude.

excess pressure, etc., are relatively large, the same thing may be said in terms of the magnitudes of the other wave parameters.

The significance of the statement above is simply that as the disturbance travels through the medium there will be a progressive change in the wave form, no matter what the wave shape may be near the source. Crests will progress faster than troughs, so that a sine wave will eventually become distorted in such a way that the portion of the graph just ahead of each crest will acquire a steeper and steeper slope, as indicated in Fig. 6–2 b.

Water waves behave in this manner when reaching shallow water. The "curling" tendency of such waves is due to just such a variation in velocity for different parts of the disturbance, although the physical reason is a different one. As the water waves move into shallow regions, the troughs are slowed down by frictional forces at the bottom, the crests continuing to advance with nearly the deep water velocity; hence the tendency for the crests to break over the troughs. In the case of longitudinal sound waves this latter effect cannot occur, since a compression would then have to move *through* a rarefaction, which is a physical impossibility. For the waves in air, the wave shape ahead of the crests will approach the vertical but the slope will never change sign. As the wave form becomes more and more of the sawtooth type, higher harmonics will become more prominent. Increased dissipation at these higher frequencies will then limit the propagation velocity and so stabilize the wave form.

Fortunately, such effects are negligible with ordinary sound waves, even after they have traveled over considerable distances. The maximum value of the condensation, s_m, rarely exceeds the order of magnitude of 10^{-3}, in which case the velocity c' differs from c, the normal velocity for small amplitude waves, by no more than 0.1 or 0.2%. If appreciable change of wave form did occur, there would be serious consequences whenever music and speech are involved; distortions of all kinds would be apparent as one moved away from the source. (As a matter of fact, "large amplitude" distortion of wave form occasionally does occur in the throats of horns fed with sound of very high intensity. This distortion is quite perceptible to the ear.)

6-6 Miscellaneous open-air effects. While dispersion is a factor of negligible importance for sound waves of the usual small amplitudes, *refraction* is perfectly possible whenever the factors determining the velocity c vary. In the open air, temperature gradients usually exist near the surface of the earth. In the daytime, temperatures generally decrease with height above the ground. Since the velocity decreases with a drop in temperature, waves leaving a source in a direction having a slight inclination above the horizontal will bend more steeply upward as they travel (Fig. 6-3). This often accounts for the inaudibility of sound to an observer at no great distance from the source; the sound waves have passed over his head. A similar effect, operating in an inverse manner, undoubtedly accounts in part for the audibility of explosions at abnormally great distances, with "skip" phenomena for certain regions much nearer to the source. Calculations from the geometry of the probable paths of sound waves indicate that, after first falling, air temperatures must rise again, and that they eventually reach a value of about 150°C at a height of 25 or 30 miles above the earth. Such a temperature gradient in the higher regions of the stratosphere would give to the sound waves a path that is concave downward, returning them to the earth at more distant points but not at intermediate positions (Fig. 6-4). A somewhat similar effect occurs when electromagnetic waves reach the ionosphere, although in this case the physical nature of the refraction is, of course, quite different.

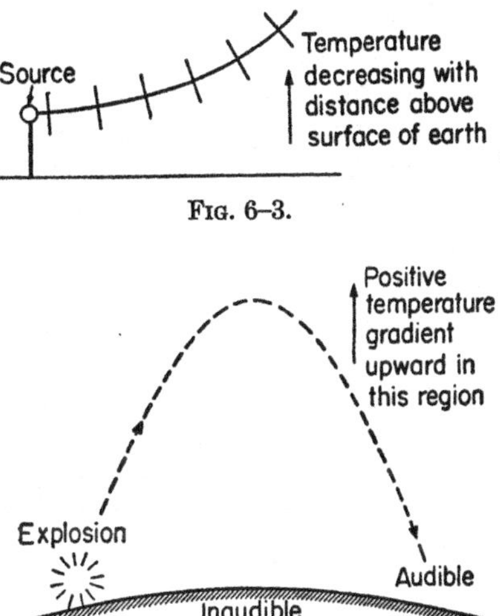

Fig. 6-3.

Fig. 6-4. Path of sound pulse from explosion heard at a great distance from the source.

Wind is also an important factor in the bending of sound waves. In passing through a region where there is a gradient in the wind velocity, the portions of the wave front that lie in the regions where the air is moving faster will move ahead of the portions lying in the slower moving air, thus rotating the plane of the wave front and so bending the path of travel. High winds are known to exist in the upper regions of the atmosphere and the bending effects referred to in the preceding paragraph are undoubtedly due, in part, to high altitude winds as well as to the existence of temperature gradients. Nearer the earth erratic behavior in the propagation of sound waves is to be expected whenever the air is not perfectly stagnant.

6–7 Acoustic focusing devices. Mirrors and lenses. High frequency sound waves reflect well from hard smooth surfaces (see Chapter 8). It should therefore be possible to construct acoustic mirrors which behave as do optical mirrors, taking due account of the important ratio of diameter to wavelength, as discussed in Chapter 4. A small source of sound, such as a whistle, placed at the focus of a parabolic reflector of diameter much greater than the wavelength, will reflect a beam of waves which are approximately parallel. To ensure small divergence, it is best to use an ultrasonic frequency (whose wavelength is therefore very short).

Since different gases transmit sound waves with different velocities, it is possible to construct an acoustic lens which will refract and focus in the manner of an optical lens. The ratio of the velocity in air to that in CO_2 is 1.28. To construct a converging CO_2 lens, the gas is forced between two circular sheets of thin rubber bound together at the edges, thus forming a double convex shape. Considerable reflection will take place at the lens surface, but enough energy will enter and leave the lens to permit focusing effects.

Recently Kock and Harvey * have built acoustic lenses whose behavior depends on somewhat more complicated principles. These lenses are of two general types. The principle of operation of the first type is easier to understand. A lens-shaped structure is built up with flat parallel metal strips whose planes are tipped with respect to the horizontal and whose shape is such as to give to the total lens a contour similar to that of a converging glass lens. Figure 6–5 is a photograph of the convex side of a plano-convex lens constructed in this manner. The theory of operation of the lens may be deduced from Fig. 6–6, which shows a vertical section of the structure. When a plane wave front strikes the flat side of the lens, the energy breaks up into segments which travel through the conduits bounded by the metal strips. The spacing of the plates is about $\frac{1}{3}$ the value of the wavelength used (the experimental frequencies ranged from 10 to 13 kc-sec^{-1}), so that the energy emerging on the other side of the lens from each conduit will diverge freely in accordance with diffraction principles. The tilt and contour of the individual plates are so chosen that for a point a on the axis of the lens the length of the various acoustic paths, as measured from the various entering points to the point a, is constant for any one of the conduits traversed. Thus the various contributions arriving at a will be in phase and this point will be the focus of the lens. The conduits supply the delay necessary to such focusing action, although in a manner different from the process of true refraction. (Such a lens will also focus short electromagnetic waves, since the metal boundaries of the conduits will also "guide" such waves.)

* *Jour. Acous. Soc. Am.* **21**, 471–481 (1949).

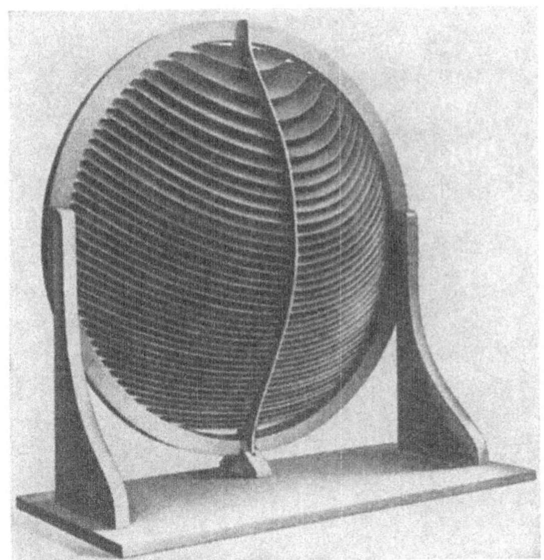

FIG. 6-5.
(*Courtesy* Bell Telephone Laboratories)

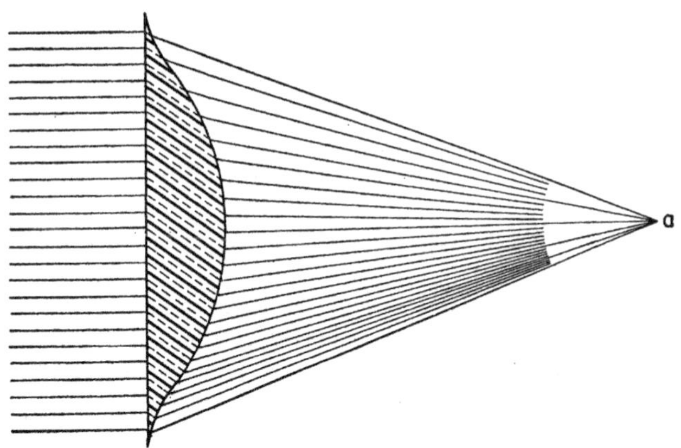

FIG. 6-6. Section of duct-type acoustic lens, showing paths of rays.

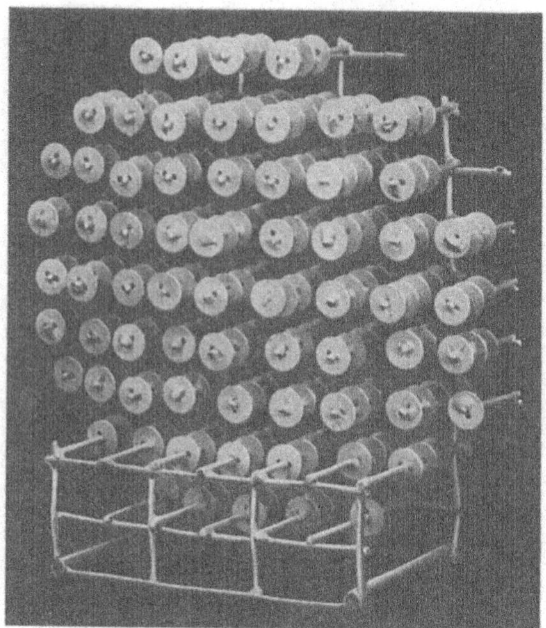

FIG. 6-7. (*Courtesy* Bell Telephone Laboratories)

A second type of lens that used "scattering" disks or strips was constructed. It had been shown by Lamb and by Rayleigh that the presence of small scattering obstacles in the path of longitudinal waves has exactly the same effect upon the velocity of wave transmission through the region as does an increase in the density of the medium itself, i.e., it decreases the velocity. It can be demonstrated that the effective index of refraction n of such a conglomerate, where the obstacles are flat disks, is given by

$$n = \left(1 + \frac{8}{3} Nr^3\right)^{\frac{1}{2}}, \tag{6-12}$$

where N is the number of disks per unit volume and r is the radius of each disk.

The photograph of Fig. 6-7 is such an obstacle lens of the converging type. For Eq. (6-12) to apply, it is necessary that the disk size be somewhat less in magnitude than $\lambda/2$; for larger sizes, resonance effects occur and the propagation velocity of the array is affected in a different manner. As with the conduit type of lens, electromagnetic as well as acoustical waves may be focused.

6-8 Attenuation of longitudinal waves in gases. As was pointed out earlier, the primary reason for the decrease in intensity of sound waves with distance from the source is the operation of the inverse square law. With a medium of infinite extent we may consider the propagation of longitudinal waves to be a nearly conservative process, so that virtually the same total energy flows per second through any sphere surrounding a small point source. That this should be exactly true is, of course, impossible, due to the existence of some attenuation factors operating both within and at the boundaries of the medium. Whenever the wave encounters a boundary, as in passing over rough ground in the open, considerable frictional and therefore dissipative effects are to be expected. It is common knowledge that sound may be heard for a greater distance over water than over land. This may be due in part to the better reflecting power of the smooth surface of the water and in part to the lesser importance of viscosity at this same surface compared with land.

As regards attenuation effects within the body of the gas itself, these have been shown to be small for travel over moderate distances. It has already been stated (Section 6-2) that some departure from complete adiabaticity begins to occur at the higher frequencies, the effect increasing with the first power of the frequency. Stokes and Rayleigh have made theoretical studies of the effect of heat radiation between regions of slightly different temperatures, and also of the losses to be expected due to viscosity within the body of the gas. These two dissipative processes were also shown to be functions of the frequency, the losses in both cases increasing at the higher values of the frequency. Experiments by Duff * and Hart ** indicate the presence of such dissipative effects, but their results are not in good agreement with theoretical computation.

There is considerable experimental evidence that at very high frequencies of the order of 10^5 to 10^6 cycles-sec^{-1}, there occurs a quite different type of dissipation, associated with vibrational resonances within molecules of a polyatomic nature. A number of workers have studied the abnormally high absorption (and hence disappearance of the energy in the form of wave motion) of CO_2 for frequencies of the order of several hundred thousand cycles per second. Dry air containing an appreciable per cent of CO_2 exhibits this property. In 1929 Sabine reported † a similar effect in air containing H_2O molecules (corroborated by Meyer ‡ and by Knudsen §). In this case the abnormal absorption occurs at frequencies well within the

* *Phys. Rev.*, **11**, 64 (1900).

** *Proc. Roy. Soc.* A, **105**, 80 (1924).

† *Jour. Franklin Inst.*, **207**, 347 (1929).

‡ *Zeits. f. techn. Physik.*, **7**, 253 (1930).

§ *Jour. Acous. Soc. Am.*, **5**, 112 (1933).

audible range (2000 cycles-sec^{-1} or over). The effect was greatest when the water molecules constituted about 0.5% of all those present. Considerable theoretical work has been done in connection with this phenomenon. In general, for such absorption to occur it is necessary for the period of the sound wave vibration to be comparable to the time required to establish thermal equilibrium between the normal and the vibrating molecules. This process should be compared and contrasted with the resonance absorption by gases of certain spectral lines. In this latter type of absorption, the absorption frequencies are exceedingly sharp as compared with the acoustic case and, being intra-atomic in nature, they occur at very much higher frequencies.

6–9 The Doppler effect. The basic principle of the frequency shift known as the Doppler effect is presented in most elementary textbooks on physics. In a wave motion taking place in a material medium like air, there are three fundamental velocities to consider if one is to predict the apparent frequency as heard by an observer. These are first, the velocity of the medium u_m, second, the velocity of the sound source u_s, and third, the velocity of the observer u_o. All of these velocities are measured with respect to the same set of axes, fixed with respect to the earth. We shall restrict our attention to the simplest case, where all these velocities are along the same straight line.

If both the source and the observer are at rest but the medium is moving, as would be the case in open air if there were a wind, the apparent frequency f' and the true vibration rate f are the same. If the wind is blowing towards the observer, the crests in the disturbance will be moving towards him more rapidly than if the air were stagnant and will be spaced farther apart than in the case of no air motion. As a result, their rate of arrival at the ear is the same as if there were no air motion at all. (This will be shown mathematically in the general formula about to be derived.) In this case, both the speed of the wave relative to the observer and the wavelength in the medium are increased, so that $f' = f$.

If both the medium and the observer are at rest but the source is moving, there will be a change in the wavelength in the medium. The crests will be closer together or farther apart, depending on whether the source is approaching or receding. In the first case, f' will be greater than f; in the second, it will be less.

If medium and source are stationary but the observer is in motion, the effect is entirely due to the motion of the observer relative to the wave. In moving in a direction opposite to that in which the wave is traveling, the observer encounters the crests more often and f' is therefore greater than f. If he is moving in the same direction, the crests will arrive at his ear less frequently and f' is less than f.

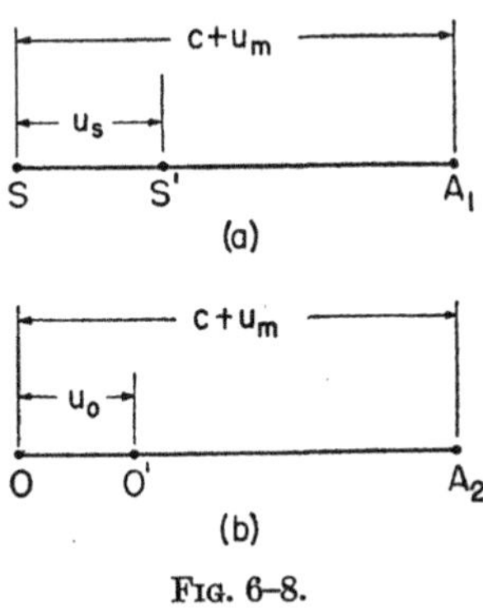

Fig. 6-8.

The general equation that takes care of all possible velocities can be set up by reference to Fig. 6-8. Let us assume all velocities to be in the same sense. If only the medium is in motion (with a velocity u_m), a crest originating at the source position (point S in Fig. 6-8a) will in one second travel a distance $c + u_m$ to point A_1, and there will be f complete cycles between S and A_1. If at the same time there is a source velocity u_s, this same number of cycles will occupy the smaller distance $S'A_1 = c + u_m - u_s$. Looking now at Fig. 6-8b, a stationary observer at point O will receive each second whatever number of cycles are contained in the distance OA_2. If, on the other hand, he is moving with a velocity u_o, he will receive only the number of cycles in the distance $O'A_2 = c + u_m - u_o$. The apparent frequency f' will then be a fraction of the vibration rate of the source f, given by

$$f' = f\frac{c + u_m - u_o}{c + u_m - u_s}. \tag{6-13}$$

This equation is in agreement with the qualitative statements made earlier in this section. It is to be noted that if u_s and u_o are both zero, the presence of u_m has no effect on the apparent frequency, but that if these velocities are not zero, the value of u_m does contribute to the numerical value of the frequency shift.

6-10 Practical importance of the Doppler effect.

Ordinarily, the velocities u_s and u_o are small fractions of the normal velocity of sound c. To see how small u_s might be and still produce a Doppler effect perceptible to the average ear, we observe that two notes on the musical scale a half tone apart bear a frequency ratio of about 16/15. Most listeners can easily distinguish a difference between these frequencies; a good musical ear can distinguish quarter tones and less. If we assume u_m and u_o in Eq. (6-13) to be zero and solve for the velocity u_s to make $f'/f = 16/15$, we find that $u_s = \frac{1}{16}c$, which corresponds to about 50 miles-hour^{-1}. As a car goes by on the street, sounding a horn of fixed frequency, there will not only be a rise in apparent frequency as it approaches, but a drop as it recedes. The car would therefore be traveling at something like half this speed if the two frequency extremes are to be in the ratio of 16/15.

Occasionally sources of waves in air move with velocities approaching or even exceeding the value of c. This is true in the case of projectiles and modern aircraft. Projectiles almost always travel faster than the air waves they produce. This behavior results in a V-shaped envelope of the air disturbance set up, very similar to the envelope of the water waves produced by a boat. Like the projectile, the boat is traveling faster than the disturbance in the medium. Much has been said in recent years of the difficulties of "piercing the sonic barrier" with airplanes, i.e., traveling at speeds approaching and exceeding the speed of sound (equivalent to about 750 miles-hour^{-1}). The physical situation may be appreciated if one considers the behavior of Eq. (6–13) for the case where u_s approaches c. This corresponds to an effective zero wavelength, i.e., all crests are coincident. If an airplane is traveling at exactly the speed c, all irregular disturbances of a pressure type set up in front of the plane, due to its motion through the medium, tend to remain for a considerable time *in the neighborhood* of the plane, contributing greatly to instability and lack of control. A good deal is known about airplane behavior at speeds greater and less than the speed of sound; *at* the speed of sound the situation is most complex and most difficult to analyze.

Some application of the Doppler effect has been made in the field of measurement. For instance, an experimental device to measure wind velocities depends for its operation on a utilization of the Doppler effect associated with a moving source of sound.

Before leaving this discussion, we might inquire into the possible importance of the Doppler effect in the production of musical sound. Since most musical sounds are complex in nature and contain more than one frequency component, the moving surface of the source may be said to be vibrating simultaneously at a high and a low frequency. The high frequency is being generated by a surface which is sometimes approaching the observer and sometimes receding from him, this oscillatory motion being due to the low frequency vibration of the surface. Thus there is a sort of "intermodulation" effect, to use the language of electronics, whereby, to the ear, the higher frequency appears to fluctuate slightly at a rate equal to that of the lower frequency. Such a Doppler effect might conceivably be a significant source of distortion in the case of a cone speaker that is called upon to reproduce simultaneously a very large variety of frequencies. Fortunately, a calculation of the magnitude of the frequency shift to be expected due to this cause shows it to be well below that detectable by the average ear, particularly in the presence of a large collection of different musical sounds. Other sources of distortion are of far greater importance, as we shall see in Chapter 11.

6-11 Transmission of longitudinal waves through liquids. The process of the transmission of longitudinal waves through liquids (in the bulk) is identical with that for gases and we have therefore only to insert the proper values of γ, P, and ρ_0 in the equation $c = \sqrt{\frac{\gamma P}{\rho_0}}$. While the density of water is about 800 times as great as that for air, the bulk modulus $\mathcal{B}$ is of the order of 15,000 to 16,000 times that for air. As a result, the velocity of waves in water is between 4 and 5 times that for air. The quantity γ is nearly unity for water and for many other liquids, so that the isothermal and the adiabatic moduli are almost the same. (In the neighborhood of 10°C, γ is 1.001 for fresh water.) For ether, however, γ is about 1.3, so that in this case $\mathcal{B}_a$ and $\mathcal{B}_i$ are quite different. For liquids, as for gases, γ is slightly sensitive to pressure variations; in the case of sea water, a variation in pressure from 100 atm to 1000 atm brings about a change in γ of approximately 1%. These pressures are extreme, but the effect is of some importance in acoustic depth sounding. Because of the relative incompressibility of liquids, the variation of density under varied conditions of pressure and temperature is of secondary importance as far as its effect on the speed of longitudinal waves is concerned. Table 6-3 gives the wave velocity in a number of common liquids.

TABLE 6-3

Liquid	*Temperature* (°C)	*Velocity* (cm-sec^{-1})
Alcohol	12.5	1.24×10^5
Ether	20.0	1.01
Mercury	20.0	1.45
Petroleum	15.0	1.33
Water, fresh	17.0	1.43
Water, sea (36 pts salinity)	15.0	1.50

6-12 Experimental measurement of c for liquids. Direct measurement of the velocity of longitudinal waves in water may be made over large distances. The velocity for fresh water is measured in lakes and rivers and the velocity for salt water in the open sea. All that is necessary is the use of electrical timing devices at the sending and receiving ends. As in the case for gases, a laboratory method for liquids is also desirable; by such a method liquids other than water may be measured under controlled conditions. It is possible to use for liquids a modification of the Kundt's tube method for gases. Instead of using dust particles to locate the nodal planes in the standing wave pattern, these planes may be located by means of a vibration detector, such as a microphone, placed against the *outside* of the tube containing the liquid. It is interesting to point out that the positions of the displacement nodes, or regions of *no motion* within the liquid are, as we shall see in the next chapter, positions of *large pressure variations* in a standing wave of the longitudinal type. These pressure variations are transmitted to the wall of the containing tube and so may be picked up by the microphone. (To ensure a velocity measurement characteristic of the elastic properties of the liquid itself, the tube walls must be very thick and fairly rigid.)

In the ultrasonic frequency range, the velocity of longitudinal waves in liquids has been measured by means of the acoustic interferometer devised by G. W. Pierce* for use with gases. A quartz crystal is made to vibrate by electrical means so as to radiate into the fluid compressional waves of a frequency of 50,000 cycles-sec^{-1} or higher. A reflecting plate is placed in the path of the waves, so that considerable energy is reflected back to the crystal. The phase of the displacement in the returning wave relative to the motion of the surface of the crystal depends in part on the total path of travel of the returning wave. If the reflecting plate is moved towards and away from the crystal, a position will be found such that the returning wave strikes the crystal surface with a particle displacement just opposite to that of the motion of the crystal surface. As a result, the crystal motion will be significantly limited, the effect being observed as an electrical change in the circuit driving the crystal. As the plate is moved progressively towards the crystal, this effect will become critical at positions $\lambda/2$ apart. In this way the value of λ may be obtained. The electrical frequency of the vibrations is easy to measure and therefore the velocity c can be computed from the relation $c = f\lambda$. This procedure is equally well adapted to the measurement of c in both gases and liquids.

6–13 Attenuation effects in liquids. Since the velocity of longitudinal waves in water is more than four times as great as it is in air, the wavelength for any given frequency is greater by the same factor. Therefore the high and low temperature regions will be farther apart in water than in air and the temperature gradients consequently smaller. In addition, the ratio of thermal conductivity to specific heat capacity is much smaller for liquids, in general, than for gases and so what heat conduction may occur to attenuate the wave will be much less important for water than for air. It can be shown also that losses of the viscous type are smaller for liquids than for gases. For these reasons, longitudinal waves will travel much greater distances through sea water than through the air. However, there are other factors that undoubtedly are of major importance. As we

* *Proc. Am. Acad.*, **60,** 271 (1925).

shall see in Chapter 8, longitudinal waves are rather efficiently reflected at the boundary between water and air (because of the great difference in the specific acoustic impedance for the two media, $\rho_0 c$) and consequently much of the energy striking the boundary is reflected back into the water. A similar effect occurs at the bottom of the water layer. As a result, instead of the spherical divergence of a medium of infinite extent, the energy is confined to a relatively thin two-dimensional layer. The intensity will therefore fall off more slowly than with the operation of the inverse square law. Other factors may accentuate this effect. For instance, it has recently been discovered that temperature gradients in the Pacific ocean area operate, through refraction, to confine the energy to so thin a layer of the sea that underwater longitudinal waves may often be picked up at distances as great as 1000 miles from the source.

More will be said regarding the transmission of waves through liquids when we come to the general subject of ultrasonics in Chapter 12.

6-14 Longitudinal waves in solids. This is a very involved branch of the subject of vibration and of acoustics and will not be discussed in any great detail in this book. We shall confine our remarks largely to a consideration of isotropic substances, like glass and metals, which are well annealed and strain-free. In such cases the elastic properties for compressional waves are the same in all directions and a much simpler state of affairs exists than in the case of nonisotropic substances such as crystals, and metals which are in a high degree of strain due to mechanical rolling, etc.

One of the important differences between fluids and solids is that only the latter offer appreciable resistance to shear. As a result, solid media can support and transmit vibrations of the transverse and torsional types. In a solid medium of infinite extent any local disturbance will ultimately give rise to both longitudinal and transverse waves. (Transverse waves in solids will be referred to again in Chapter 7.) Considering only the longitudinal waves set up in an isotropic medium of infinite extent, the velocity of propagation depends on the density ρ_0 and on the two elastic constants λ and μ (known as the Lamé constants), according to the equation

$$c = \sqrt{\frac{\lambda + 2\mu}{\rho_0}}, \tag{6-14}$$

where the effective elastic factor is therefore $(\lambda + 2\mu)$.

Our chief concern with longitudinal wave propagation in solids will be for samples of limited dimensions, in particular for bars whose cross-sectional area is uniform and where such areas have dimensions small compared with the length of the bar and with the wavelength associated with the disturbance. In this case the velocity for longitudinal waves may be expressed in terms of Young's modulus. This may be readily shown by an analysis very similar to that used for plane waves in air.

Figure 6–9 shows the effect of a small longitudinal distortion upon an infinitesimally thin slab of a bar. The total force F elongating a bar originally of length l is

$$F = YS\frac{\Delta l}{l},$$

where Y is Young's modulus, Δl is the change in length, and S is the cross-sectional area. (The latter is assumed to change by a second-order amount during the deformation.) For the thin slab of Fig. 6–9, we have

$$F = YS\frac{\partial \xi}{\partial x}. \tag{6–15}$$

Since the value of F varies continuously along the bar in the presence of the wave and since it is the *difference* in the forces acting at the two faces of the slab which produces its acceleration, we need

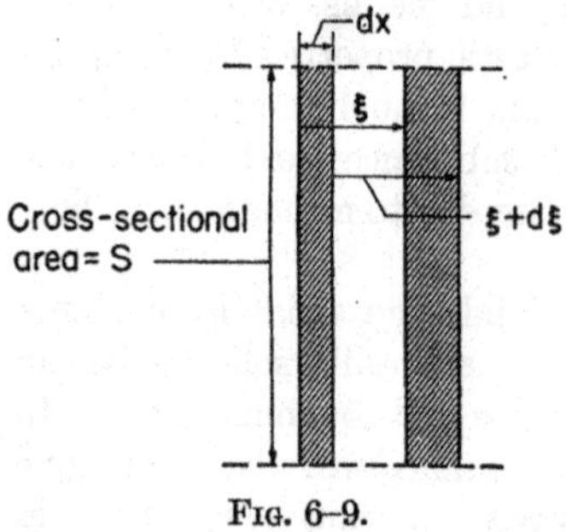

$$dF = \frac{\partial F}{\partial x}\,dx = YS\frac{\partial^2 \xi}{\partial x^2}\,dx. \tag{6–16}$$

By Newton's second law, we have

$$YS\frac{\partial^2 \xi}{\partial x^2}\,dx = \rho_0 S\,dx\,\frac{\partial^2 \xi}{\partial t^2}$$

or

$$\frac{\partial^2 \xi}{\partial t^2} = \frac{Y}{\rho_0}\frac{\partial^2 \xi}{\partial x^2}. \tag{6–17}$$

Fig. 6–9.

Since this differential equation is in the wave form, we may conclude that the velocity of wave propagation c is

$$c = \sqrt{\frac{Y}{\rho_0}}. \tag{6–18}$$

To make use of this result, it is not necessary to assume that the bar is isotropic, as long as we measure Young's modulus for the long dimension of the bar. It is interesting to point out that since the elastic constant Y is in general smaller than the effective elastic constant $\lambda + 2\mu$ for the case of a medium of infinite extent, the velocity along the bar is smaller than would be the case for the same material in large bulk. This is because in the latter case there is *lateral* as well as longitudinal resistance to deformation. In the case of the bar of small cross section, the lateral resistance is small (and may be neglected), so that the effective "stiffness" is thereby reduced.

When the above assumptions of dimensions are not valid, Eq. (6-18) may not be used. As we shall see later, a vibrating quartz crystal is a very common source of longitudinal waves of ultrasonic frequency. In this case, the material is anisotropic and the wave velocity within the crystal is a function of the orientation of the crystalline axes and also of the particular shape of the bounding surfaces, and therefore no single formula will suffice.

6-15 The measurement of c as a means of studying the elastic properties of solids. For isotropic solids there are two important elastic constants, the Lamé constants λ and μ, already mentioned. These constants are simply related to the velocity of longitudinal waves in the materials and also to the velocity of torsional waves. Both these types of waves may easily be set up experimentally and the measurement of these two velocities enables computation of the important elastic constants. With the more complex situation that obtains with anisotropic solids, there may be 20 or more significant constants. These also are related to the various velocities of wave propagation. The measurement of c is therefore a very useful tool for determining more fully the nature and behavior of the solid state.

6-16 Dissipation within solids. Because of the high heat conductivity of solids, and of metals in particular, dissipation of energy by heat conduction is of greater importance than for gases or liquids, and for many solids, the losses due to viscosity phenomena are even more important. Rubber is a good example. The sources of internal dissipation are, in general, exceedingly complex for solids. In well-annealed polycrystalline metals, heat flow of a resonant type accounts for a considerable fraction of the dissipation, under certain conditions.* In addition, the presence of internal strains (imperfections in the crystal structure along crystal planes) is important. All of these effects are usually called loosely "internal friction." Much experimental and theoretical work has been done in this field. The study is important from a practical point of view because of its relation to the well-known failure of metallic structures under the influence of constant vibration.

---

* Zener, *Phys. Rev.* **52,** 230 (1937) and **53,** 90 (1938). Also Randall, Rose, and Zener, *Phys. Rev.* **53,** 343 (1939).

PROBLEMS

1. The phenomenon of dispersion is observed for longitudinal waves in air only at very high frequencies. Give the reasons for this statement.

2. Would the velocity of sound waves in the open air be expected to vary as the square root of the barometric pressure? Explain.

3. The velocity of longitudinal waves is being measured in a closed cylindrical tube immersed in a temperature-controlled bath. How will the velocity be found to vary with the temperature? Compare with the situation obtained in the open air.

4. An acoustic piston producing plane waves in open air is oscillating at a constant frequency and with constant amplitude. Compare the intensity in the beam at $-10°C$ with the intensity at $30°C$.

5. Assuming that the effective velocity of a wave of large amplitude is given by $c' = c(1 + s)^{[(\gamma+1)/2]}$, (a) find the value for the condensation s in air at 0°C and 76 cm of Hg, for which the effective velocity c' exceeds the small amplitude velocity by 1%. (b) Find the corresponding *intensity level* at which this occurs.

6. The apparent frequency of an automobile horn when the car is approaching a stationary observer is 10% higher than when the car is moving away. There is no wind. Find the velocity of the car, assuming it to be constant. Take the velocity of sound waves in still air to be 331 m-sec^{-1}.

7. A sound source of frequency 1000 cycles-sec^{-1} is mounted on the end of a horizontal bar rotating about a vertical axis. The source is 2 feet from the axis. An observer in the plane of rotation of the bar hears a periodic frequency shift, due to the Doppler effect. Assuming he can detect the presence of two apparent frequencies if their ratio is 20/19 or greater, find the minimum speed of rotation in rpm at which a frequency shift might be detected.

8. A loudspeaker cone is vibrating simultaneously at the frequencies 50 and 1000 cycles-sec^{-1}. If the amplitude of motion of the lower frequency is 1.0 mm, find the maximum Doppler shift in the higher frequency, which might occur, for an observer sitting directly in front of the loudspeaker.

CHAPTER 7

STATIONARY WAVES. VIBRATING SOURCES. MUSICAL INSTRUMENTS

7-1 Introduction. The fundamental relations for plane and for spherical waves developed in Chapters 2 and 3 have assumed that a disturbance, once set up, travels out from the source an indefinite distance. This picture of a medium infinite in extent is useful for any fundamental description of the physics of wave propagation since, at any one point in space, we are concerned with waves traveling in one direction only. For sound sources radiating into the open air, and with few obstacles to reflect or scatter the energy, the medium may be considered virtually infinite in extent. However, when sound waves strike hard, relatively rigid structures, an appreciable fraction of the incident energy may be deflected and perhaps returned in the direction of the source. In the region where this occurs there will be two wave trains moving in opposite directions, each contributing to the deformation of the medium.

Under certain conditions this situation may give rise to *stationary* or *standing* waves, with a whole new set of features quite foreign to waves of the unidirectional type. In this case the "pattern" of the deformation in the medium remains fixed in space, with no evidence at all of propagating crests or troughs. It is with this general phenomenon that we shall be mainly concerned in this chapter.

Stationary waves may occur in any medium having definite boundaries. In air, such waves may be of primary importance within a room, where the medium is confined by the surrounding walls. Wave reflection, and the consequent production of a standing wave pattern, may sometimes take place without an actual change of medium. This is the case when waves traveling down a cylindrical pipe reach an end open to the surrounding air. In this case the reflection is associated with the change in the acoustic impedance as the wave passes from the region within the pipe to the region of free space beyond it. This phenomenon is very similar to that taking place in an electrical transmission line whenever the line characteristics change abruptly and we shall have more to say about it later.

Sound, musical or otherwise, originates in many cases from the vibration of some solid of limited dimensions, such as a stretched string, a metallic bar or plate, etc. (A few sources, like the organ pipe and the siren, obviously do not fall into this classification.) These elastic solids are set into vibration by deforming the material either by a direct contact force or, as in the case of sources like the radio loudspeaker cone, by forces of an electromagnetic or similar nature. As we saw in Chapters 2 and 3, a deformation of an elastic medium (in the earlier discussion a gas) gives rise to a wave disturbance traveling with a speed determined by the elastic and inertial properties of the medium. One would expect that waves of some sort would follow the deformation of a stretched string, an elastic plate, or the like, these waves moving within the material of the solid itself. Such waves do occur and in the case of solids may be either of a longitudinal or of a transverse type. In either case, these disturbances are sure to be reflected at the boundaries of the solid, so that the conditions are right for the production of stationary waves.

Stationary waves are inevitably present in vibrating sound sources whose rigidity is anything short of the infinite rigidity assumed for an ideal acoustic piston (only at low frequencies can any practical source be assumed to be equivalent to an acoustic piston). In the general subject of the vibration of extended bodies (rather than of particles), the variety of standing wave patterns that may occur is of considerable interest for its own sake. The study of these patterns is also important, however, because of the part they play in the radiation of sound waves by such a vibrating body. For any one variety of standing wave in the sound source there will be a particular set of surface motions, these surface motions being, of course, the cause of the longitudinal waves in the air. In this way the whole character of the resulting sound waves will be significantly affected by the wave motion *within the source itself*. Waves in the source and waves in the outside medium are intimately connected.

The transverse motion of a stretched string furnishes the simplest example of waves in a solid having fixed boundaries, since a string can be considered a body of one dimension only and since the stresses in a string are of a particularly simple type. We shall consider first the physical properties of the string that make possible wave propagation. Later we shall investigate the effect of reflection for a string of limited length, leading to the production of stationary waves. The stationary wave equations will prove very useful in the discussion of vibrations of strings and of air columns. While the physical picture is different for air columns, the mathematics of string vibrations may be used with very little modification.

7–2 The ideal string. Our ideal string is of uniform mass per unit length and is under tensile stress only, even when deformed. This implies that the string is perfectly flexible, so that no bending moments are ever called into play. The particles of the string are free to move in a direction transverse to the long dimension of the string under the action of restoring forces that are due, as will be seen presently, to the inclination of the string on either side of such particles. (Any *longitudinal* motions of the particles may be ignored in view of the great stiffness of the string for deformations of that type.) The extent of the transverse displacement of any string particle will be assumed small enough so that the tension may be considered to remain constant in magnitude while the motion is taking place. No forces of a dissipative nature exist.

7-3 The differential equation. We shall assume, as we did for plane waves in air, a very general type of deformation. Figure 7-1 represents the particular shape of the string in some local region at a given instant of time. A segment of the string of differential length dx is acted upon by two forces only, the tensions F_t at either end. These forces will not, in general, make the same angle ϕ with the x-axis, but because of the curvature of the string will differ by a small amount as shown. As a result, there will be a small net force in the transverse or y-direction. By Newton's second law, we have

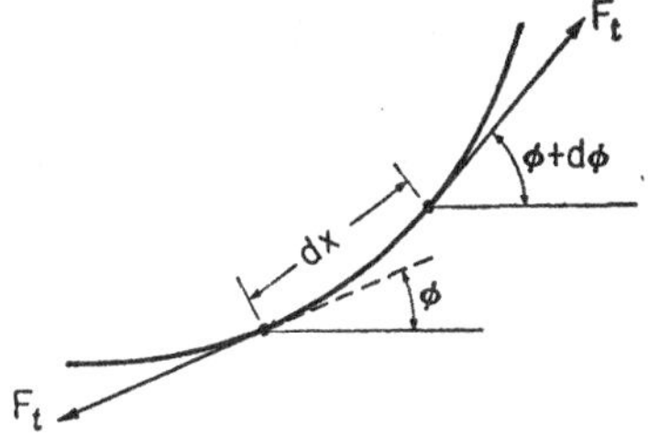

FIG. 7-1. Essential geometry for the deformation of an ideal stretched string.

$$F_t \sin(\phi + d\phi) - F_t \sin\phi = \sigma_l\, dx\, \frac{\partial^2 y}{\partial t^2}, \tag{7-1}$$

where σ_l is the mass per unit length of the string.

The net force on the left may be rewritten

$$F_t \sin(\phi + d\phi) - F_t \sin\phi = F_t d(\sin\phi) = F_t \frac{\partial}{\partial x}(\sin\phi)\, dx. \tag{7-2}$$

Remembering that the angle ϕ is never very great if the transverse string displacements are small, we may, with negligible error, replace $\sin\phi$ by $\tan\phi$. Since $\tan\phi$, the slope, is equal to $\frac{\partial y}{\partial x}$, the net force on the string segment may be written

$$F_t \frac{\partial}{\partial x}(\sin\phi)\, dx = F_j \frac{\partial}{\partial x}(\tan\phi)\, dx = F_t \frac{\partial^2 y}{\partial x^2}\, dx.$$

Equation (7-1) then becomes

$$F_t \frac{\partial^2 y}{\partial x^2}\, dx = \sigma_l\, dx\, \frac{\partial^2 y}{\partial t^2}$$

or

$$\frac{\partial^2 y}{\partial t^2} = c^2 \frac{\partial^2 y}{\partial x^2}, \tag{7-3}$$

where

$$c = \sqrt{\frac{F_t}{\sigma_l}}. \tag{7-4}$$

7-4 The solution. Equation (7-3) is identical in form with the equation for plane waves in air. We may therefore write down at once its general solution,

$$y = f(ct \pm x), \tag{7-5}$$

where y in this case refers to a *transverse* displacement of a string particle, whereas ξ in the plane wave solution specified a longitudinal displacement, The interpretations of Eqs. (2-13) and (7-5) are identical as far as wave propagation is concerned. Any local deformation of the string will immediately start two waves, one moving in the $+x$ direction, corresponding to $y = f(ct - x)$, and one moving in the $-x$ direction specified by $y = f(ct + x)$. (Functions of the form $f(x \pm ct)$ are also solutions.) The speed of travel will be the same, i.e., $c = \sqrt{F_t/\sigma_l}$. If the string is very long, so that possible reflection at the ends need not be considered, there are no restrictions placed upon the form of the function in Eq. (7-5) as long as it contains the argument $(ct \pm x)$. The disturbance may be periodic but it is not necessarily so.

The graphs of Fig. 2-3 may also be used for the case of a string, where they have the following particularly simple interpretation. If we plot in cartesian coordinates the string displacement y as a function of x at any one fixed instant of time, the resultant graph will be a virtual photograph of the string, with the wave shape "frozen" upon it. The graph *is* the string shape. In the case of longitudinal waves, it will be remembered, a vertical y-coordinate represented a particle displacement *along the horizontal x-direction.*

7-5 The string of limited length. For a string of limited length our solution must satisfy the conditions imposed at the boundaries, as well as the differential equation. Let us suppose that at the position $x = 0$ the string is attached to a support of infinite rigidity. (The rigidity of actual supports is often sufficiently great to justify this assumption.) The boundary condition at this point is that y must be zero at all times. Equation (7-5) may therefore be written

$$y = f_1(ct + x) + f_2(ct - x), \tag{7-6}$$

where f_1 and f_2 are not necessarily the same functions. If, however, the functions f_1 and f_2 are taken to be of the same form but of opposite sign, we may write, for the point $x = 0$,

$$0 = f(ct) - f(ct). \tag{7-7}$$

This is an obvious identity and the boundary condition is satisfied. The motion of the rest of the string will then be given by

$$y = f(ct + x) - f(ct - x), \tag{7–8}$$

implying two wave trains *of exactly the same wave form* traveling in opposite directions.

7–6 Reflection at one end of a string. The graphical interpretation of Eq. (7–8) is interesting and important. From a mathematical point of view, the two wave trains exist to the left of the point $x = 0$ as well as to the right. Physically, of course, there can be no string vibrations except where x is positive. In Fig. 7–2a are drawn the two oppositely moving wave disturbances represented by $f(ct + x)$ and $-f(ct - x)$ at some particular time t. In the figure the solid line represents the disturbance moving to the left and the broken line the disturbance moving to the right. Only to the right of the support do these disturbances have any real significance. To satisfy the boundary condition it is necessary that the ordinates due to the separate disturbances always be equal and opposite at $x = 0$ (from Eq. (7–7)). At other positions along the string this is not at all necessary, although at certain instants of time it may be true. The actual displacement of any particle of the string at any position x will be the algebraic sum at that instant of the particular ordinates associated with

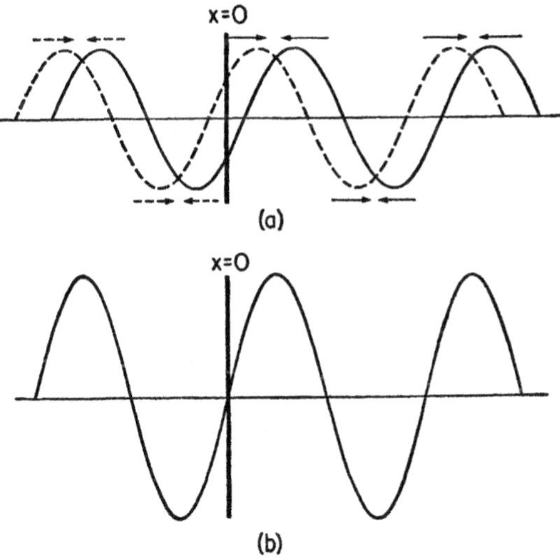

FIG. 7–2. Reflection of transverse waves at a rigid boundary.

the separate disturbances, whatever they may be. In Fig. 7–2b is shown the graphical sum of the solid curve and the broken curve for the time chosen. This sum curve represents the actual shape of the string.

The situation to the right of the string support may be summarized as follows. A train of waves is approaching the support and a second train of exactly the same wave form is simultaneously moving away from the support. This is the ordinary notion of *reflection.* In this case, where the support is taken to be completely rigid, the reflection is perfect. The energy in the reflected disturbance must then necessarily be identical to that in the incident disturbance. The maintenance of the wave form after the reflection is consistent with such a conservation of energy.

The process of reflection might have been approached somewhat differently. It is possible to follow along the string a small pulse of a highly localized nature moving towards the support position $x = 0$. As the deformation reaches the support, causing the string to pull, say, upward upon it, the support (being completely unyielding) reacts upon the string in such a way as to throw the string particles *downward.* In this way the arriving and departing pulses may be seen to involve ordinates of opposite algebraic sign at the support position, as was apparent from our more formal mathematical analysis. The dynamics of the situation at the support, as far as the string particles are concerned, is exceedingly difficult to follow in any quantitative way. It therefore seems preferable to use the approach of the preceding paragraphs. This is one instance, not at all uncommon in physics, where physical reasoning must bow to formal mathematics as the more useful tool.

If the string support is not infinitely rigid, the reflection process is modified. Obviously, we may no longer use the boundary condition that y is at all times zero at the position $x = 0$. We shall have more to say later in this chapter on the effect of yielding supports; this matter is also related to wave absorption at boundaries as discussed in Chapter 8.

7–7 Simultaneous reflection at both ends of a string. For a string of limited length, reflection will, in general, take place at both ends. It is this double reflection that makes possible, indeed *necessary*, the periodic motion of a string, so important to stringed instruments as sources of musical sound. Consider a pulse of some sort near one end of the string, traveling along it to the other end and there being reflected back towards the first end. The total time T for the pulse to return to its starting point will be $2l/c$, where l is the length of the string and c is the velocity of wave propagation. After a time T, therefore, a particle of the string will repeat its original motion. In this manner the motions of all particles of the string will necessarily be periodic. It will turn out that with certain initial deformations of the string the period of its motion may be less than $2l/c$ but it is never greater. In the next section we shall see just what the period of the string motion may be, consistent with both end conditions.

7-8 Vibrating string fixed at both ends. In view of the expected periodic motion of the string, let us take as the solution to the wave equation two periodic functions of the same form but of opposite sign, in order to satisfy the boundary condition at $x = 0$:

$$y = y_m \sin\left[\frac{2\pi}{\lambda}(ct + x) + \alpha\right] - y_m \sin\left[\frac{2\pi}{\lambda}(ct - x) + \alpha\right], \qquad (7\text{–}9)$$

where y_m is the amplitude of motion in each of the two waves and α is a phase angle. Equation (7–9) may then be rewritten

$$y = y_m \sin\left[\left(\frac{2\pi ct}{\lambda} + \alpha\right) + \frac{2\pi x}{\lambda}\right] - y_m \sin\left[\left(\frac{2\pi ct}{\lambda} + \alpha\right) - \frac{2\pi x}{\lambda}\right]$$

Expanding the sines of the sums and differences of two angles and collecting terms, we find that

$$y = 2y_m \sin\left(\frac{2\pi x}{\lambda}\right)\cos\left(\frac{2\pi ct}{\lambda} + \alpha\right). \qquad (7\text{–}10)$$

At this point we may introduce the boundary condition for the right-hand end of the string. Since this end is fixed, when we insert in Eq. (7–10) the value of $x = l$, it is necessary that y remain zero for all values of the time. For this to be true, the value of λ must be restricted by the relation

$$\frac{2\pi l}{\lambda} = n\pi$$

or by

$$\lambda = \frac{2l}{n}, \qquad (7\text{–}11)$$

where n is any integer. When the value of λ given by Eq. (7–11) is inserted into Eq. (7–10), we obtain

$$y = 2y_m \sin\left(\frac{n\pi}{l}x\right)\cos\left(\frac{n\pi c}{l}t + \alpha\right). \qquad (7\text{–}12)$$

This is one form of the *stationary wave equation.* It applies in this case to a string stretched between rigid supports, but with different variables it may be used for other vibrating bodies as well, as we shall see later, provided that the boundary conditions are analogous to those assumed for the string.

7–9 Interpretation of the stationary wave equation. According to Eq. (7–12), there are an infinite number of *possible* periodic motions, depending on the value of the integer n. The frequencies are restricted to the values given by

$$f = \frac{\dfrac{n\pi c}{l}}{2\pi} = \frac{nc}{2l} = \frac{n}{2l}\sqrt{\frac{F_t}{\sigma_l}}, \tag{7–13}$$

and hence constitute a harmonic series of the Fourier type. As with the simple progressive type of wave, y is a periodic function of both x and t, but in this case there is no evidence of traveling crests and troughs. The amplitude of motion of the different particles of the string varies along the string with the value of x. For positions where $(n\pi/l)x$ is an even integral multiple of $\pi/2$, the amplitude is zero, since the sine function in Eq. (7–12) is zero. These points are called the *nodes*. Between the nodes, at positions where $(n\pi/l)x$ is an odd integral multiple of $\pi/2$, the amplitude is a maximum and is equal to $2y_m$. These points are called the *antinodes* or loops.

Between an adjacent node and antinode the string particles have amplitudes intermediate between zero and $2y_m$. On opposite sides of a nodal position the particles are, at any one instant, moving in opposite directions. This may be seen by examining the algebraic sign of the expression $\sin\left(\frac{n\pi}{l}x\right)$ in the neighborhood of a value of x for which the angle is some even integral multiple of $\pi/2$ (a nodal point). If this sign is positive on one side of a node it will, at the same instant, be negative on the other, indicating a relative phase of 180° for the particle motions.

The distance between the nodes (or the antinodes) will depend on the period of vibration and hence on the value of the integer n. Let us suppose the integer n to be unity. Since a node will occur at intervals along the string whenever the angle $(n\pi x/l)$ changes by an amount π, the corresponding change in x will be equal to l. Hence for $n = 1$, the spacing of the nodes is the length of the string itself, i.e., there are only two nodes, at the ends. From Eq. (7–11), the length of the string will be $\lambda/2$. This is the simplest mode of vibration, where f_f, the so-called "fundamental" frequency, is given by

$$f_f = \frac{c}{2l} = \frac{c}{\lambda}. \tag{7–14}$$

If $n = 2$ $(f = 2f_f)$, the angle $(n\pi x/l)$ will increase by π when x increases by an amount $l/2$, and there will be a node in the middle of the string as well as at the ends. The length of the string is then equal to the wavelength.

If $n = 3$, the change in x need be only $l/3$, etc. The various possible patterns for the first few integral values of n are given in Fig. 7-3. The 180° phase relationship mentioned above is also indicated.

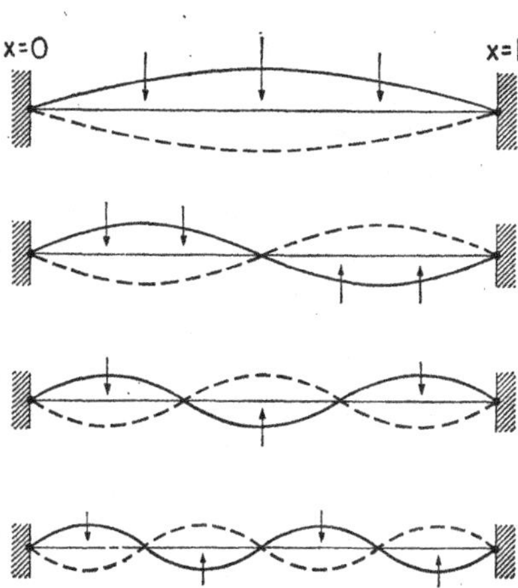

FIG. 7-3. Several possible stationary wave patterns for a string stretched between rigid supports.

By far the clearest idea of the nature of stationary waves is gained if a vibrating string is viewed under stroboscopic light. Such a light source is regularly periodic in character and the electrical circuit is usually such that the frequency may be smoothly adjusted over quite a wide range. If the illumination frequency is set to be the same as that of the string while the latter is executing one of its possible modes of vibration as given by Eq. (7-12), the string will appear to stand still, since it will be visible only when its various particles are in the same position in space. If the illumination frequency is slightly higher or lower than that of the string, the latter will appear to vibrate very slowly, in the manner peculiar to stationary wave behavior. The successive changes in shape, the different particle amplitudes, and the phase relationships may then be clearly seen. (Similar effects may occur when the ratio of string frequency to illumination frequency has certain values differing from unity.) It is regrettable that no such simple demonstration is possible with unidirectional waves along a string, since with a finite length the pattern is also complicated by reflections.

7-10 Other end conditions. Both ends free. When stretched strings are used on musical instruments, the slight yielding of the supports is an important factor in the radiation of musical sound by the instrument as a whole. From the standpoint of the possible modes of vibration of the string itself, however, the assumption of complete rigidity at the points of attachment predicts a set of stationary wave patterns in close agreement with experiment. The quite different boundary condition that one end of the string be *completely* unrestrained is of interest. Such a boundary condition is never encountered in musical instruments, since it would then be impossible to maintain any tension in the string. We consider this highly artificial situation only because of its mathematical importance in the later discussion of stationary air waves in pipes. As mentioned earlier, the mathematics developed for the string is directly applicable to the resonance behavior of air in pipes. Such columns of air may be considered "free" to move at an open end, the elastic factor in this case being present along the air column without the necessity for any "pull" at the ends. We consider the free-end condition for the string because it is easier to observe than in the air column.

A simple way of demonstrating the free-end condition for a body like a string (and therefore observing its physical effect experimentally), is to hang a heavy flexible rope in a vertical position. The lower end will be virtually unrestrained and the tension will be maintained by the weight of the rope itself. The transverse vibrations of such a vertical rope are affected in an important way by the fact that the tension is a variable up and down the rope, being a maximum at the top and approaching zero at the bottom. All transverse waves will therefore travel with a speed which is a function of the position along the rope. Despite this major complication, the condition at the bottom is that of a "free end." If the upper end is moved back and forth with a periodic transverse motion, there will be some frequency or frequencies which will set up standing waves along the rope. A close observation of the lower end will reveal the physical effect of the free-end condition. The particles of the rope will there have a *maximum* amplitude of motion (as compared with particles farther up the rope). The motion of a heavy rope will usually be slow enough to show that the *slope* of the wave shape close to the lower end is zero, i.e., the end portion always remains vertical and therefore parallel to the long dimension of the rope (Fig. 7–4). Using y and x, respectively, to specify the transverse and longitudinal directions relative to the rope, the mathematical boundary condition at a perfectly free end is that $\partial y/\partial x$ must always be zero, rather than that y must be zero.*

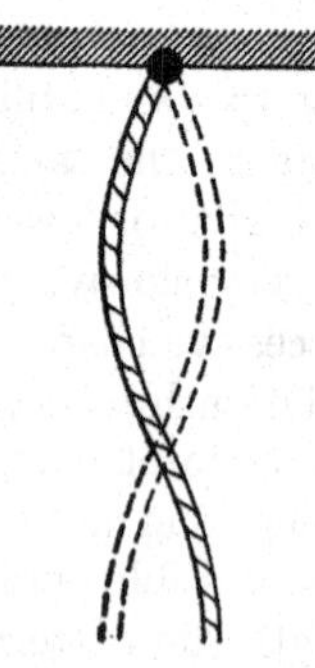

FIG. 7–4. Simplest mode of vibration of a heavy rope hung vertically from a rigid support.

* The mechanism of free-end reflection may be considered somewhat as follows. At a rigid boundary the transverse force acting on the support upon the arrival of an incident wave or pulse is proportional to the slope of the string just in front of the boundary (it is proportional to $F_t \tan \phi$ for small slopes). If we imagine the mass and stiffness of the support to decrease and approach zero, this transverse force must go to zero also. Since T is assumed constant, the slope must therefore approach a zero value at the boundary as we approach the completely free-end condition.

Let us return to the general solution for waves along a string (Eq. (7–5)). If we suppose that the end at $x = 0$ is free, we must in this case select two functions f which are the same in form and of the *same* algebraic sign. For if the equation for y is then written

$$y = f(ct + x) + f(ct - x), \tag{7–15}$$

it is seen that upon performing the differentiation $\partial y/\partial x$ and setting $x = 0$, we obtain

$$\frac{\partial y}{\partial x} = f'(ct) - f'(ct) = 0. \tag{7–16}$$

The boundary condition is thus seen to be satisfied. If a periodic form for the two waves is chosen as in Eq. (7–9), but with a positive sign in front of the second part of the solution, it is easy to show by the same trigonometric process that y may be written

$$y = 2y_m \cos\left(\frac{2\pi x}{\lambda}\right) \sin\left(\frac{2\pi ct}{\lambda} + \alpha\right). \tag{7–17}$$

As in the procedure for the two fixed ends, we now introduce the condition that the end at $x = l$ is also completely free. If we differentiate y partially with respect to x, we have

$$\frac{\partial y}{\partial x} = -2y_m \frac{2\pi}{\lambda} \sin\left(\frac{2\pi x}{\lambda}\right) \sin\left(\frac{2\pi ct}{\lambda} + \alpha\right). \tag{7–18}$$

Inserting the value $x = l$, we see that for $\partial y/\partial x$ to be zero for all values of the time, the following relation must hold:

$$\frac{2\pi l}{\lambda} = n\pi$$

or

$$\lambda = \frac{2l}{n}, \tag{7–19}$$

where n is any integer. This will be recognized as exactly the same restriction upon wavelength as was imposed by the condition that both ends be fixed. Therefore the *possible* frequencies of vibration are exactly the same as for fixed ends and Eq. (7–13) applies equally well for either set of conditions. Inserting λ from Eq. (7–19) into Eq. (7–17), the complete stationary wave equation may be written

$$y = 2y_m \cos\left(\frac{n\pi}{l}\, x\right) \sin\left(\frac{n\pi c}{l}\, t + \alpha\right). \tag{7–20}$$

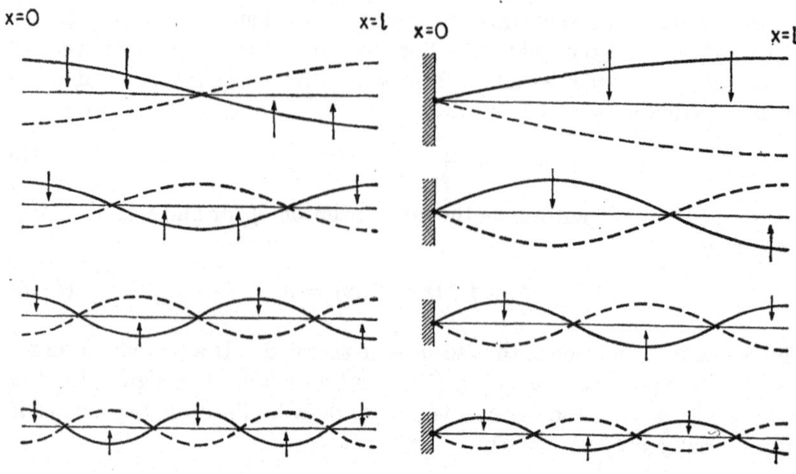

FIG. 7–5. Several possible stationary wave patterns for a hypothetical stretched string whose ends are completely free to move transversely.

FIG. 7–6. Several possible stationary wave patterns for a hypothetical stretched string fixed at one end and perfectly free to move at the other.

While the possible frequencies are the same as for the string with fixed ends, the stationary wave patterns do look different, as seen in Fig. 7–5 (compare with Fig. 7–3). At the free ends there are always *antinodes*. The simplest mode of vibration involves a wavelength which is just twice the length of the string, with a single node at the center.

7–11 Vibrating string, one end fixed, one end free. This particular set of end conditions, like the case just discussed, where both ends were free, is of interest primarily in connection with air waves in pipes. Let us assume that the left end is fixed and the right end is free. In this case the stationary wave equation becomes

$$y = 2y_m \sin\left(\frac{n'\pi}{2l}\,x\right)\cos\left(\frac{n'\pi c}{2l}\,t + \alpha\right), \tag{7-21}$$

where n' is an *odd* integer only. The possible frequencies of vibration are given by

$$f = \frac{n'c}{4l} = \frac{n'}{4l}\sqrt{\frac{F_t}{\sigma_l}}. \tag{7-22}$$

Each of the higher modes of vibration is an *odd multiple* of the fundamental lowest frequency. In Fig. 7–6 are shown a few of the simpler standing wave patterns

7-12 Initial conditions. Throughout the previous discussion we have been discovering *possible* modes of vibration. This does not at all imply that a string *necessarily* executes any particular one of these modes, nor must all modes exist simultaneously. Just how the string will move is determined *by the initial way in which the string is deformed*, as well as by the special boundary conditions for certain positions along x. In other words, since y is a function of both x and t, we must satisfy boundary conditions for both variables.

It must be freely admitted that to state precisely the initial time conditions for the vibration of strings on musical instruments is a very difficult matter, impossible in most cases. The most common modes of excitation are plucking, as in the banjo, the harpsichord, and occasionally the violin; striking, as in the case of the piano; and bowing, a process peculiar to the violin and related instruments. The initial conditions are progressively more complicated, in that order. While numerous analytical statements have been attempted, few may be said to accurately describe the situation. In addition, the project hardly seems worth-while from a practical point of view since, as we shall see presently, the resonance vibrations set up in the body of a musical instrument color the quality of the radiated sound fully as much as do the vibration properties of the string itself. In a few cases, however, the initial conditions are fairly simple. We shall briefly consider one such case, to indicate the procedure.

It will be noted that a phase angle α was included in Eq. (7–12) for the vibrations of a string fixed at both ends. Let us consider to begin with that all possible modes *may* exist simultaneously as well as separately (by the superposition principle). Then we may write for y,

$$y = \sum\left[A_n \sin\left(\frac{n\pi}{l}\,x\right)\cos\left(\frac{n\pi c}{l}\,t + \alpha_n\right)\right], \tag{7–23}$$

where A_n and α_n are the amplitudes and phase angles, respectively, associated with each mode. This equation may be put in the form

$$y = \sum\left[\sin\frac{n\pi}{l}\,x\left(b_n \cos\frac{n\pi c}{l}\,t + a_n \sin\frac{n\pi c}{l}\,t\right)\right], \tag{7–24}$$

where

$$\left.\begin{aligned} a_n &= -\,A_n \sin\alpha_n \\ b_n &= A_n \cos\alpha_n. \end{aligned}\right\} \tag{7–25}$$

and

The problem now is to determine coefficients of the form a_n and b_n in such a way as to completely satisfy the initial conditions. This process constitutes essentially a Fourier analysis, for at the time $t = 0$,

$$y\Big|_{t=0} = \sum\left(b_n \sin\frac{n\pi}{l}x\right). \tag{7-26}$$

If, in addition, an initial velocity condition is imposed,

$$\dot{y}\Big|_{t=0} = \sum\left(a_n \frac{n\pi c}{l} \sin\frac{n\pi}{l}x\right). \tag{7-27}$$

If we can now state y as a function of x, and $\dot{y}$ as another known function of x at the time $t = 0$, the coefficients in the cosine and sine series of Eq. (7–24) can be determined by use of the formulas developed in Chapter 1. We then have the complete picture for the subsequent motion of the string.

Example. The initial conditions for a plucked string can be stated rather simply. Assume that just before release the string is given the initial shape shown in Fig. 7–7. Let us suppose, as in the diagram, that the string is plucked at the center. Then at the time $t = 0$ we may express the function y as

$$y = \frac{2h}{l}x, \text{ for } 0 < x < \frac{l}{2},$$

and

$$y = \frac{2h}{l}(l - x), \text{ for } \frac{l}{2} < x < l,$$

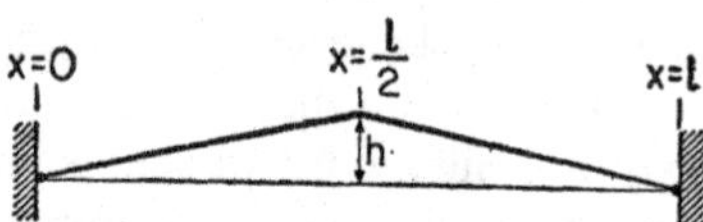

FIG. 7–7. The geometry for a string plucked at the center.

where l is the length of the string and h is the displacement at the center. In addition, let us picture the string as released from a state of complete rest, i.e., at time $t = 0$, $\dot{y} = 0$, everywhere on the string. This means, from Eq. (7–27), that all the coefficients of the form a_n must be zero, so that the sine series (Eq. 7–24) must be completely missing. To determine the coefficient of a typical cosine term, we make use of Eq. (7–26) and Eq. (1–16), that is,

$$\begin{aligned} b_n &= \frac{2}{l}\left[\int_0^{l/2} \frac{2h}{l}x \sin\left(\frac{n\pi x}{l}\right)dx + \int_{l/2}^{l} \frac{2h}{l}(l - x)\sin\left(\frac{n\pi x}{l}\right)dx\right] \\ &= \frac{8h}{n^2\pi^2}\sin\frac{n\pi}{2}. \end{aligned}$$

Plainly, the coefficients are zero for *even* values of n and consequently only the *odd* harmonic multiples in the harmonic series will be present after the string is plucked. Their amplitudes as determined from the expression above for b_n turn out to be in the ratio of $\frac{1}{(1)^2}, \frac{1}{(3)^2}, \frac{1}{(5)^2}$, etc., diminishing rapidly as the frequency rises.

7–13 Other initial conditions. General considerations. The example just given will suffice to show the general procedure for an analysis of this type. Since for most modes of excitation the lowest frequency involves the greatest amplitude, it will usually be the most prominent among the various string vibrations and will, in general, be the most prominent to the ear and so determine the characteristic "pitch." The amplitudes of the higher harmonics, so important to the quality of the emitted sound, are completely determined by the mathematics. It should not be concluded, however, that the intensity ratio as observed by the ear will bear any simple relation to these relative amplitudes. The vibration of the body of the musical instrument has an enormous effect upon the radiated sound, as we shall see later in the chapter, and alters radically the harmonic content in the transmitted sound wave.

For a string which is *struck*, as in the case of the piano, the initial conditions are more complicated, because of the added forces exerted by the hammer over a finite time. To represent this situation analytically is very difficult if one desires to take account of the transient period of the hammer contact. In addition, with a sharp blow (as indeed also with a violent plucking) the small amplitude vibration assumed for the string is easily exceeded and, as for waves in air, different portions of the disturbance will travel with different speeds. One thing may be noted, however, from the result for a plucked string. Any harmonic whose stationary wave pattern involves a node at the point of initial plucking or striking is always absent, e.g., the even harmonics were not present in the problem just presented. A piano string is struck by the hammer at a point about $\frac{1}{8}$ of the distance from one end. This discourages the 7th and 9th harmonics, whose combination is particularly dissonant to the ear.

A further effect of the hammer is associated with the softness of the felt face. Since the blow is made less abrupt and is less highly localized because of the felt, the harmonics above about the 6th are reduced in prominence in the general interests of tone purity. The hardness of the felt in pianos differs considerably and corresponding quality differences in tone are readily recognizable.

Parenthetically we might stress the point of view of the physicist in contrast to the opinion of many musicians as to the exact role played by the performer in determining the *quality* of a single note struck on a piano. Jeans and others have pointed out the purely mechanical linkage between the key and the hammer, whose characteristics are largely beyond the control of the musician. As a matter of fact, oscillograph records of notes produced by a musician's finger and by the falling of a weight (where the strength of the blows is identical) are practically indistinguishable. The pianist, of course, must possess many technical skills relating to the manner in which a *sequence* of notes is played, but to the physicist the existence of any such skill in the striking of a *single* note is pure fiction.

7–14 Bowing. Relaxation oscillations. The excitation of a violin string by bowing is too complicated to discuss in any mathematical detail. As the gut is slid over the string, its rosin-coated surface alternately holds to the string and then lets go. As a result, a disturbance of a periodic nature travels down the string and stationary waves are quickly set up. The phenomenon occurring at the portion of the string in contact with the bow is mainly a function of local elastic and frictional forces only. Nevertheless, the vibrations occurring over the string as a whole are of frequencies characteristic of conventional free string motion, which indicates that the inertial as well as the elastic properties of the string must enter into the total phenomenon. The bow, serving as a driving agent, merely supplies enough energy, through a mechanism involving rather weak coupling, to maintain the string vibrations at constant amplitude. Raman* has been able to develop a reasonably satisfactory mechanical theory of the bowing process, although not complete in all details.

The phenomenon occurring at the contact point of a bowed string is of some importance outside the field of musical sound. Vibrations of this general type are called "relaxation" vibrations and, as has been suggested above, are of a frequency dependent only upon elastic and frictional factors, with inertia often playing a very minor part. For there to be the alternate slipping and nonslipping referred to, the maximum static friction must be regularly exceeded, with short intervening periods during which a lower sliding friction is in operation. A frictional force which periodically changes sign will, mathematically speaking, bring about such a state of affairs. Van der Pol† and others have studied equations involving friction of this type and have obtained solutions in approximate agreement with experiment. The *period* of motion, as so determined, is proportional to the *ratio* of a frictional coefficient to an elastic constant. Many examples of this type of vibration may be cited. The squeaking of a door hinge, the flapping of a flag in the wind, and the scraping of chalk on a blackboard are a few. The operation of the so-called sweep circuit in an oscilloscope, where a capacitor alternately charges and discharges, is an example of a relaxation oscillation of an electrical nature. In this case the period of the oscillations depends on the ratio of capacitance to electrical resistance.

7–15 Vibration of membranes. Stationary waves in two dimensions. A membrane may be considered as a two-dimensional flexible string. It consists of a thin sheet of elastic material under uniform stress in a direction

---

* *Ind. Assoc. for Cult. of Science*, **15,** 1–158 (1918). Also *Phil. Mag.* **38,** 573–581 (1919).

† *Phil. Mag.* **2,** 978 (1926); also **6,** 763, (1928).

tangent to its surface and of uniform mass per unit area. A circular drumhead is essentially a membrane, in this case held bound at its circumference. The mathematics of a square membrane, although this shape is seldom used, is simpler than that of a circular membrane. In cartesian coordinates, the differential equation for the motion of a membrane may be written (almost intuitively) as

$$\frac{\partial^2 z}{\partial t^2} = \frac{F_s}{\sigma_s}\left(\frac{\partial^2 z}{\partial x^2} + \frac{\partial^2 z}{\partial y^2}\right), \tag{7–28}$$

where x and y are measured in the plane of the membrane and z is a displacement perpendicular to this plane. The quantity F_s is the stress across unit length in the plane of the membrane (defined as for liquid surface tension) and σ_s in this case is the mass per unit area.

If periodic deformations having circular symmetry are assumed (and this will obviously be the case when a drumhead is struck initially at the center), this equation may be transformed to read

$$\frac{\partial^2 z}{\partial r^2} + \frac{1}{r}\frac{\partial z}{\partial r} + Az = 0, \tag{7–29}$$

where r is a radial coordinate and A is a constant related to the inertial and elastic factors as well as to the frequency of any simple harmonic motion which may take place.

The solution of the differential equation (7–29), assuming a steady state symmetrical stationary wave pattern, must, of course, take account of the boundary conditions. For a drum, z must be zero when r is equal to the radius of the drumhead. We will simply summarize the results of the mathematics. The possible frequencies do *not* constitute a harmonic series. The numerical relations between the first few higher frequencies and the fundamental frequency are given in Table 7–1.

TABLE 7–1

Frequency	f_1	f_2	f_3	f_4	f_5	f_6
Relative frequency	1.0	1.59	2.13	2.29	2.65	2.91

In Fig. 7–8a, b, c are shown the instantaneous shapes, together with the nodal lines for three symmetrical modes of vibration. For comparison, Fig. 7–8d shows a possible mode of vibration which does not have circular symmetry. Such a mode might be excited by striking the drumhead off-center.

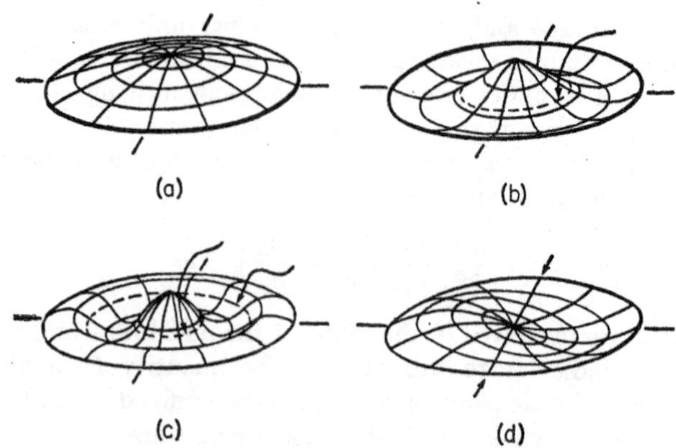

FIG. 7–8. A few of the possible modes of vibration of a stretched circular membrane. Arrows point to nodal lines. (*After* Morse)

The prominence and number of the possible modes of vibration which will exist with any practical kind of excitation will depend on the initial conditions, just as for a string. To determine the exact nature of the vibration, the procedure is similar to that employed for the string, although somewhat more difficult. Drumheads are often stretched over the tops of rigid airtight containers (as, for instance, the kettle drum of the orchestra) and the presence of the trapped air significantly increases the stiffness of the drumhead and hence raises the natural resonance frequencies.

7–16 Longitudinal stationary waves in bars. We have seen in Section 6–14 that plane longitudinal waves travel down a bar of uniform cross section with a velocity $c = \sqrt{Y/\rho_0}$. This velocity is quite independent of frequency, as is the case with waves of small amplitude in air and with transverse waves of small amplitude along a stretched string. Since the differential equation for the waves in a bar is identical in form with the equation for waves along a string, we may take over the solution and write

$$\xi = f(ct \pm x).$$

For a bar of limited length, one may impose boundary conditions at the ends which are completely analogous to the end conditions for the string, remembering that the particle displacement ξ for the bar is longitudinal. This will lead quite naturally to a stationary wave equation of the same type as Eq. (7–12). If the bar could be clamped between infinitely rigid supports, the particle displacement ξ at the ends would always be zero. Equation (7–12) would then apply exactly, except that y would be replaced by ξ. The possible periodic motions are then given by Eq. (7-13), that is,

$$f = \frac{nc}{2l},$$

where c in this case is $\sqrt{Y/\rho_0}$.

Since a solid bar is itself so rigid, it is difficult to provide supports of greater rigidity, and hence the end conditions assumed above are hard to realize in practice. The free-end condition may be achieved quite readily, however, by supporting the bar on strings placed some distance in from the ends. The stationary wave patterns given in Fig. 7-5 then apply equally well here. To reduce the energy dissipation at the supports to a minimum, the strings should be located at the nodes. For the simplest mode of vibration, where there is a single node at the center, a knife-edge clamp should be used (as in Kundt's original experiment, Fig. 6-1).

The resonance frequencies associated with longitudinal waves in a bar constitute a harmonic series, as they do for the string. Such frequencies are always much higher for the bar, however, since the velocity of wave propagation is much greater than for waves along a string. For a metal like steel, Y is 2×10^{12} dynes-cm^{-2}, ρ_0 is about 7.8 gm-cm^{-3}, and therefore $c = \sqrt{Y/\rho_0}$ becomes 5.0×10^5 cm-sec^{-1}. For a stretched string on a typical musical instrument, c might be of the order of 10^4 cm-sec^{-1}. Since $f = c/2l$ for the fundamental mode of either the string or the bar (ends either both fixed or both free), the lowest frequency to which a bar will resonate might be roughly fifty times that of a string of the same length.

7-17 Transverse waves in bars. Bars are more easily set into transverse than into longitudinal vibration, since they are more yielding to deformations of that type. Most of the annoying vibrations in the frame of a car, an airplane, etc., are of this kind. Vibrations of the transverse type are of some importance in the production of musical sound, particularly in instruments having reeds, like the saxophone, and in the xylophone, whose bars are struck transversely.

Unlike the ideal string, a bar resists bending. Therefore even without longitudinal stress, flexure will give rise to restoring forces. These forces are of two general kinds over any cross section of the bar, as indicated in Fig. 7-9. There is a transverse "shearing" force F_{sh} and a couple or "bending moment" M_b, the latter being the result of the compression and the elongation of opposite sides of the bar. The shearing force and

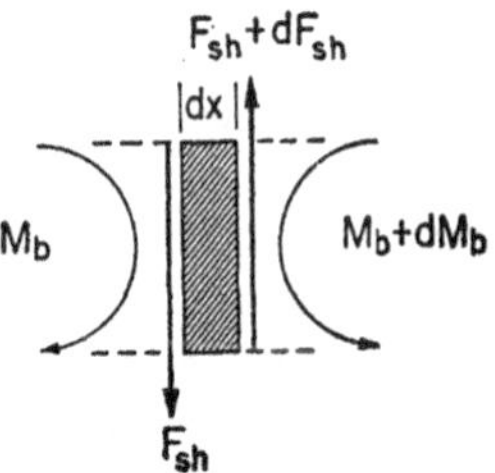

FIG. 7-9. Forces and moments acting on a thin slab of differential thickness in a bar which has been deformed.

the bending moment will, in general, vary along the bar in the presence of a deformation, as indicated in the figure. On the basis of this illustration one may express Newton's second law for the transverse translation and for the rotation of a thin slab the area of whose face is the cross-sectional area of the bar. We shall not go into the details of this analysis, but shall merely summarize the results.* One may finally obtain a differential equation relating the transverse displacement y for any point on the bar to the position x along the bar and to the time t. This equation, for small amplitudes, becomes

$$\frac{\partial^2 y}{\partial t^2} + A' \frac{\partial^4 y}{\partial x^4} = 0, \qquad (7\text{–}30)$$

where A' is a constant related to the elastic and inertial properties of the bar and to the shape of the cross section. This is a fourth order differential equation and its solution presents difficulties, some of which may be best surmounted by the use of graphical methods. Upon examination, the solution implies the possibility of wave motion, but the speed of travel of these waves turns out to be a function of the frequency, higher frequencies traveling faster than lower ones. For a bar of limited length, stationary waves are possible at certain discrete frequencies. The higher frequencies do not bear any simple integral relationship to the fundamental frequency and hence do not comprise a harmonic series. Table 7–2 gives the relationships for a few possible modes (assuming free-end conditions).

TABLE 7–2

Frequency	f_1	f_2	f_3	f_4
Relative frequency	1.0	2.76	5.40	8.93
Number of nodes	2	3	4	5

The simplest mode involves the existence of two nodes, as indicated in Fig. 7–10. The wooden bars of the xylophone are supported horizontally on two strings located at these nodal points, a little less than $\frac{1}{4}$ the way in from the ends (actually a fraction equal to 0.224).

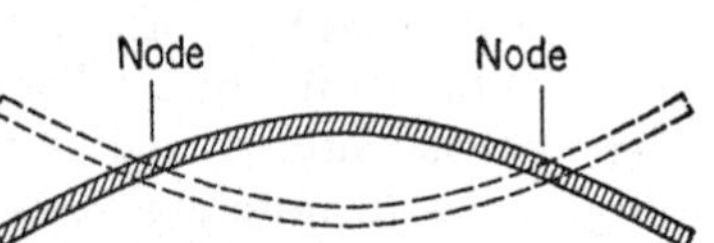

Fig. 7–10. Simplest mode of vibration of an elastic bar.

* A good discussion of the mathematics of this problem is given in H. Lamb, *Dynamical Theory of Sound*, Edward Arnold & Co. (1931).

If a bar is clamped at one end and the other end is left completely free, the possible modes consistent with these end conditions are somewhat different. The number of nodes (including the one at the clamped end), together with the frequency relationships, for a few of the various modes are given in Table 7-3. Other frequencies may be discovered for other combinations of end conditions.

TABLE 7-3

Frequency	f_1	f_2	f_3	f_4
Relative frequency	1.0	6.27	17.5	34.4
Number of nodes	1	2	3	4

7-18 The tuning fork. We should not leave the subject of bar vibrations without mention of the tuning fork. Such a fork is really a bar bent in the shape of a letter U. The bending can be shown both experimentally and theoretically to bring about a closer spacing between the two nodes which are characteristic of the simplest mode of vibration of a straight bar, free at the ends. Compare Fig. 7-11 with Fig. 7-10 for the straight bar. If a tuning fork is struck with only a moderate blow, so that the amplitude of the prong motion is small, frequencies higher than the fundamental will be of negligible prominence. In addition, since the drop in amplitude per cycle is about the same for all frequencies, the higher modes will disappear sooner than the lower ones. Also, the various possible frequencies are rather widely spaced, as was pointed out for the straight bar, and the upper frequencies are therefore of little consequence within the audible range. As a result of this general behavior, the tuning fork is an excellent source of pure sinusoidal vibrations of a single frequency. As mentioned in Chapter 3, it is a very poor direct radiator of sound waves, since it possesses double source action. Its radiation efficiency may be greatly enhanced by touching its stem to a table top or other plate of large surface area. The area of this secondary radiating surface, together with the self-baffling action of the plate for its own dipole components, will add much to the audibility of the sound.

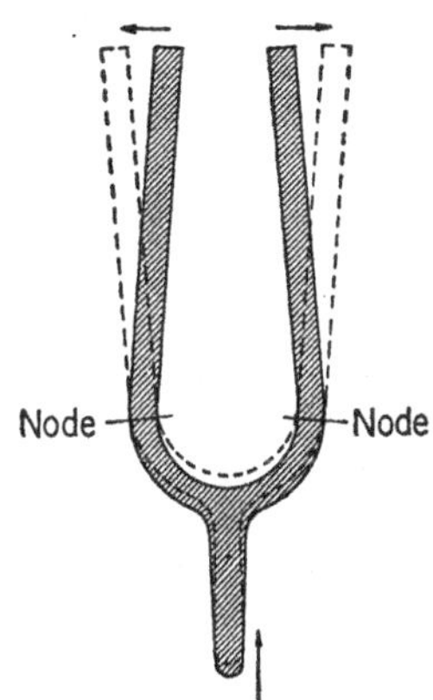

FIG. 7-11. Vibration of a tuning fork.

7–19 The vibration of plates. A plate is, in a sense, a bar with two large dimensions and it compares to a membrane as does a bar to a string. Transverse deformation of a plate is always accompanied by shearing forces and bending moments, just as in the bar, except that in the plate there is the additional complication that waves may travel in *any* direction parallel to its surface. The mathematical difficulties connected with the analysis of the vibrations of a plate are great, and we shall only indicate the general nature of the motion. For a circular plate, like the diaphragm of a telephone receiver, the possible symmetrical modes of vibration are quite similar in appearance to those of a circular membrane, as shown in Fig. 7–8. For a square plate, symmetrical modes of vibration of a different variety may occur. Most of us have seen demonstrated the classic sand patterns known as Chladni plate patterns. In this experiment a heavy plate is set into transverse vibration by means of bowing or, more effectively, by electromagnetic means. Sand sprinkled on the plate will then gravitate towards the nodal regions. The variety of patterns obtainable is very great. A few are shown in Fig. 7–12.

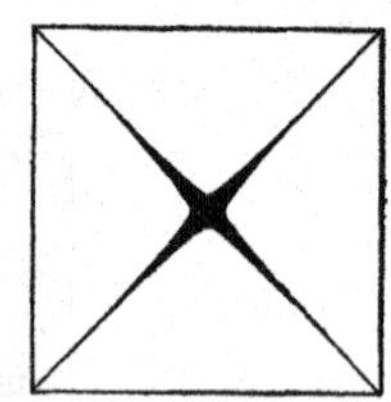

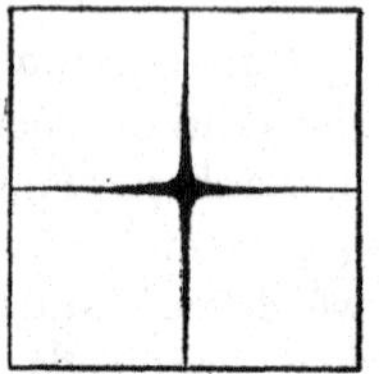

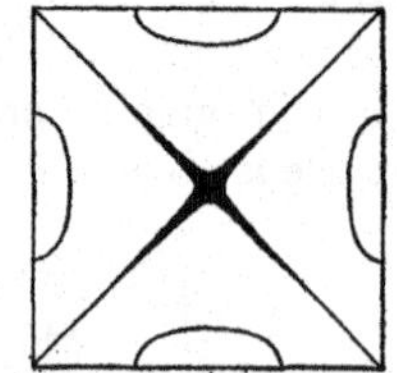

FIG. 7–12. Typical Chladni sand patterns for a square plate.

We should note that in general, as for a bar, the various possible frequencies for a plate are not related by integers and are relatively *far apart* as compared with those for a simple string. This latter fact is of some practical importance in telephone receiver design. Utilization of the fundamental mode is desirable from the standpoint of efficiency. If the diaphragm is designed to resonate in the middle of the useful frequency range, the higher resonance frequencies for the most part will be outside the important part of the audible spectrum.

7–20 Stationary air waves in pipes. Resonance effects in enclosed bodies of air may be of two general types, the first of which was introduced in our discussion of the Helmholtz resonator in Chapter 5. It was carefully noted there that for the resonance formula developed to be valid, any dimension of the enclosed volume must be *small* compared with the wavelength of the radiated sound. The enclosed air then acts as a simple spring. Mention was also made of the possibility of higher resonance frequencies associated with *stationary wave phenomena*. Resonances of this latter kind occur when the dimensions of enclosed bodies of air are *comparable to or larger than the wavelength.* The various modes of vibration are very similar to those for strings and for bars executing longitudinal motion and the mathematics is almost identical, as we shall see.

As an example of stationary waves in air, the vibration of a cylindrical air column of limited length is of the most general interest, since such a geometry is basically that of the organ pipe and other of the wind instruments. As a matter of fact, the analysis to be given does not require a circular cross section. The cross-sectional shape of wooden organ pipes is usually square and their behavior does not differ significantly from that of pipes having circular cross sections.

7-21 Vibrations in a pipe closed at both ends. The particular manner in which an air column may be set into vibration will be discussed in Section 7-27. At this point we shall simply assume that some disturbance in the air has started plane waves traveling along the axis of the pipe. If both ends of the pipe are closed with rigid plates, such waves will be reflected, the phase relations at the boundaries being identical with those for a string having fixed ends. For the air column, ξ must be zero at the ends. As a consequence, the incident and the reflected wave must be just 180° out of phase at those positions. The number of steady periodic motions consistent with the condition that the ends of the column be nodal points may be determined from the stationary wave equation

$$\xi = 2\xi_m \sin\left(\frac{n\pi}{l}x\right)\cos\left(\frac{n\pi c}{l}t + \alpha\right). \tag{7-31}$$

(This equation is Eq. (7-12) for the string rewritten in terms of the air particle displacement ξ.) Just as for the string, the frequencies are given by

$$f = \frac{nc}{2l}. \tag{7-32}$$

The diagrams of Fig. 7-3 for the string may be taken over bodily for the column of air, if we are careful to interpret ordinates as *longitudinal* air displacements.

It should be understood that what has been said above refers strictly to the displacement variable ξ in the wave. A sound wave involves other parameters, such as the important pressure and density variables, and the stationary wave picture for these latter variables is quite different from that for the displacement. For instance, to see just what the pressure situation is at the ends of the pipe, let us operate upon Eq. (7–31). This expression is the exact equivalent of that for two oppositely traveling waves. For a plane wave, the relation between p and ξ is

$$p = -\mathcal{B}\frac{\partial \xi}{\partial x}. \tag{7-33}$$

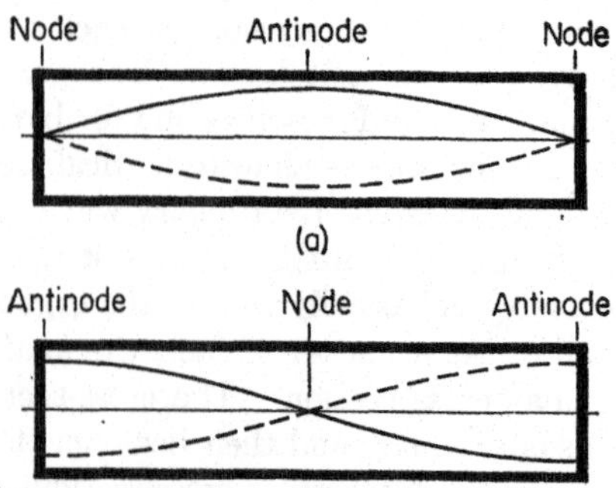

FIG. 7–13. Stationary wave pattern, pipe closed at both ends, (a) for *displacement*-fundamental mode, (b) for *pressure*-fundamental mode.

Hence the acoustic pressure in the pipe, as a function of x and t, is, from Eqs. (7–31) and (7–33),

$$p = -\ 2\xi_m\mathcal{B}\frac{n\pi}{l}\cos\left(\frac{n\pi}{l}x\right)\cos\left(\frac{n\pi c}{l}t + \alpha\right). \tag{7-34}$$

It will now be seen that with various integral values of n the end positions (where $x = 0$ and $x = l$, respectively) become pressure *antinodes*, since at these points the cosine expression has its maximum value of unity. The same thing may be said of density, since the condensation and the acoustic pressure are linearly related. It is therefore correct to call the end points either nodes or antinodes, provided one is careful to specify the proper parameter. The diagrams of Fig. 7–13 emphasize this distinction.

7–22 Vibration of an open organ pipe. Organ pipes are either open at both ends or open at one end and closed at the other. The former is called an "open" pipe, the latter a "closed" pipe. In this section we shall discuss the open pipe. Reflections of longitudinal air waves may take place at an open end as well as at a closed end. From a qualitative physical viewpoint the process is understandable if one looks at the pressure parameter in the wave. When a pulse associated with a positive value of the excess pressure p reaches an open end, it suddenly leaves the confines of the tube walls and enters the less restraining region of free space. Its abrupt exit leaves behind a pulse of rarefaction which propagates back down the tube and constitutes the reflected component. Such a conception carries with it the notion of a phase reversal of the pressure component of the disturbance upon reflection. We may verify this conclusion in a more convincing manner if we consider the *displacement* feature of the wave.

The last layers of air near an open end of a pipe are under somewhat smaller restraining forces than are layers well within the pipe, since the air beyond the end is unconfined and the medium can move laterally as well as axially. As far as the particle displacement ξ is concerned, we may consider the boundary condition at an open end as somewhat analogous to the free-end condition assumed earlier for a string or for a solid bar vibrating longitudinally. If this picture is essentially correct, we may adopt as the expression for ξ in an organ pipe open at both ends, the form of the string solution given by Eq. (7-20), that is,

$$\xi = 2\xi_m \cos\left(\frac{n\pi}{l}x\right)\sin\left(\frac{n\pi c}{l}t + \alpha\right), \tag{7-35}$$

which shows the amplitude of the variation in ξ to be a maximum at the ends. These points therefore are *displacement antinodes*. If, however, we are interested in the *pressure* at the ends, by making use of Eq. (7-33) relating ξ to the acoustic pressure p, we find that

$$p = 2\xi_m \mathcal{B}\frac{n\pi}{l}\sin\left(\frac{n\pi}{l}x\right)\sin\left(\frac{n\pi c}{l}t + \alpha\right). \tag{7-36}$$

This equation indicates that the end points for an open pipe are *pressure nodes*, since when $x = 0$ or l, p is zero for all values of the time.

This result is quite consistent with the qualitative description of pulse behavior. If a pulse involving *positive* acoustic pressure returns as a pulse of *negative* pressure, the incident and reflected pulses at the open end are out of phase and hence the end point is located at a pressure node.

The various facts regarding the two kinds of end conditions in an air column are summarized in Table 7-4.

TABLE 7-4

	"Displacement" wave		"Pressure" wave	
	Closed end	Open end	Closed end	Open end
Phase change at reflection	180°	none	none	180°
	node	antinode	antinode	node

7-23 Reflection and acoustic impedance. The reflection at an open end is not complete. The fact that wave energy does leave the pipe (since the sound is audible) is evidence enough that with each reflection only some fraction of the energy incident at an open end returns down the tube. When the perimeter of the pipe end is small compared with the wavelength, the reflection is more complete. This follows from the discussion of diffraction in Chapter 4 and from certain features of horn behavior as presented in Chapter 5. The open end of a pipe may be considered as a possible source of waves radiating into space. If the perimeter is small compared with the wavelength, the energy will radiate with great spherical divergence and the specific acoustic impedance z_s at the pipe end will have a relatively large imaginary or reactive component characteristic of a sphere of small radius (small, that is, compared with the wavelength). Little real power will then leave the pipe and reflection and stationary waves will be encouraged. Since practical organ pipes *do* have cross sections of these relative proportions, the degree of reflection is quite adequate to maintain stationary waves of large amplitudes. Some escape of energy takes place, however, because the real component of the acoustic impedance, while relatively small at an open end, is never zero.

We shall have more to say in the next chapter about the possible effect of a partial reflection at the end upon the wave pattern within the pipe. For the present we shall assume the reflection at an open end to be practically complete and so deal with a simpler picture.

7–24 Frequencies of vibration of a "closed" organ pipe. The boundary conditions for a closed pipe correspond to those for a string with one end fixed and one end free, and have already been discussed in Section 7–11. The possible modes of vibration of the air column will be given by

$$f = \frac{n'c}{4l}. \tag{7–37}$$

Only odd integers may be used for n', it will be remembered, and hence only the odd harmonics will be possible for a closed organ pipe under the ideal conditions assumed. The actual harmonics present when an organ pipe is sounded will depend upon how it is excited. This point will be considered when we discuss musical instruments.

7–25 General features of stringed instruments. The violin. As everyone knows, a violin is much more than a stretched string, and wind instruments are not, in general, simple cylindrical tubes. The design of musical instruments is the outgrowth of hundreds of years of experimentation, with until recently little or no careful scientific analysis. Empirical study over a long period will often achieve the desired results and the construction of musical instruments is a good example of this sort of development. To deal completely with this subject would require a book in itself. We can here emphasize only some of the essential physical principles involved.

A violin is a thin-walled box of unique shape, with the strings stretched tightly across the so-called "bridge" (Fig. 7–14). The four strings, sometimes of gut and sometimes of metal, are all of the same length (measured between the bridge and the upper clamping point) but are of different linear mass density and are under different tensions. As a result, the fundamental modes differ. The "open" string resonances cover a range of somewhat over an octave. The fundamental modes can be raised by shortening the string, i.e., by "stopping" the string with the finger. The range of the instrument may in this way be extended to about four octaves.

It is important to recognize that the sound waves produced by a violin originate almost exclusively with the vibrations of the body of the instrument, not with the vibrations of the strings themselves. A vibrating string, as mentioned in Chapter 3, is a linear array of double sources whose dipole components are almost coincident. The rate of dissipation of the vibrational energy by the radiation of sound waves is practically zero. Most of the energy supplied by the bowing action is transmitted to the body of the instrument through the bridge, the latter being set into motion by the periodic shortening and lengthening of the string associated with the stationary wave pattern. The amplitude of the motion at the top of the bridge is longitudinal and is smaller than that of the average transverse displacement of the string. The ends of the string can therefore still be considered fixed as far as stationary waves are concerned. Vibrations are transmitted to the bottom of the instrument by a wooden rod called the "sounding post," which extends from the belly downwards (*SP* in Fig. 7–14). The sound waves originate largely with the vibration of the top and the bottom of the violin. These areas are enormously larger than the surface areas of the strings and the dipole components are separated in space by a much greater acoustical distance than are those of the strings. For these reasons the body of the instrument is an efficient radiator.

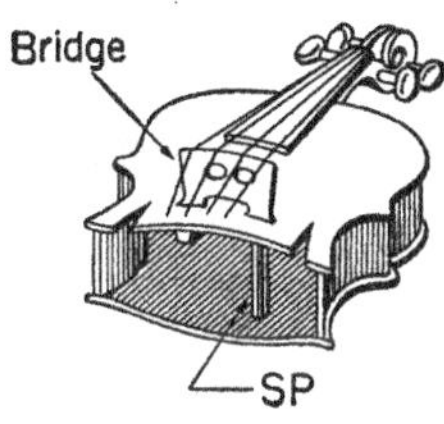

FIG. 7–14. Cross section of the body of a violin.

The harmonic content of a note played upon a violin is a complex function of a number of different factors. The most obvious are the manner of excitation (bowing or plucking); the position on the string where the excitation occurs (although this is apparently of minor importance); the complex modes of vibration of the various sides of the instrument body, which are impossible to predict on any analytical basis; and the nature and shape of the air cavity within the instrument, which may sometimes act as a Helmholtz resonator and sometimes, for the higher frequencies, may act in the manner of a pipe closed at one end. The wooden body possesses internal dissipation qualities due partly to the presence of joints and partly to the nature of the wood itself. Such dissipation is greater for the higher frequencies than for the lower, and so contributes in an important way to the strength or weakness of the upper harmonics.

Many experimental studies of the harmonic structure of the sound from a violin have been undertaken. Figure 7–15 shows the relative prominence of the various harmonics present for the four strings, each vibrating as an "open" string (using its maximum length). Each graph is different. Note that although for the G and E strings the lowest frequency is not the most intense, the characteristic pitch is still usually associated with the lowest frequency. This results from an interesting property of the ear, discussed further in Chapter 9. If presented with a more or less complete harmonic series, the ear apparently will associate the whole set with the mathematical fundamental, even though the latter may be weak or missing altogether.

Except for the factor of size, the other members of the violin family are of similar construction and behave in a similar fashion. Each of the numerous stringed instruments has, of course, a musical quality all its

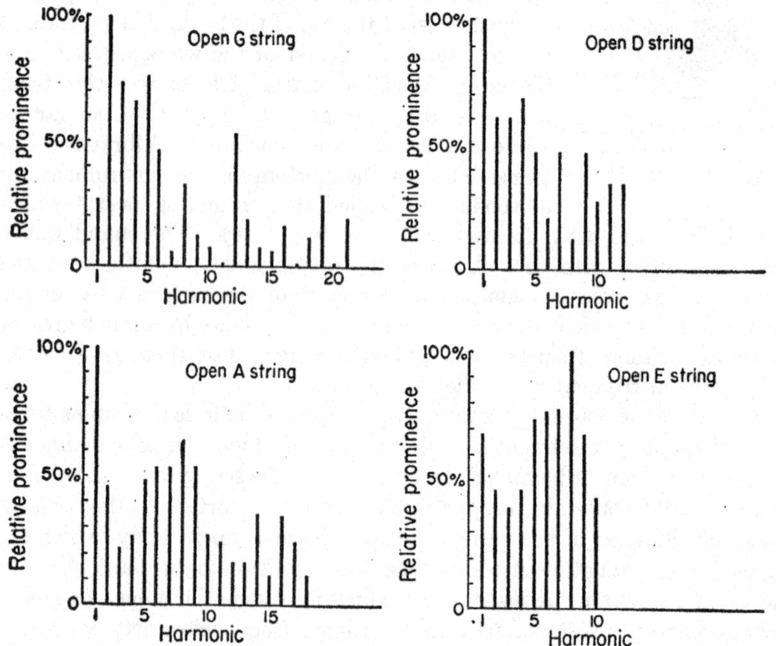

FIG. 7–15. Harmonic content of the sound from a violin when each of the four open strings is bowed. (*After* Culver)

own. In some cases a plucked note will sound brilliant, in others somewhat muffled. This effect is associated with the degree of attenuation of the upper harmonics. The attenuation is rather large, for instance, for the banjo and is less for the guitar. A string of gut has greater attenuation than a string of metal. Stringed instruments have been carefully studied in recent years and summaries of the significant factors affecting tone are readily available.*

7-26 The piano. The piano, certainly the most widely played of all the string instruments, is somewhat simpler to analyze from an acoustical point of view than is the violin. As with the violin, the vibrating strings of the piano are not direct radiators of sound waves. Instead, the vibrations of the strings are transmitted through the string supports to a massive sounding board, analogous to the light wooden box of the violin. This radiating plate has very broad resonance properties and is capable of vibrating at all the fundamental frequencies to which the strings are tuned and also at a great many of the harmonics which the hammer blow inevitably excites. Since the area of the sounding board is large, it is an efficient radiator for even the very low frequencies. Because of the great physical size of the piano, strings of considerable length may be used and this makes possible fundamental modes of vibration of much lower frequency than in the violin. By "loading" the bass strings with wound copper wire the wave velocity may be decreased (since σ_l in Eq. (7–4) is thus increased) and this makes possible still lower fundamentals. The range of fundamental frequencies for the piano (over seven octaves) is wider than that of any other instrument except the pipe organ.

The radiated sound from a piano is rich in harmonic content, particularly from the strings of low register. This is a feature of piano music that delights the listener and is of much concern to the manufacturers of small pianos of the spinet type. To achieve compactness, these pianos have strings which are shorter than normal in the low register and these must be under less tension, or more heavily loaded, or both, in order to resonate at the proper frequencies. These shorter strings will execute rather large amplitude vibrations when struck and the harmonics will usually be more prominent than in the case of a piano of conventional size. Since the sounding board is also smaller, there is danger that it may not resonate at the lowest frequencies and also that it may be a less efficient acoustic radiator. As a result of all of these factors, the fundamental associated with the very low notes is often almost completely missing to the ear, the second or even the third harmonic being far more prominent. One way of

* For interesting reading, see Culver, *Musical Acoustics*, Blakiston Co. (1947).

surmounting these difficulties is to utilize air cavity resonances but these are usually too sharp and an unnatural enhancement of certain notes often results. The building of a small piano with tonal qualities equal to that of a large one is still an unsolved problem.

7–27 The wind instruments. Excitation of an organ pipe. The wind instruments form a second large and important group of musical instruments. These employ resonating columns of air, either of uniform cross-sectional area, like the organ pipe, or with a flare, as in the horns. The manner in which the air column is set into vibration differs with the different instruments, although there is a certain similarity among them all. We shall look first at the organ pipe.

Figure 7–16 shows an organ pipe of the simple "flue" type in cross section. During excitation a steady stream of air is blown against the rigid wooden (or metal) lip. At very low velocities the stream of air divides, part entering the pipe and part flowing to the outside. As the velocity of the stream is increased, however, turbulence sets in near the edge of the lip and the stream begins to swing first to one side and then to the other. Rotational motion of the air results and small vortices may form with each swing. As shown in the figure, these vortices break off at the sides of the lip. One set is dissipated outside the pipe. The other set, inside the pipe, travels along the axis of the pipe with approximately regular spacing if the stream velocity is maintained constant. These regularly spaced regions of disturbance represent *pulses* of a periodic nature and, being pulses, may be thought of in terms of a Fourier harmonic series. Some one of these harmonics is apt to be close to a natural pipe resonance and so the air column may begin to vibrate in a sinusoidal manner. A variation of the stream velocity may result in a variation in the spacing of the vortices and a consequent favoring of one or another of these modes of vibration. The whole process is somewhat analogous to the thermionic tube generator of electrical oscillations. In both the organ pipe and the vacuum tube circuit a nonlinear element (the lip mechanism for the pipe and the vacuum tube characteristic for the circuit) soon transforms a steady flow of energy into a flow of periodic nature.

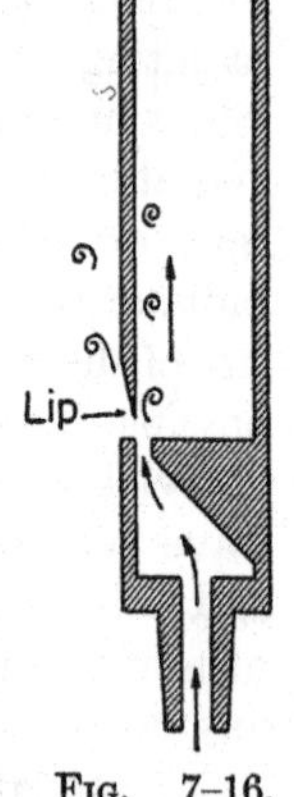

FIG. 7–16. Pulse formation during the blowing of an organ pipe.

The interesting phenomenon of "singing" telephone wires is related to the excitation mechanism just described. When wind velocities are sufficient, small whirlpools will form and break off on the leeward side of a wire exposed to the stream. As they break off a mechanical reaction is exerted on the wire, which then may begin to vibrate at one or more of its natural modes.

7-28 Wind instruments of the reed type. Some organ pipes and many of the smaller wind instruments are equipped with reeds near the air inlet. The reed (of metal, bamboo, or some other suitable material) is set in vibration by a steady stream of air blown past it. The process is very similar to the pulse formation just described. In some organ pipes the reed (considered as a bar clamped at one end) is tuned so that its fundamental corresponds to the fundamental or other desired mode of the organ pipe. In the smaller reed instruments, such as the clarinet, the reed is tuned to a frequency considerably higher than the fundamental mode of the air column. The frequency of the reed is controlled by the natural frequencies of the air column for the lower notes, since the coupling between the reed and the air column is rather close. Such frequencies are far removed from the natural reed frequencies. At higher modes of vibration the natural reed frequencies are more of a controlling factor and the reed resonances therefore considerably enhance the higher harmonics of the instrument, giving to it its characteristic tonal quality. Ports, opened and closed with keys, are provided to give the single tube of the instrument flexibility as regards frequency. These keys provide a musical scale based on the fundamental for the tube length and determined by the particular position of a port. It is also possible to accentuate harmonics by opening a special port with the so-called "speaker" key. Other subtleties of fingering and blowing make the instrument increasingly flexible.

Each of the wind instruments has its own special features of control and its own peculiar musical quality, determined by the harmonic structure of the sound waves produced. It is interesting to note that in the clarinet, discussed above, the driving end (where the air is blown in) is virtually a node, instead of the antinode that might be expected. The amplitude of the air vibration is so small at this end compared with that at the mouth of the instrument that once stationary waves are set up the driving end is in effect a point of no motion at all. As a result, the harmonics usually present are those shown to be characteristic of the "closed" organ pipe, i.e., the odd ones only. For other of the wind instruments, such as the flute, the resonances are those for a pipe open at both ends, where both even and odd harmonics may be present. This is true for the oboe, whose tube is conical in shape rather than cylindrical. Such a conical tube may be shown to resonate like an open organ pipe and so has a complete harmonic series of vibrational modes.

7-29 Wind instruments as radiating sources of sound. With the exception of the open organ pipe, wind instruments are almost exclusively "single" acoustical sources. (An open organ pipe is a double source, and the two sources may be in phase or out of phase, depending upon the mode of vibration. In any case, the two sources are rather widely separated in space and cancellation effects are small.) The mouth of a wind instrument is the virtual source of the waves. The propagation in surrounding air is approximately directional for the higher frequencies but much less so for the lower, since the mouth size then becomes smaller in comparison to the wavelength. For the air column to have resonance properties it is absolutely necessary that there be considerable spherical divergence from the mouth of the instrument, since such divergence is associated with a complex form for the acoustic impedance at the mouth and hence with reflection back into the tube. A musical instrument must, of course, radiate energy to be useful, but it must also resonate (through wave reflection). The dimensions of the mouths of all the wind instruments are such that both radiation and reflection can occur. It is interesting to note that in this case a satisfactory design has been achieved by purely empirical means.

PROBLEMS

1. The length of a stretched string A is twice that of string B. The tension in A is twice the tension in B, but the total mass of A is the same as that of B. Find the ratio of the fundamental frequency of A to that of B.

2. Show that the periodic motion of a string fixed at one end and completely free at the other is given by the equation

$$y = 2y_m \sin\left(\frac{n'\pi}{2l}x\right)\cos\left(\frac{n'\pi c}{2l}t + \alpha\right),$$

where n' may be only an odd integer.

3. A stretched string of length l is pulled to one side a distance h, at a position $\frac{1}{4}$ the distance from one end. The string is then released from rest. (a) Find the subsequent frequencies of vibration. (b) Are any frequencies in the harmonic series missing? If so, discuss the connection between such frequencies and the position of plucking.

4. From the stationary wave equations for p and for ξ, determine the specific acoustic impedance at the end of a resonating organ pipe for (a) a closed end (infinitely rigid cap), and (b) an open end (assumed perfectly "free").

5. Consider stationary air waves in a pipe, assuming perfect reflection at the ends. (a) Show, by considering the product $p\dot{\xi}$, that there is no average net flow of power at any point along the axis of the tube. (b) Discuss the specific acoustic impedance at various points along the axis.

6. A vibrating tuning fork held in the hand is almost inaudible. If held over the end of a pipe of the proper length for resonance, the sound may be clear and loud. (a) Why does this occur, since the fork is the source of energy in both cases? (b) Is any violation of the conservation of energy implied in the above phenomenon? Explain.

7. For what resonant frequencies will an open pipe 5 feet long radiate as a double source of the type considered in Chapter 3?

8. (a) For a progressive transverse wave on a string, compare the instantaneous distribution of potential and kinetic energy along the direction of wave travel with the energy distribution for a progressive longitudinal wave in air. (b) For a wave on a string, draw a graph of the total energy (potential plus kinetic) as a function of x, and compare with the graph for a longitudinal wave (Fig. 2–5).

9. A closed organ pipe 2 feet long and of cross-sectional area 50 cm^2 is resonating at its fundamental frequency. If the amplitude of motion of the air at the open end is 0.2 cm, find the average acoustic power radiated (Section 3–7).

10. (a) Show that the bridge of a violin will vibrate at twice the string frequency. (b) Suggest a physical picture of the energy transfer which does not require a final frequency doubling, i.e., for which the body of the violin vibrates with the same frequency as that of the string.

11. The paper cone of a loudspeaker sometime generates sound waves which are *sub*harmonics of the frequency of the axial driving force (applied at the apex of the cone). With a driving frequency, for instance, of 1000 cycles-sec^{-1} there may appear a 500-cycle note as well. This is due to the fact that the cone is not completely rigid, so that flexure may occur. Draw a physical picture to explain the appearance of such subharmonics.

CHAPTER 8

REFLECTION AND ABSORPTION OF SOUND WAVES

8-1 Introduction. In the preceding chapter we have seen how a more or less complete wave reflection at the end of a string, a bar, or an air column led to the formation of a stationary wave pattern in front of the boundary. Two general types of end condition were considered, the immovable or fixed end, and the end perfectly free to move, with no restraining force of any kind. In the first instance the fixed end was necessarily a nodal point, as far as displacement of the particles of the medium was concerned. Interpreting this situation in terms of reflection phenomena, we saw that at such a node the incident and the reflected waves are just 180° *out of phase*. For the free end the conditions were right for an antinode; at an antinode the incident and the reflected waves are exactly *in phase*.

Only in special types of practical problems will these extreme conditions of restraint or freedom apply with any degree of accuracy. An elastic string stretched between rigid supports constitutes a system where the end conditions are close to ideal. On the other hand, an air column in an open pipe is under some restraint at the ends, due to the presence of the surrounding air, and to call the ends "free" is only an approximation. In general, all practical boundary conditions involve partial freedom (or partial restraint) and under such conditions we have no right to make use of conclusions as to phase, etc., that are the outgrowth of mathematics suited only to the two limiting cases.

It is the purpose of this chapter to examine more closely the reflection phenomena for longitudinal waves when the boundary conditions are intermediate between the two extremes. The problem is of great practical interest in connection with the acoustical behavior of rooms, since in such enclosures sound waves impinge upon wall surfaces which are rarely perfectly rigid. The exact extent and manner of the "yielding" in the presence of acoustic pressures affect the reflection process in an important way and this in turn, as we shall see later, largely determines the suitability of the room for speech or music.

8-2 Reflection of longitudinal waves at a boundary between two ideal elastic media, each infinite in extent. As is usual in all branches of physics, we shall begin with a relatively simple situation and later, when we have developed some basic principles, proceed to the more complicated problems. Consider first two media, each having elastic and inertial properties, continuous and isotropic in nature and separated by a plane boundary, as in

Fig. 8–1. The regions to the right and to the left of the boundary are to be considered infinite in extent, so that no reflections may occur except possibly at the boundary of separation. For the medium at the left, the density and wave velocity are ρ_1 and c_1, respectively; for the medium at the right they are ρ_2 and c_2. The density ρ, as used in this chapter, will always refer to the average undisturbed value. No dissipative forces are present in either medium.

Let us assume that a periodic plane wave advances toward the boundary from the left with exactly normal incidence. (The wave front will then be parallel to the plane of the boundary.) Assuming that reflection and also transmission into the second medium may take place, we have then to consider the presence of three separate wave trains in the neighborhood of the boundary, the incident wave and the reflected wave to the left of the boundary, and the transmitted wave set up in the second medium, to the right of the boundary. In Fig. 8–1 these wave components are labeled respectively i, r, and t.

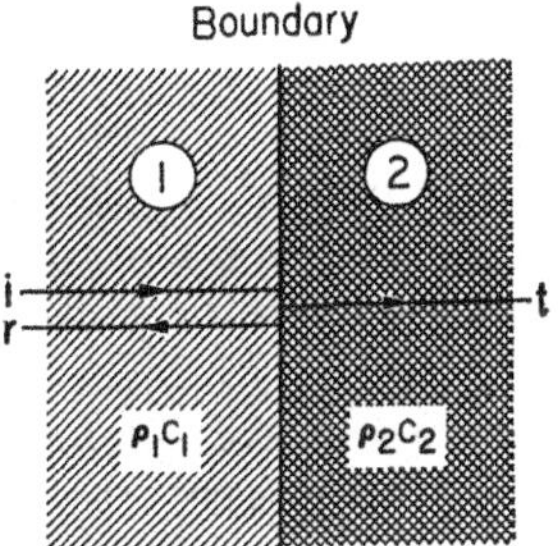

FIG. 8–1. Partial reflection of plane waves at a boundary between two different media.

As we approach the boundary plane from either side, it is necessary that the last layer of each medium, located at the boundary, be identical in motion to the other and be in a region of identical pressure. In mathematical terms, both *particle velocity* and *acoustic pressure* must be mathematically continuous across the boundary. (This notion of "continuity" across a boundary is used in other branches of physics. It will be recalled that the normal component of induction, B, is continuous across a boundary between two magnetic media.) To the left of the boundary, both the incident and the reflected waves contribute to the total particle velocity (the latter is the vector sum of the two contributions). To the right of the boundary there is only the transmitted wave to consider. Making use of i, r, and t as subscripts to indicate the three separate wave components, we may therefore state that

$$\dot{\xi}_i + \dot{\xi}_r = \dot{\xi}_t \tag{8–1}$$

and

$$p_i + p_r = p_t. \tag{8–2}$$

The instantaneous particle velocities in Eq. (8–1) are written arbitrarily as positive, the positive direction being the direction of travel of the incident wave. Since $\dot{\xi}$ is periodic in each case, the algebraic sign will vary with the time. At this point there is nothing to indicate the relative phases of $\dot{\xi}_r$, $\dot{\xi}_t$, and $\dot{\xi}_i$. Such phase relationships will appear, however, in the course of this analysis.

The acoustic pressure p associated with each wave component is connected with the corresponding particle velocity by the plane wave relations. A comparison of Eqs. (2–19b) and (2–19e) shows that

$$p = \frac{\mathcal{B}}{c}\dot{\xi}$$

or

$$p = \rho c\dot{\xi}. \tag{8–3}$$

As was pointed out in Section 2–13, the particle velocity is *in phase* with the pressure for a wave traveling in the *positive* x-direction but is *out of phase* by 180° for a wave traveling in the *negative* x-direction. If we introduce into Eq. (8–2) the relationship given by Eq. (8–3), we must therefore use the negative sign for the reflected wave. The expression for the continuity of pressure may then be written

$$\rho_1 c_1 \dot{\xi}_i - \rho_1 c_1 \dot{\xi}_r = \rho_2 c_2 \dot{\xi}_t. \tag{8–4}$$

It is now a simple matter to discover the exact relationship among the three particle velocities at the boundary. If we eliminate $\dot{\xi}_t$ between Eqs. (8–1) and (8–4), we can obtain the ratio of $\dot{\xi}_r$ to $\dot{\xi}_i$, that is,

$$\frac{\dot{\xi}_r}{\dot{\xi}_i} = \frac{\rho_1 c_1 - \rho_2 c_2}{\rho_1 c_1 + \rho_2 c_2}. \tag{8–5}$$

If, on the other hand, we eliminate $\dot{\xi}_r$, we will have the ratio of $\dot{\xi}_t$ to $\dot{\xi}_i$,

$$\frac{\dot{\xi}_t}{\dot{\xi}_i} = \frac{2(\rho_1 c_1)}{\rho_1 c_1 + \rho_2 c_2}. \tag{8–6}$$

It should be noted that the ratio of the particle displacements is identical to the ratio of the velocities. Therefore we may replace $\dot{\xi}_r/\dot{\xi}_i$ by ξ_r/ξ_i, and $\dot{\xi}_t/\dot{\xi}_i$ by ξ_t/ξ_i.

8–3 Relative magnitudes of the particle velocities. Equations (8–5) and (8–6) describe the boundary conditions for two elastic media. It will be observed that the ratio of $\dot{\xi}_r$ to $\dot{\xi}_i$ and of $\dot{\xi}_t$ to $\dot{\xi}_i$ in each case is a function of the specific acoustic impedance characteristic of each medium in the presence of progressive plane waves. Equations (8–5) and (8–6) give the magnitudes of the particle velocities in the reflected and in the transmitted waves, relative to the incident wave. Whenever ρc for the first medium differs from that for the second, there will be a reflected wave and also a transmitted wave. When ρc is the same for both media, there is only a transmitted wave. If we use for $\dot{\xi}_r$ and $\dot{\xi}_t$ the root-mean-square values

$(\xi_r)_{\text{rms}}$ and $(\xi_i)_{\text{rms}}$, we may readily compute the relative *intensities* of the three wave components. Using the same subscripts for the intensity I as for the particle velocities, we may state that

$$\frac{I_r}{I_i} = \frac{\rho_1 c_1 (\xi_r)^2{}_{\text{rms}}}{\rho_1 c_1 (\xi_i)^2{}_{\text{rms}}} = \left(\frac{\rho_1 c_1 - \rho_2 c_2}{\rho_1 c_1 + \rho_2 c_2}\right)^2 = \left(\frac{(z_s)_1 - (z_s)_2}{(z_s)_1 + (z_s)_2}\right)^2, \tag{8-7}$$

and

$$\frac{I_t}{I_i} = \frac{\rho_2 c_2 (\xi_t)^2{}_{\text{rms}}}{\rho_1 c_1 (\xi_i)^2{}_{\text{rms}}} = \frac{\rho_2 c_2}{\rho_1 c_1}\left(\frac{2\rho_1 c_1}{\rho_1 c_1 + \rho_2 c_2}\right)^2 = \frac{(z_s)_2}{(z_s)_1}\left(\frac{2(z_s)_1}{(z_s)_1 + (z_s)_2}\right)^2. \tag{8-8}$$

Since no dissipative factor is present, it is necessary that $I_i = I_r + I_t$. It is easily verified algebraically that Eqs. (8–7) and (8–8) are consistent with this statement of the conservation of energy.

8–4 Relative phases. In Eq. (8–5) the ratio of ξ_r to ξ_i may be either positive or negative, depending on the relative magnitudes of $\rho_1 c_1$ and $\rho_2 c_2$. If $\rho_1 c_1$ is greater than $\rho_2 c_2$, the ratio is positive, indicating that ξ_r and ξ_i, both sinusoidal quantities for disturbances of a simple periodic nature, are always in phase. On the other hand, if $\rho_1 c_1$ is less than $\rho_2 c_2$, the ratio is negative. For ξ_r and ξ_i to be always directly opposite in algebraic sign indicates a 180° phase relationship and since $\rho_1 c_1$ and $\rho_2 c_2$ are real numbers, no phase relationship other than this is ever possible.

From the phase relationship just described, we may conclude that in a case where ξ_r is in phase with ξ_i, the two waves will reinforce each other just in front of the boundary, as they do for a completely free end of a string. When there is a phase reversal, on the other hand, the incident and the reflected waves will partially cancel. The cancellation will not be complete because the maximum value of ξ_r will, in general, be less than the maximum value of ξ_i and so we cannot properly call the boundary a true nodal point. As a matter of fact, because of the partial reflection the whole character of the stationary wave pattern in front of the boundary is different from that of a string attached to rigid supports.

8–5 Magnitudes and phases of the acoustic pressures. The pressure in the three wave components under discussion is of especial interest. If we make use of the relation $p = \rho c \xi$, we may write Eq. (8–1) in terms of the pressure rather than in terms of the particle velocity. Combining this equation with Eq. (8–2), we find that

$$\frac{p_r}{p_i} = \frac{\rho_2 c_2 - \rho_1 c_1}{\rho_1 c_1 + \rho_2 c_2} = \frac{(z_s)_2 - (z_s)_1}{(z_s)_1 + (z_s)_2} \tag{8-9}$$

and

$$\frac{p_t}{p_i} = \frac{2\rho_2 c_2}{\rho_1 c_1 + \rho_2 c_2} = \frac{2(z_s)_2}{(z_s)_1 + (z_s)_2}. \tag{8-10}$$

It will be noted that the acoustic impedances in the numerator of Eq. (8–9) are interchanged as compared with Eq. (8–5) for the particle velocity ratio. The ratio of the *magnitudes* of the pressures p_r and p_i is identical with the ratio of the magnitudes of the particle velocities. This is to be expected, since the intensity ratio for plane waves in the same medium must be proportional to the square of either the velocity ratio or the pressure ratio.

The phase relationships for pressures, however, are reversed as compared with the relationships for particle velocities. When ρ_1c_1 is greater than ρ_2c_2, p_r is 180° *out of phase* with p_i, and when ρ_1c_1 is less than ρ_2c_2, p_r is in phase with p_i and thus the conditions for reinforcement of ξ_r and ξ_i are just right for the partial cancellation of p_r and p_i. This is in line with the results of our discussion of stationary waves along strings or in pipes, as presented in Chapter 7. The ratio of p_t to p_i is always positive, as is the ratio of ξ_t to ξ_i, so that the pressures in the transmitted wave and in the incident wave are always in phase. The ratio of the pressure *magnitudes* differs from the ratio of the velocity magnitudes because of the part ρc plays in the equation connecting p and ξ.

We must emphasize again that these simple phase relationships, 0° or 180° as the case may be, apply only to the particular physical system described at the beginning of this analysis. We shall presently consider boundary conditions which are more complicated.

8–6 Practical implications. The formulas developed in the previous sections apply well to such a pair of elastic media as air and water. For air under normal conditions, ρc approximates 42 cgs units. For fresh water, under the same conditions, ρ is 1.0 and c may be taken as 143,000, both in cgs units. Therefore ρc for water is 1.43×10^5 cgs units, a figure about 3400 times that for air. As a result, plane waves impinging normally upon a boundary between air and water will, by Eq. (8–7), be almost completely reflected, whether the waves are incident on the air side of the boundary or on the water side. If the waves do not strike with normal incidence, we need consider only the normal component of the particle velocities in the incident wave, since the component parallel to the boundary surface remains virtually unaffected. As in the reflection of light waves from a polished surface, the angle of reflection will equal the angle of incidence, with an intensity in the reflected wave practically equal to the intensity in the incident wave. A longitudinal wave started within a relatively thin layer of water such as a shallow lake will therefore remain confined to the water by a process of internal reflections.

We may apply our reflection criteria to the problem of a metal bar, vibrating longitudinally and surrounded by air. Instead of considering the bar to be completely free to move at one end, let us consider the end of the bar as a boundary plane between two different media, steel and air. We must be very careful, as usual, that the assumptions back of any equations we use are at least approximately realized. To be sure that we are dealing with plane waves throughout, let us picture the bar as being relatively short and of large cross section. A short bar will have high frequency modes of vibration (steady state stationary wave modes) and the large area of its end, since the wavelength in air will be short, will ensure plane wave propagation in the surrounding medium. We may now apply the findings of the previous sections.

Suppose a single train of waves is set up in the bar by a blow at one end. When such a train of waves reaches the opposite end, it will encounter a boundary between steel, for which ρc is about 3.9×10^6 cgs units, and air, for which ρc is 42 cgs units. The difference is very great, so that, according to Eq. (8–7), almost complete reflection will take place. Since in this case $\rho_1 c_1$ is greater than $\rho_2 c_2$, we may conclude that upon reflection there will be no phase shift in ξ (or, conversely, that there will be a 180° shift for p). Therefore this end of the bar becomes a velocity or displacement antinode, corresponding to a pressure node. These results are in agreement with those of the earlier more direct physical argument.

The value of the product ρc for liquids is much closer to that for solids than is the value for gases as compared with liquids, or gases as compared with solids. Therefore when plane waves reach boundaries between liquids and solids the reflection is less complete than in the cases we have been discussing. If we consider the two media steel and water, we find that for waves passing from steel to water,

$$\frac{I_r}{I_i} = \left[\frac{(\rho c)_{\text{steel}} - (\rho c)_{\text{water}}}{(\rho c)_{\text{steel}} + (\rho c)_{\text{water}}}\right]^2 = \left[\frac{3.9 \times 10^6 - 1.43 \times 10^5}{3.9 \times 10^6 + 1.43 \times 10^5}\right]^2 = 0.86.$$

Therefore 14% of the incident energy will pass into the water. It should be noted that this result is obtained for a wave traveling in either direction. This is also true, it will be recalled, for optical reflections.

Example. Plane longitudinal waves, passing first through water, strike a boundary between water and ice with normal incidence. Assume each medium to be infinite in extent. Compute the magnitudes of the following ratios:

$$\frac{\xi_r}{\xi_i}, \frac{\xi_t}{\xi_i}, \frac{I_r}{I_i}, \frac{I_t}{I_i}.$$

Compute also the relative phases for the particle velocities and for the acoustic pressures.

$$\rho_{\text{water}} = 1.0 \text{ gm-cm}^{-3}. \qquad c_{\text{water}} = 1.43 \times 10^5 \text{ cm-sec}^{-1}.$$
$$\rho_{\text{ice}} = 0.92 \text{ gm-cm}^{-3}. \qquad c_{\text{ice}} = 3.8 \times 10^5 \text{ cm-sec}^{-1}.$$

$$(\rho c)_{\text{water}} = 1.0 \times 1.43 \times 10^5 = 1.43 \times 10^5 \text{ gm-cm}^{-2}\text{sec}^{-1}.$$
$$(\rho c)_{\text{ice}} = 0.92 \times 3.8 \times 10^5 = 3.49 \times 10^5 \text{ gm-cm}^{-2}\text{sec}^{-1}.$$

$$\frac{\xi_r}{\xi_i} = \frac{(1.43 - 3.49) \times 10^5}{(1.43 + 3.49) \times 10^5} = -0.42.$$

$$\frac{\xi_t}{\xi_i} = \frac{2(1.43) \times 10^5}{(1.43 + 3.49) \times 10^5} = 0.58.$$

$$\frac{I_r}{I_i} = (0.42)^2 = 0.177.$$

$$\frac{I_t}{I_i} = 1 - \frac{I_r}{I_i} = 0.823.$$

The relative phases are:

ξ_r is 180° out of phase with ξ_i.
ξ_t is in phase with ξ_i.
p_r and p_t are both in phase with p_i.

If we are interested primarily in enhancing the transmitted component, i.e., in obtaining the maximum energy in the *second* medium, we should obviously select two media for which ρc is nearly the same. As $\rho_2 c_2$ approaches the value of $\rho_1 c_1$ the value of I_r approaches zero (Eq. (8–7)), and for the transmitted wave I_t approaches I_i, as one would expect. For a steel bar immersed in water the two values of ρc are still quite different but less so than for a steel bar in air. The vibrations of a freely oscillating bar immersed in water will very quickly die out, due to the rapid loss of energy at the ends. The bar in air will continue to vibrate for a much longer period of time and it is evident that its rate of decay is controlled more by internal dissipative forces than by radiation effects. This property is of interest in the production of underwater longitudinal waves, where the problem of introducing maximum acoustical power into the medium is of primary concern (Chapter 12).

8–7 The effect of partial reflection upon the stationary wave pattern. For the stationary wave patterns considered in Chapter 7 we assumed perfect reflection at the boundaries. If any appreciable fraction of the energy incident at the boundary leaves the first medium, the reflected wave will have a lower amplitude than the incident wave and the stationary wave pattern will be altered.

Let us consider the reflection of a longitudinal wave at a boundary between two media where the product ρc for the first medium is less than for the second medium. The wave is approaching the boundary from the right, as in Fig. 8–2. As we have seen, there will be a 180° phase shift as far as ξ or $\dot{\xi}$ is concerned, and the incident and reflected waves will therefore partially cancel at the reflection boundary. The periodic solution to the differential equation for plane waves may then be written, as before, in two parts of the same form, except that in this case the amplitude of the reflected wave $(\xi_m)_r$ is less than that of the incident wave $(\xi_m)_i$.

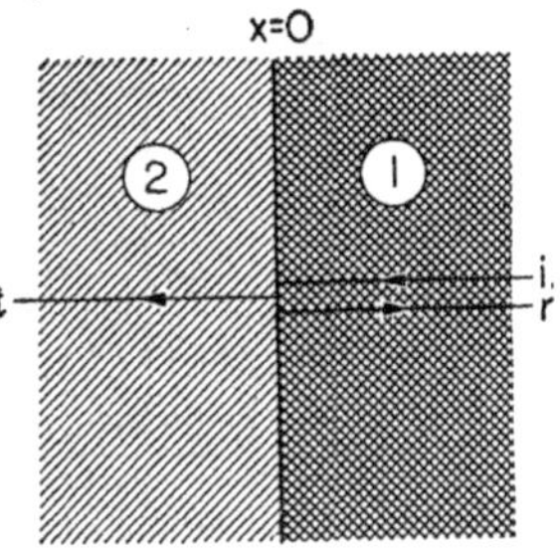

Fig. 8–2. Direction relationships for partial reflection of plane waves.

$$\xi = (\xi_m)_i \cos \frac{2\pi}{\lambda}(ct + x) - (\xi_m)_r \cos \frac{2\pi}{\lambda}(ct - x). \tag{8–11}$$

The negative sign for the reflected wave is necessary to satisfy the condition at the boundary, where $x = 0$.

If we now expand the cosine functions, considered as the cosines of the sums and differences of angles, and collect terms, we obtain

$$\xi = [(\xi_m)_i - (\xi_m)_r] \cos \frac{2\pi c}{\lambda} x \cos \frac{2\pi c}{\lambda} t - [(\xi_m)_i + (\xi_m)_r] \sin \frac{2\pi c}{\lambda} x \sin \frac{2\pi c}{\lambda} t. \tag{8–12}$$

Comparison of this equation with the several stationary wave equations of Chapter 7 will show that the complete expression on the right represents *two* separate sets of stationary waves. The frequencies associated with each are the same but the amplitude of motion at the antinodes is different for the two patterns, being the sum of the separate wave amplitudes in the one case and the difference in the other. The two sets of antinodes do not occur for the same value of x, since in one case a cosine function is involved and in the other a sine function. The antinodes of one set appear halfway between the antinodes of the other or, in other words, at the other's nodal points (Fig. 8–3a, b). The two component stationary wave motions are 90° out of phase with each other in respect to time as well as position, since one involves a sine function of t and the other a cosine function.

The two graphs of Fig. 8–3a and b do not represent the conditions at the same instant of time, and consequently cannot be directly added to obtain the complete picture. Consideration of these two patterns, however, will show the envelope of the particle motions along x to be something like the graph of Fig. 8–3c, where the amplitude at point a_1 is $[(\xi_m)_i + (\xi_m)_r]$ and the amplitude at point a_2 is $[(\xi_m)_i - (\xi_m)_r]$, there being no true nodal positions at all.

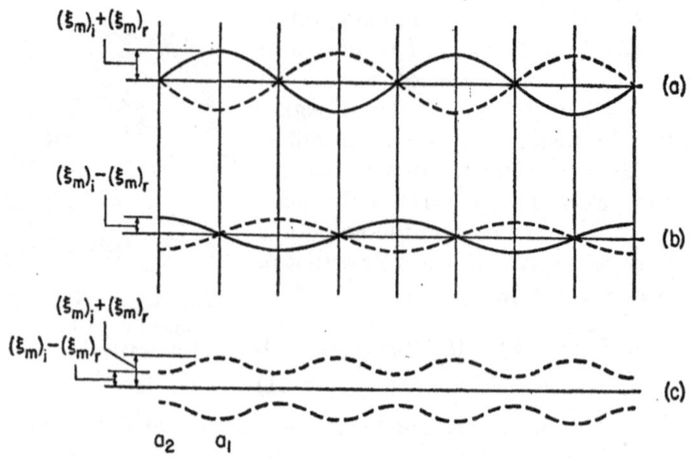

FIG. 8-3. Stationary wave patterns resulting from partial reflection of plane waves.

A similar pattern will occur for acoustic pressure, but with the maximum and minimum points interchanged, for reasons discussed earlier. As we shall show later, a convenient stationary wave method of measurement of acoustic impedance is based on pressure measurements taken in a pattern of this type.

8-8 The absorption coefficient. The absorption coefficient for a boundary between two media may be defined as the ratio of the acoustic power transmitted through a given area of the boundary to the power incident upon the same area. For *normal incidence of plane waves* at a boundary between two perfectly elastic media of infinite extent, we can, by Eq. (8-8), define a coefficient α_n as

$$\alpha_n = \frac{I_t}{I_i} = 1 - \frac{I_r}{I_i} = \frac{\rho_2 c_2}{\rho_1 c_1}\left(\frac{2\rho_1 c_1}{\rho_1 c_1 + \rho_2 c_2}\right)^2. \tag{8-13}$$

We are interested in acoustic absorption coefficients primarily for problems arising indoors, where we can in no sense assume only two media of infinite extent. A room has limited dimensions and, what is even more important, the wall itself is usually a complex structure made up of comparatively thin laminations, each component of which may have different physical and acoustic properties. More often than not the wall is constructed of panels of limited area, rigid at the edges and relatively flexible near the center. The nature of the surface material may change radically from one portion of the wall to another, due to curtains, windows, wood trim, etc. Finally, to mention one other important complication, although sound waves originating within the room *may* reach a wall from predominantly one direction (rarely the normal to its plane), more often than not energy will arrive from practically *all* directions because of multiple reflections at other walls.

In view of the remarks above one would expect to use Eq. (8-13) with great caution and then only for certain special sets of conditions. There are, as a matter of fact, several different kinds of absorption coefficients, not easily defined. Before giving these definitions we shall look more carefully into the relation between acoustic impedance and sound absorption.

8-9 Specific acoustic impedance of a boundary surface. The specific acoustic impedance in the path of a unidirectional plane wave is ρc. For spherical waves radiating outward from a pole, z_s is in general complex, its value approaching ρc only at distant points. In the presence of two oppositely moving waves such as exist in front of a boundary, the value and nature of the acoustic impedance may be still different, if we consider it to be the ratio of a total pressure to a total particle velocity, each of which is dual in nature. To understand the procedure for determining the impedance under these conditions, we shall consider a number of simple special cases.

8-10 Both media perfectly elastic and infinite in extent. We shall define the *normal specific acoustic impedance* z_n *for a boundary surface* as the ratio of the instantaneous *total* acoustic pressure at the surface to the *total* particle velocity, assuming a plane longitudinal wave to impinge upon the surface with normal incidence. For the system of two elastic media of infinite extent used so far in this chapter,

$$z_n = \frac{p_i + p_r}{\xi_i + \xi_r}. \tag{8-14}$$

Because of the continuity of both pressure and particle velocity across the boundary, z_n may also be written

$$z_n = \frac{p_t}{\xi_t}. \tag{8-15}$$

In this case, therefore, $z_n = z_s$, where z_s, as defined, is the specific acoustic impedance characteristic of a free plane wave moving through the second medium only. Since ρc is a real positive number, z_n is comparable to an electrical impedance which is a pure resistance.

8-11 Boundaries for which z_n is reactive. The process of reflection of plane waves at the end of a cylindrical tube filled with air can be as readily considered as that at an infinite plane. As will be remembered, plane waves started down cylindrical tubes of not too small cross section (so as not to introduce dissipation) remain essentially plane. If a rigid flat plate is placed at the end of the tube there will be no transmitted wave and z_n will be practically infinite. This follows from Eq. (8–14), since ξ_r will in this case be equal and opposite to ξ_i. Suppose, however, that instead the tube is closed by a thin circular plate mounted on a spring, allowing axial motion (Fig. 8–4). Now consider the effect of an acoustic pressure due to the arrival of a plane wave. The total force over the face of the plate due to this pressure is Sp, where S is the area of the plate. This force is applied to the plate-spring system, whose *mechanical impedance* z_m is reactive and is equal to $-j(K/\omega)$, where K is the elastic constant of the spring (inertia and friction in the plate-spring system are neglected). The velocity of the plate must be the same as that of the air immediately in front of the plate and ξ will then be

$$\xi = \frac{Sp}{z_m} = \frac{Sp}{-j(K/\omega)}, \tag{8–16}$$

and the specific acoustic impedance at the surface may be computed as

$$z_n = \frac{p}{\xi} = -j\frac{K}{S\omega}. \tag{8–17}$$

Since this is a purely imaginary quantity, there is no real power, on the average, delivered to the plate at the boundary. Just as for the case of a capacitor terminating a transmission line, instantaneous power is alternately fed to the plate-spring system and then returned to the acoustic line. Equation (8–17) shows that at the boundary the pressure is 90° out of phase with the particle velocity.

If we assume the plate to have considerable mass and the spring upon which it is mounted to be very weak, with a very small elastic constant, the mechanical impedance of the plate-spring system will be predominantly

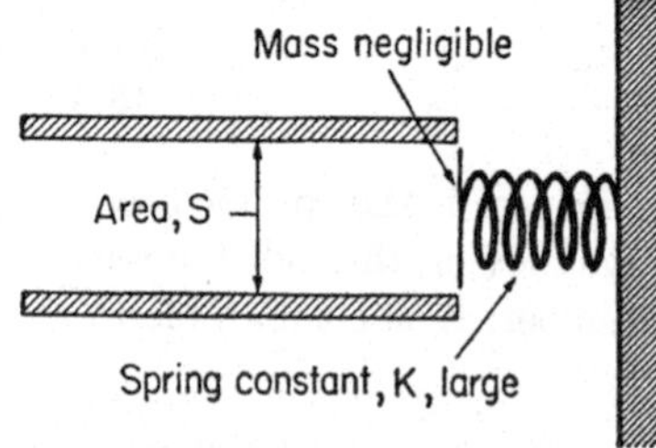

FIG. 8–4. A pipe terminated by a mechanical reactance having the properties of capacitance (compliance).

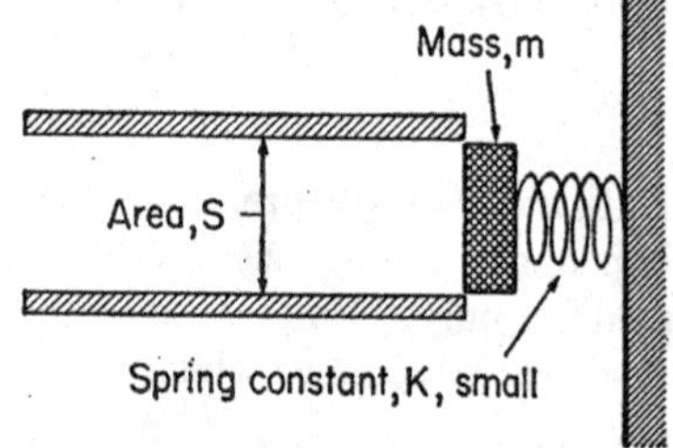

FIG. 8–5. A pipe terminated by a mechanical reactance having the properties of inductance (mass).

inertial (Fig. 8–5). By an argument similar to that just presented, we may conclude that the specific acoustic impedance at the boundary is

$$z_n = j\frac{\omega m}{S}, \tag{8–18}$$

where m is the mass of the plate.

If both elasticity and inertia must be considered, and if in addition some frictional forces are present when the plate moves, we may conclude that z_n, in general, is of the form

$$z_n = R + jX. \tag{8–19}$$

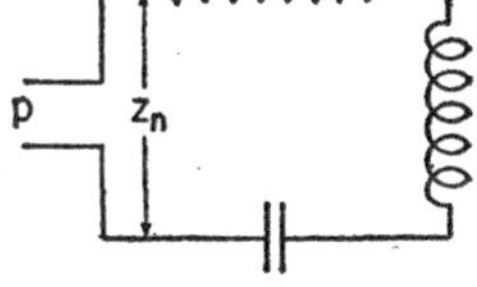

FIG. 8–6. Equivalent "network" for the normal specific acoustic impedance at the end of a pipe.

The imaginary part of z_n may be positive or negative (indicating either a lag or lead of ξ relative to p), depending on the relative values of the constants. At the frequency for which the plate-spring system resonates, ξ will be in phase with p and its root-mean-square value will be p_{rms}/R. The acoustic impedance z_n may be considered the equivalent of a series *LRC* circuit, as in Fig. 8–6, where the impressed potential is the acoustic pressure.

8–12 Specific acoustic impedance at positions of discontinuity in the tube cross section. If the medium in a tube is air, effects similar to those just discussed are obtained whenever there are changes in the tube cross section or when there are partial obstructions. In Fig. 8–7 is shown one possible type of closure for a cylindrical tube. The system to the right of the plane *aa* is very similar in geometry to the Helmholtz resonator described in Chapter 5. In the region just to the left of the air channels the ratio of the acoustic pressure to the particle velocity, averaged over the total cross section of the main tube, will be determined by the inertial, elastic, and dissipative constants of the system to the right of the plane at *aa*. The air free to move in the several air channels will have mass, as in the neck of the Helmholtz resonator. The larger volume of air to the right of the channels will supply an elastic factor and friction along the channel walls will add the dissipative factor. As a result, one would expect the specific acoustic impedance at *aa* to be complex, its exact form determining the numerical and phase relationships between p and ξ at that position. Depending upon the exact geometry, the impedance may be predominantly inductive, capacitive, or resistive. (Each of the special cases can be described on the basis of Fig. 8–7.)

In the examples given in this and the preceding section, z_n is definitely a function of the frequency in the incident wave. Only for the system discussed in the early part of this chapter, i.e., the two elastic media of infinite extent, is the value of z_n independent of frequency.

8–13 Specific acoustic impedance at the surface of absorbing materials. We are now in a better position to understand the absorption process at the actual wall surfaces of a room or other enclosure. Every portion of the wall surface of a room may be considered to offer to the incident waves a certain acoustic impedance. We shall presently show that the quantitative extent of the reflection as well as the phase relationships are directly related to the specific acoustic impedance z_n, which in general is complex in nature. All wall materials are to some extent yielding. Some, like cloth and felt, are highly porous and have a structure (on a scale approaching the microscopic) similar to that shown in Fig. 8–7. There is clearly no possibility of a direct computation of z_n in terms of the geometry of the channels, the mass of the filaments, etc. The obvious recourse is to experiment. The most common method of measuring z_n experimentally is based on observation of a stationary wave pattern. A resume of several methods will be given in Chapter 10.

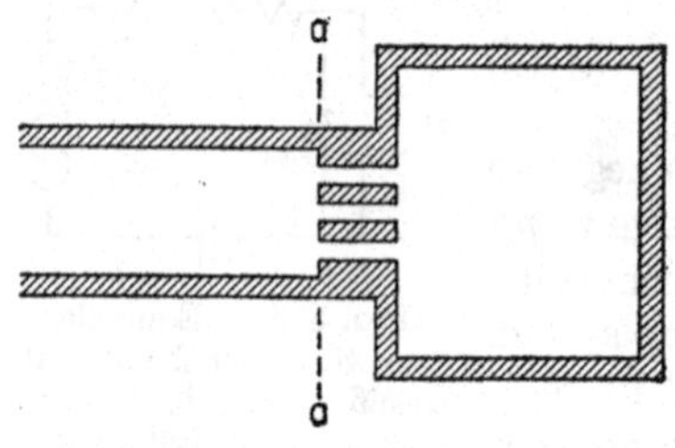

FIG. 8–7. One type of closure for tube containing air.

8–14 The relation between z_n and the absorption coefficient for plane air waves of normal incidence. Returning now to the absorption coefficient defined in Section 8–8, we shall show its important relationship to the normal specific acoustic impedance of the absorbing surface. Consider the total pressure p and the total particle velocity $\dot{\xi}$ at the surface. By Eqs. (8–1) and (8–2), we may say that

$$p = p_i + p_r \tag{8–20}$$

and

$$\dot{\xi} = \frac{p_i}{\rho c} - \frac{p_r}{\rho c}, \tag{8–21}$$

where p_i and p_r respectively represent the pressures in the incident and the reflected wave at the surface, and ρ and c refer to air. The negative sign in the second equation is due to the direction of the wave velocity, as explained earlier.

Since we cannot assume p_i and p_r to be the same in magnitude or in phase, we shall consider each to be a complex quantity. Therefore when we compute z_n we shall expect to find it complex, the relationship to the pressures being

$$z_n = \frac{p}{\xi} = \rho c \left(\frac{p_i + p_r}{p_i - p_r}\right). \tag{8-22}$$

The ratio of p_r to p_i as a complex quantity may be stated in the polar form:

$$\frac{p_r}{p_i} = M\epsilon^{j\theta}, \tag{8-23}$$

where the modulus M represents the *numerical* ratio of the maximum values of the two pressures, and θ is the relative phase angle (see the discussion on complex quantities in Chapter 5). Combining Eq. (8-22) with Eq. (8-23), we find that

$$z_n = R + jX = \rho c \frac{1 + M\epsilon^{j\theta}}{1 - M\epsilon^{j\theta}}. \tag{8-24}$$

By a general property of complex quantities, it is necessary that the real parts on each side of this equation be equal and also that the imaginary parts be equal. By rewriting the complex parts of the right-hand side in the form $a + jb$ and collecting terms according to the rules of complex algebra, one may therefore obtain two independent equations. These equations essentially relate M and θ to R and X. Knowing R and X, it is possible to find M and θ or, conversely, knowing M and θ, it is possible to find R and X.

On the basis of the above mathematics, which we have here only outlined, the normal absorption coefficient α_n may be determined, once z_n is completely defined. The absorption coefficient α_n is given by

$$\alpha_n = 1 - \frac{I_r}{I_i} = 1 - \left[\frac{(p_m)_r}{(p_m)_i}\right]^2 = 1 - M^2. \tag{8-25}$$

There are other mathematical and graphical tricks which may be developed to shorten the actual computation of the absorption coefficient (see Morse, *Vibration and Sound* and Beranek, *Acoustic Measurements*). In principle, the notion of absorption is the same for all types of surfaces. The fact that z_n must often be considered complex is simply a computational detail.

8-15 Other absorption coefficients. The absorption coefficient α_n, often called the free-wave coefficient, is only one of several coefficients, each defined somewhat differently. The coefficient already discussed is perhaps the most fundamental from a physical point of view but it is not necessarily the most directly useful. The *Sabine absorption coefficient* α_s is probably the most important historically and until recently has been the one most commonly used. This coefficient is the ratio of absorbed acoustic power to incident power when the latter is arriving simultaneously from *all* directions. For surfaces that react only locally to the incident wave, the absorption coefficient is independent of the angle of incidence. For surfaces that transmit transverse waves well and rapidly, the coefficient is a function of the incident angle. The quantity α_s is the mean of a random assortment of angles and it may be computed on the basis of laboratory measurements or it may be determined from the reverberation properties of "live" rooms. It is often less than α_n. Table 8–1 lists values of the Sabine coefficient for various materials.

TABLE 8–1*

VALUES OF THE SABINE COEFFICIENT α_s

Frequency (cycles-sec^{-1})	128	512	1024	4096
Acoustic plaster	0.30	0.50	0.80	0.50
Brick wall, unpainted	0.02	0.03	0.04	0.05
Carpet, lined	0.11	0.37	0.34	0.24
Curtains, light	0.04	0.11	0.25	0.30
Curtains, heavy	0.10	0.50	0.80	0.75
Floor, concrete	0.01	0.02	0.02	0.02
Floor, wood	0.05	0.03	0.03	0.03
Temcoustic F2 (when attached to plaster or concrete)	0.33	0.54	0.52	0.42
Ventilator grill	0.50	0.40	0.35	0.25
Human body, seated (assuming an exposed body area of 9 ft^2)	0.17	0.42	0.56	0.50

* Based on *Bull. Acous. Materials Assoc.* **7,** 1940.

It is to be noted that α_s is a function of the frequency, the values usually being somewhat lower at the lower frequencies. This dependence upon frequency is to be expected, from the complex nature of specific acoustic impedance. It is unfortunate that the middle and high frequency bands of the audible spectrum suffer greater absorption than do the low frequencies, since the former are the important components of speech. A recent solution to this problem will be mentioned presently.

A third group of absorption coefficients, called "chamber" absorption coefficients, are characteristic of certain particular chambers of definite size and shape where special conditions of intensity distribution, etc., lead to special values of α. For each such coefficient, the conditions of measurement must be described.

8-16 Use of panel resonance. By the use of carefully selected panel resonances the relatively low values of the absorption coefficient at the lower frequencies, characteristic of most wall surfaces, may be replaced by values considerably higher. An elastic panel may be so constructed and supported that it will resonate at some particular low frequency. At the resonance frequency the average value of z_n at the surface will be low, since the reactive part will be zero and only the resistive component will remain. At this particular frequency, z_n takes on a value much closer to the ρc of free plane waves in air and greater absorption is encouraged. In order to absorb a *band* of frequencies, the wall structure may be broken up into a large number of panels of assorted resonant frequencies (Fig. 8-8). In this way the wall becomes virtually a "band-pass filter," similar to a filter of the electrical variety.

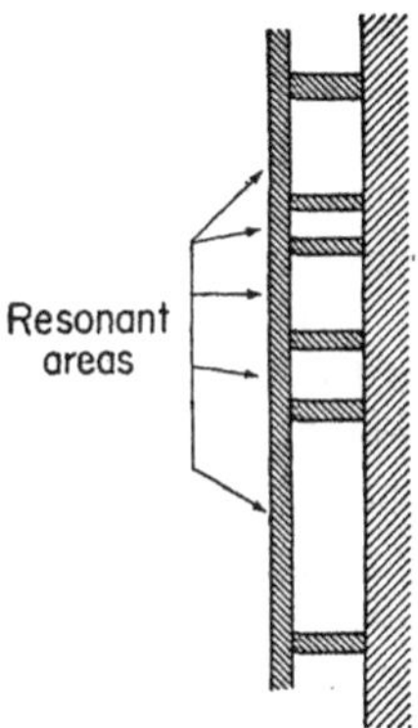

FIG. 8-8. The use of resonating panels to improve the low frequency absorption by walls.

8-17 Absorbing "layers." The effect of thickness. If laboratory measurements of the value of z_n or α_n are made at the surface of an actual sample of absorbing material, the results will be determined in part by the particular backing used. If the sample is fastened to wood, plaster, or any other material which is itself a good reflector, internal reflection at the boundary between the sample and the backing may return much energy to the air. A complete circuit analogy can be set up and the general results for any type of lamination can be predicted. In general, a surface material of high absorption coefficient and as thick as construction will permit is the most desirable. It is equally desirable that the material *attenuate* as rapidly as possible what energy does cross the surface boundary, in order to reduce the internal reflection just referred to. Porous materials supply such attenuation through viscosity and heat conduction effects along the minute ducts. The fibers of such materials themselves move in the presence of a wave, and the resultant internal friction also increases attenuation. All these factors, together with the important effect of the thickness of the material upon the emerging intensity of the internally reflected wave, are automatically taken into account in the measurement of z_n and of α_n. Measurements should be taken with a thickness of sample and a type of backing characteristic of the use to which the material is to be put.

8–18 Good reflectors and good absorbers. We have thus far stressed the quantitative features of the reflection and absorption of sound waves at boundaries of various sorts. Now let us consider what constitutes a good reflector for plane waves. Plainly, this would be a boundary at which the acoustic impedance is widely different from that characteristic of free plane waves in air. For many substances such as liquids or solids of the hard plaster or cement type, z_n is much greater than ρc and is often largely real in character. To compute the actual fraction of the incident energy which will be absorbed, we must know the value of z_n (or α_n) and then proceed as in Section 8–14. Conversely, a good absorber must be a poor reflector. The quantity z_n at the surface should approach as closely as possible the numerical value of ρc for plane waves in air and should be real, so that a maximum fraction of the power in the incident wave may cross the boundary and never return.

A "perfect" reflector or a "perfect" absorber is quite unattainable. As a matter of fact, such a boundary would be highly undesirable, as we shall point out when we consider room acoustics in Chapter 12. By understanding the basic factors controlling the degree of reflection or absorption, we can usually design surfaces and materials which fulfill the necessary requirements.

PROBLEMS

1. (a) Find the expression for the ratio of the condensation s in the reflected wave to the condensation in the incident wave (for normal incidence). (b) What is the ratio of s in the transmitted wave to s in the incident wave?

2. A sound wave in air of pressure level 10 db strikes, with normal incidence, a boundary between the air and a second medium of large extent for which $c = 84$ cgs units. Find (a) the rms pressure in the reflected wave, (b) the rms pressure in the transmitted wave, and (c) the phase relationship at the boundary between these two instantaneous pressures and the instantaneous pressure in the incident wave.

3. If sound waves in air strike the surface of a second medium of large extent with normal incidence, what must be the value of z_s for the second medium such that one-half of the incident energy returns to the air?

4. Will an open window allow sound waves of all frequencies to pass through undiminished? Give reasons for your answer.

5. A cylindrical tube is closed by a flat plate of negligible mass mounted on a spring, as in Fig. 8–4. The spring constant is 10^4 dyne-cm^{-1}; the cross-sectional area of the tube is 15 cm^2. A longitudinal wave in the tube, of frequency 100 cycles-sec^{-1}, impinges upon the plate, imparting to it a velocity of rms value 6 cm-sec^{-1}. Find the rms acoustic pressure at the surface of the plate.

6. A viscous force of friction is added to the plate-spring system of problem 5. This force is 20 dynes per unit velocity (in cm-sec^{-1}). (a) Find the acoustic pressure at the plate required to impart to the plate the same rms velocity as in problem 5, that is, 6 cm-sec^{-1}. (b) Determine the average power delivered to the plate-spring system under these conditions.

7. A cylindrical tube of diameter 4 cm is closed by a structure similar to that shown in Fig. 8–7, there being a large number of small air channels. The average value of z_n over the cross section of the tube is measured at the position *aa*, using waves of a certain frequency, and is found to be $z_n = 15 + j5$. (a) What is the phase angle between the pressure and the particle velocity just in front of the position *aa*? (b) If the rms pressure at the position *aa* is 3 dyne-cm^{-2}, what is the average rate of loss of wave energy at the boundary *aa*?

8. The maximum pressure in the reflected wave in the air just in front of an absorbing surface is one-half the value in the incident wave, and p_r lags p_i by a phase angle of 45°. Making use of Eqs. (8–23) and (8–24) and using the mathematical procedure suggested in Section 8–14, find the real and imaginary parts of the acoustic impedance z_n at the boundary. (b) Compute the absorption coefficient α_n.

9. A plane sound wave strikes with normal incidence an isolated small disk of absorbing material whose diameter is small compared with the wavelength. Will the value of z_n at the surface of the disk under these conditions be the same as if the disk were at the end of a close-fitting cylindrical tube along which are traveling plane waves? (Compare with problem 4.)

CHAPTER 9

SPEECH AND HEARING

9–1 Importance of the subjective element in acoustics. The primary interest of human beings in the subject of sound arises because of the acoustic equipment in the possession of every normal person, the voice apparatus and the hearing mechanism. Since these two pieces of equipment make possible one of the most important channels of communication between individuals, it is important that we understand, as far as possible, their physical structures and modes of operation. Physical structure can be determined by dissection, but the complete physics of the production of speech and of the hearing process is another matter and, especially in the case of hearing, many uncertainties still exist. Apart from the localized complexities of the mechanisms themselves, there are links with the psychological whose exact nature is difficult to discover by the usual procedures of experimental physics.

It will be the purpose of this chapter to describe briefly the voice and hearing mechanisms and to summarize the essential physics of their operation, as far as it is known today. Many of the peculiarities of hearing will be presented descriptively, with no attempt at explanation.

9–2 The vocal apparatus. The energy associated with speech or with the singing voice originates with the forcible expulsion of air from the lungs. This steady stream of air may be looked upon as a "carrier" of energy, just as is the steady stream of air entering an organ pipe. In order for there to be audible sound, there must be a periodic variation in velocity (and hence in pressure). This necessary "modulation" is brought about in two basic ways, the type of sounds so produced being called, respectively, *voiced* sounds and *unvoiced* or *breath* sounds.

The voiced sounds include the vowels of ordinary speech and the tones which predominate in the singing voice. The fundamental modulating organ is the larynx, across which are stretched two membranelike bands. These are the vocal cords. In the production of voiced sounds, air is forced through a rather narrow slit between the bands, whose flexibility allows them to yield under the pressure of the air stream. The result is a widening of the slit, with a resultant rush of air and a consequent drop in the pressure. The membranes then return to their original positions and the phenomenon is repeated. The action of the vocal cords is thus seen to be fundamentally that associated with a relaxation oscillation. Their behavior is similar to that of the reeds of certain musical instruments, that is, motion induced in this manner converts a steady flow into a periodic one, and so makes sound waves possible.

As with any oscillation of the relaxation type, one would expect the generation of a fundamental frequency along with a large number of very prominent higher harmonics. The value of the fundamental frequency, and hence of the range of frequencies covered by the harmonic series, may be varied considerably by voluntary control of the tension of the vocal cords. This primarily accounts for the range in basic pitch of the speaking or singing voice. The numerous resonating cavities both above and below the larynx have a large number of assorted natural frequencies which do not necessarily bear a whole number relationship. Such resonances contribute in a very important way to the frequency content in the emitted sound and, in addition, their presence accounts for the existence of certain inharmonic frequencies.

It is possible for the vocal equipment to give rise to sound without use of the vocal cords. Such sounds are called breath sounds. A steady forcible exhaling of the breath will produce a hissing sound like that of escaping steam. This is also a result of relaxation effects due to the turbulence set up in the flow of air around the numerous irregularities along the vocal tract. An analysis of this type of sound reveals a band of practically continuous frequency coverage largely confined to the upper portion of the audible frequency range. As we shall see presently in connection with speech, it is the existence of this type of breath sound that makes whispering possible.

A third type of sound results from a combination of voiced and breath sounds. Such sounds as "zee" belong to this classification.

9-3 The speech process. A steady-state analysis of the frequency content of the various intoned vowel sounds has been made by Fletcher.* Figure 9-1 shows the relative amplitudes of the harmonic terms for several such sounds. Although the lowest frequency gives the characteristic pitch recognized by both the speaker and the listener, higher frequencies are more often than not of greater prominence (just as with some of the notes played on the violin). For the sounds "oo" and "oh," for instance, the most prominent frequencies are the third and fifth harmonics, respectively, even though the fundamentals are identical (Fig. 9-1a and b). For the sound "ee" (Fig. 9-1c), there are important harmonic frequencies as high as 4300 cycles-sec^{-1}, the series from about 2500 cycles up being fully as prominent as the group below 2500 cycles.

* Fletcher, *Speech and Hearing*, D. Van Nostrand Co. (1929), pp. 51-55.

The variation in the harmonic content of the different vowel sounds is brought about by a deliberate variation of the size and shape of the constrictions along the vocal tract, the position and shape of the tongue, the shape of the mouth opening, etc. In a rather complicated manner this alters the nature and extent of the various resonances along the way, thus controlling the relative prominence of the numerous frequencies set up at the larynx.

A single intoned vowel does not convey speech information. The process of speech may be described somewhat as follows.* As has been pointed

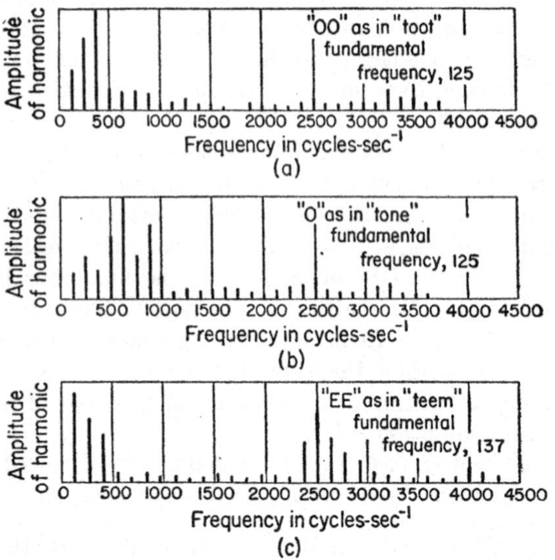

Fig. 9–1. Relative prominence of frequency components in intoned vowel sounds.

out, the production of audible sound in the vocal tract is a process of modulation impressed upon the carrier motion, in this case the steady velocity of the expelled air. A voiced sound originating in this manner and containing a steady-state mixture of frequencies may now itself serve as a carrier for speech information contained in a second type of superimposed modulation associated with word formation. The steady sound wave may be compared to the carrier wave generated by a radio transmitter. In the latter case the wave form is sinusoidal and of course is of a frequency above

* For a fuller discussion of this view of speech, see Dudley, "The Carrier Nature of Speech," *Bell System Jour.* **19**, 496 (1940).

audibility, whereas in the case of a steady voiced sound the wave form is complex, since there are many frequency components which lie within the audible range. Both carriers are nevertheless similar in that no information is conveyed by their mere steady-state existence. In the electromagnetic wave the information is impressed upon the carrier in several ways, the most common types of modulation being change of amplitude and of frequency. The complicated actions of the vocal tract described above accomplish the same thing in the acoustic wave. For speech, the modulation rate, i.e., the time rate of change from one speech sound to another (as from one vowel to another vowel) is very low, the effective frequency for this variation being far below audibility. The amplitude type of modulation seems to be more important for conveying information, although frequency modulation of the carrier plays an important part, for instance, in the speech attribute known as inflection.

Just as the collection of frequencies set up by the vocal cord action may serve as the carrier for speech modulation, so may also the breath sound referred to earlier. It is this carrier whose frequency composition resembles that of a continuous spectrum, which is so essential to the production of the consonant sounds. Like the voiced sound, the breath sound may be modulated by altering the configuration of certain features of the vocal tract. In actual speech, containing vowels and consonants, use is made either alternately or simultaneously of both voiced and breath sound carriers. In whispering, the breath sound alone serves as the carrier. It would thus seem that the exact nature and characteristics of the carrier are of much less importance than the modulation imposed by the speech information. This has, of course, been found to be true for a large variety of electrical carrier waves.

9-4 The vocoder. An interesting speech-synthesizing circuit known as the *vocoder* demonstrates the carrier nature of speech. By means of electrical filter circuits the original carrier components are removed from a certain sequence of speech sounds. The modulation containing the information is retained and then impressed electrically upon an artificial carrier. For the voiced type of carrier an electrical oscillator of the relaxation type may be used. For the breath variety of carrier, a gas-filled tube will supply a "hisslike" output. When the low frequency modulation carrying the speech information is impressed upon either carrier, intelligible speech sounds result.

One of the interesting outgrowths of the studies of Dudley on the carrier nature of speech is that almost *any* carrier sound wave of mixed frequency content may be used. Even orchestral music may be used to transmit speech information, if it has a reasonably constant intensity level, and provided it is modulated at the proper rate, determined by the speech characteristics.

9–5 Energy distribution in speech as a function of frequency. Many measurements have been made of the energy distribution among frequencies involved in ordinary speech. Most of the earlier data must be credited to the Bell Telephone Laboratories, whose interest in the subject is quite

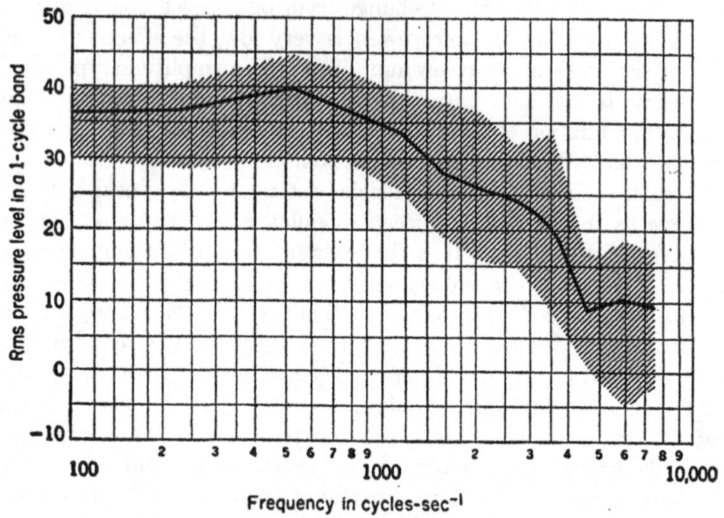

FIG. 9–2. Pressure-frequency distribution in normal speech.

understandable. Crandall found that for the average male voice the energy peak lies in the neighborhood of 120 cycles-sec^{-1}; for the average female voice the peak occurs at about twice this frequency. The many harmonics above these frequencies are associated with very low relative energy.

Some recent data along these lines, recorded at Harvard University,* are represented in the graph of Fig. 9–2. The voices of seven different men were studied, each one speaking the sentence: "Joe took father's shoe bench out; she was waiting at my lawn." This quaint and decidedly unliterary sentiment was chosen because of the wide variety of representative speech sounds it contains. The horizontal axis of the graph is fre-

* Rudmose et al., *Jour. Acous. Soc. of Amer.* **20,** 507 (1948).

quency, plotted on a logarithmic scale. The ordinates represent rms sound pressures, each point on the curve signifying the average pressure level over a band width of one cycle for the time during which the test sentence was being spoken. (The measurements involved the use of electrical filters and integrating circuits.) The central curve within the shaded area is an average for the seven voices. The limits of the shaded area are drawn so as to include all the separate curves representing individual characteristics and therefore to give some idea of the spread among the test subjects. As may be seen, there is a progressive falling off in sound pressure above 800 to 1000 cycles-sec^{-1}, indicating that by far the greater part of the energy in speech is carried by the lower frequencies. Since the vertical scale of the graph corresponds to the pressure level, it may be considered an intensity scale. It is interesting to note that the average total acoustic power in the typical speech of a single individual was of the order of 30 microwatts (1 microwatt = 10^{-6} watt). This value assumes hemispherical distribution.

9–6 Intelligibility of speech as related to frequency band width. Although most of the energy of speech is concentrated in the lower frequency regions, it must not be concluded that the upper frequencies are of negligible importance for purposes of communication. The higher frequencies constitute the main carrier for the important parts of the speech message associated with consonant production; speech with the consonant sounds removed is notoriously difficult to understand.

Much data on the essential factors governing the intelligibility of speech have been reported in the technical journals, and standard test sounds, syllables, words, and sentences have been devised. The tests and the method of analysis of the results are described by Fletcher.* Interesting experiments have been performed wherein certain frequencies and bands of frequencies within the speech range have been suppressed by means of electrical filters. Data were taken on the observed articulation of certain sounds and syllables (the portion correctly recognized) and also on the intelligibility of sentences (the portion representing correct thought reception). All the data point to the primary importance of the upper frequencies. In fact, the use of a filter designed to eliminate all frequencies below 500 cycles-sec^{-1} still permits an articulation score of 98%. For some of the consonant sounds the cutoff may be placed as high as 1500 cycles-sec^{-1} with negligible effect upon the articulation. Frequencies up to at least 3000 cycles-sec^{-1} are important; frequencies higher than this are of

* Fletcher, *Speech and Hearing*, D. Van Nostrand Co. (1929).

decreasing importance to articulation and intelligibility. (Much higher frequencies must be included, however, for facsimile reproduction of speech quality and musical detail.)

9–7 Miscellaneous voice properties. To the general results summarized in the preceding sections may be added other significant information regarding the voice. Male voices are usually richer in harmonic content than are female voices. This is primarily because the lower fundamental frequencies of the male voice involve harmonics which are in general multiples of the fundamental and must therefore be more closely spaced in the frequency spectrum. Despite the differences in fundamental pitch and harmonic content that distinguish between individuals of the same sex and also between men and women, the vowel sounds discussed in Section 9–3 are each characterized by certain invariant clusters of closely spaced frequencies. The vowel sound "ah," for instance, has a strong group of frequencies in the neighborhood of 900 cycles-sec^{-1} for practically all individuals, the long "ee" sound almost invariably contains two clusters of frequencies in the neighborhood of 750 and 1600 cycles-sec^{-1} respectively, and for the long "a" sound there are two groups near 500 and 2500.

9–8 Artificial voices. Tests involving human subjects are always laborious and liable to errors of the statistical type. In an effort to obtain more standardized test conditions, artificial voice mechanisms have been developed. The sound source for an artificial voice may be a more or less conventional loudspeaker unit of appropriate size. In front of the vibrating diaphragm is placed an acoustically designed structure simulating the essential impedance properties of the mouth and the mouth opening and of comparable size (to supply acoustic resistance and acoustic reactance). Since the diffraction properties of the human head play an important part in determining the sound distribution pattern, it is often desirable to surround the artificial voice mechanism with a life-sized model of the head. The use of this standardized equipment facilitates testing of equipment such as transmitter microphones, for instance.

9–9 The hearing process. The process of hearing may be said to take place in three stages. A portion of the wave front is first intercepted by the opening of the outer ear, which funnels the energy through the auditory canal to the eardrum, separating the inner from the outer ear. At the eardrum, the acoustical energy is transformed (partially) to the mechanical energy of vibration of the membrane. The second stage consists of the transmittal of this vibrational energy, through the interconnection of several levers, to a second membrane which lies at the entrance to the liquid-filled cochlea, a complex structure within which lie the sensation detectors. The final stage is the translation of the physical stimuli of these detectors, brought about indirectly by the pressure variations set up within the cochlear fluid, into a definite nerve message to the brain.

9-10 The structure of the ear. An anatomical drawing of the ear structure is apt to be confusing because of the wealth of nonessential detail. Figure 9-3 is designed to emphasize the functional features of the various parts of the ear. The outer, visible part is vestigial in size

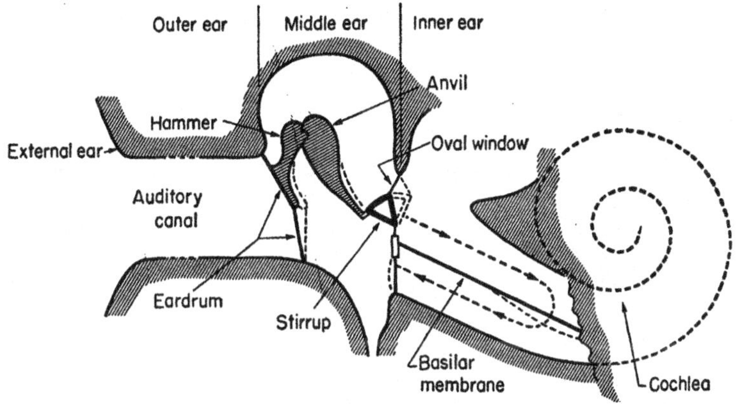

FIG. 9-3. Schematic diagram of the ear (scale distorted). (*After* Stevens and Davis)

and shape and adds little to the collecting power of the auditory canal. (In many animals the external part of the ear is large and mobile and is more important acoustically.) The auditory canal, about 3 cm in length, may be considered as a pipe closed at the inner end by the eardrum. The calculated fundamental resonance of such a pipe occurs at approximately 2700 cycles-sec^{-1}, not far from that at which the average ear is most sensitive. The sensitivity of the ear, however, falls off much more slowly with frequencies lower or higher than 2700 cycles-sec^{-1} than would be expected on the basis of a pipe resonance.

The eardrum (tympanic membrane) is stretched tightly across the inner end of the auditory canal and separates the outer from the middle ear. The middle ear contains three leverlike bones that serve to transmit the motion of the eardrum to the oval window, a second membrane which, in part, separates the middle from the inner ear. The first of these bones, the hammer, is attached to the eardrum and to the second bone, the anvil, which in turn is pivoted to the stirrup. One side of the stirrup is attached to the oval window. (It is interesting to compare the mechanical structure of the middle ear to the linkage system employed in the pickup head of the old-style "acoustical" phonograph.) The structure of the middle ear, taken as a system of closely coupled parts, has very broad resonances, due partly to the damping effect of tissue at the joints and partly to the loading supplied by the liquid on the inner ear side of the oval window. The response may therefore be described as approximately aperiodic in character.

The structure of the inner ear has excited great interest among research workers. The oval window, connected to the stirrup, is one of two membranes closing the end of the important spiral-shaped cavity called the cochlea. The second membrane, the round window, completes the closure of the cavity but is not directly connected to the stirrup. Figure 9-4 is a diagrammatic side view of the cochlea. Extending longitudinally along the turns of the spiral are two membranes, the basilar membrane and the membrane of Reissner. The latter is very thin and flexible and apparently of secondary importance. Figure 9-5 is a cross section of one of the turns of the cochlea, showing the nearly central position of the basilar membrane. Close to the oval window and to one side of the basilar membrane is the entrance to the semicircular canals, which play no direct part in the hearing process.

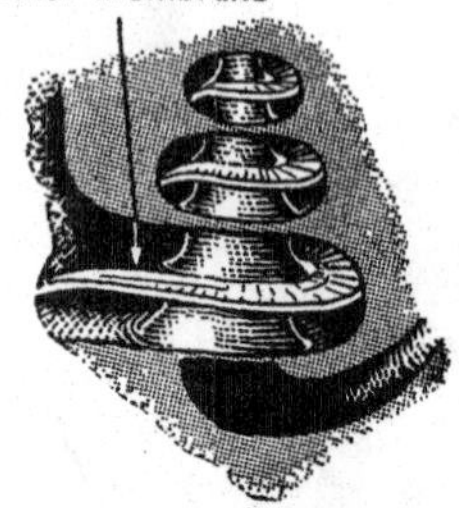

FIG. 9-4. Section of the cochlea, parallel to the axis.

The regions on each side of the basilar membrane contain liquid. Any motion of the stirrup, such as that caused by the entry of sound waves into the outer ear, will vary the pressure on the basilar membrane near the oval window, with a resultant flexure of the membrane itself (Fig. 9-3).

9-11 The organ of Corti. Mounted along the center of the basilar membrane is the *organ of Corti*, a structure which has been the subject of much controversy among physiologists. The most prominent feature of this complex structure is the *arch of Corti*, a basic framework upon which are supported clusters of hair cells from which cilia project into the liquid above the membrane. As the basilar membrane moves in response to the pressure changes occurring at the oval window it carries with it the whole organ of Corti. For some time it was thought that the arch of Corti was the sensory organ that detects the vibrations set up in the inner ear. It is now known that the auditory nerves run from the hair cells and that the latter are the true organs of sensation. There are five rows of hair cells along the organ of Corti, totaling about 23,000 in number. The bundles of auditory nerve channels emerge from the center of the cross section of the cochlea, as shown in Fig. 9-5.

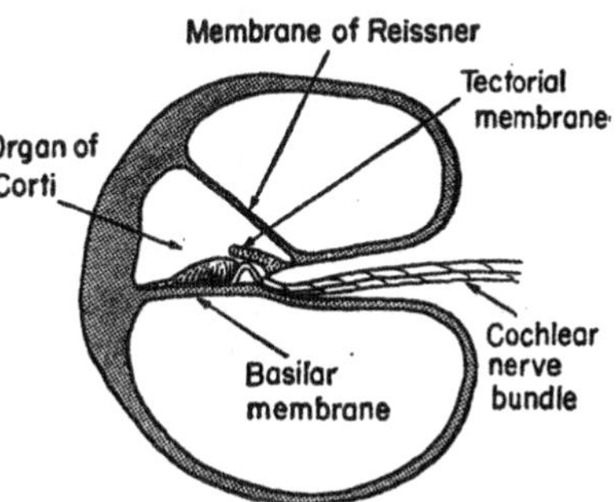

Fig. 9-5. Transverse section of one turn of the cochlea.

9-12 Mechanical properties of the cochlea. Resonance theory of Helmholtz. A physicist, familiar with resonance phenomena for vibrating bodies, is naturally tempted to picture the frequency-sensitive portion of the ear as made up of a large number of individual units, each tuned to a different frequency in the audible spectrum. Helmholtz, who was certainly part physicist, first believed that the rods or arches of Corti were the resonators, and that the vibration of the basilar membrane was the source of excitation of the arches of Corti. In a later development of his theory, he considered the fibers of the basilar membrane itself to be the group of resonators. These transverse fibers are shorter near the base of the cochlea than near the apex. There are about 10,000 membrane fibers in all; their varying lengths suggest the frequency distribution associated with a stringed instrument like the harp.

9-13 Other resonance theories. Numerous critics voiced their objections to the theory of Helmholtz, even during his lifetime. The mechanical behavior of an individual resonator remained obscure. The stretched fibers of the basilar membrane appear incapable of resonance over the wide range required for the ear, especially near the upper frequency limit. By the end of the 19th century there were, for these and other reasons, few supporters of the Helmholtz theory.

A number of more recent resonance theories have been suggested. Among several theories of a similar nature may be mentioned that of Ewald, proposed in 1898. Instead of concentrating upon a distributed set of resonators, Ewald considered the basilar membrane as a whole to be the single resonator for *all* frequencies, its response being much like that of a single stretched string. Many difficulties arose, however, because of the complexity of the stationary wave patterns which might result in accordance with this theory.

In 1928 Békésy began to report upon a series of experiments with mechanical models of the cochlea containing a thick membrane simulating the basilar membrane. (Ewald had done some experimenting with models, also.) When excited with sound waves, this membrane was set into a rather complicated motion which involved vibrations of the sound frequency. In addition, a sort of modulation effect occurred, in the nature of a traveling envelope of the vibrations. The peak of this vibration envelope, however, always occurred for any one frequency at the same point, as measured along the long dimension of the membrane. For low frequencies this peak occurred near what would be the apex of the cochlea. For high frequencies its position was near the base. For intermediate frequencies there were corresponding intermediate positions for the peak. Békésy later performed similar experiments with preserved specimens of human and animal cochlea, and observed motions of the same type, in agreement with the results of his study of models. These experiments would indicate that there are certainly mechanical resonant effects within the cochlea, apparently associated with waves moving longitudinally along the liquid-filled structure, with the maximum mechanical response occurring, as in Helmholtz's theory (but for quite different reasons), at definite positions for any one particular frequency.

Various theoretical treatments of the behavior of such a liquid-filled tube having flexible sides have been given in recent years. Among the more recent are those of Ranke* and Zwislocki.† Both formulations predict cochlear resonance very similar to that found experimentally by Békésy.

9–14 The organs of sensation. One of the interesting aspects of the behavior of the cochlea concerns the existence of alternating potentials within the cochlear fluid in the presence of sound stimulation. These so-called cochlear potentials follow closely the wave form characteristic of the original sound disturbance. Since these potentials appear to be nonexistent in the cochlea of animals with normal ear structure except for the absence of the hair cells, it may be concluded that these hair cells are the source of the cochlear potentials. Because these cells move with the motion of the basilar membrane, it seems reasonable to suppose that they are distorted in shape as the upper ends of the cilia press periodically against the tectorial membrane immediately above the hair ends. The alteration in the state of polarization of the outer cell surfaces which results from this distortion is presumed to be the source of the potentials.

Since the terminations of the auditory nerve fibers are entwined around the lower ends of the hair cells, it seems likely that, either directly or

* O. F. Ranke, *Jour. Acous. Soc. of Amer.* **22,** 772 (1950).

† Zwislocki, *Jour. Acous. Soc. of Amer.* **22,** 778 (1950).

indirectly, the potentials developed in the hair cells are responsible for the nerve impulses associated with the hearing process.

9-15 Frequency perception. As has been seen, the early theories of Helmholtz and others assumed that the mechanical stimulation of certain restricted areas within the cochlea accounted for a definite sense of frequency peculiar to that area. Later study has shown the presence of a type of true resonance phenomenon, but it is so widely spread along the cochlea that the perception of frequency at a specific "point" seems most unlikely.

In contrast to the point theories described above, Rutherford, Meyer, and Wrightson, among others, advanced the so-called "telephone" theory. By this theory the mechanical system within the cochlea is supposed always to be stimulated more or less as a whole. The mechanism is considered, like a telephone, simply to serve as a relay to transmit stimuli to the nervous system, where the true frequency discrimination takes place.

A point of view which, in the main, concentrates upon the stimulation of the auditory nerve fibers has been described by Wever.* This concept presumes that for the lower range of frequencies the nerve fibers as a whole are stimulated (as in the older "telephone" theories), but because of the "relaxation times" of the individual fibers, each such unit may "fire" (i.e., deliver its impulse) at time intervals corresponding to, say, five cycles of the audible vibration. Because of the large number of such fibers there will always be some ready to discharge at each peak of the vibration stimulus and hence the frequency of occurrence in the total nerve message will correspond to that in the sound vibration (Fig. 9-6). This response of the nerve fibers in groups has been called by Wever the "volley principle." While for the lower frequencies there is seen to be no particular localization of sensation along the cochlea, there is good evidence from the study of impaired hearing that sensitivity to frequencies above about 2000 cycles-sec^{-1} is concentrated over the first three-quarters of a turn near the base of the cochlea (the region sensitive to the very highest frequencies lies near

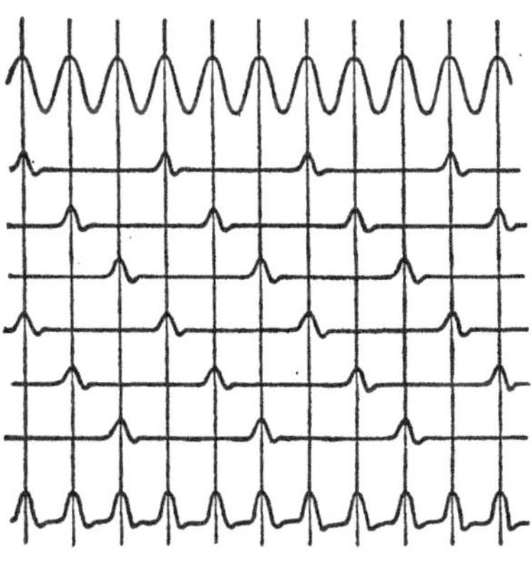

FIG. 9-6. Volley principle in nerve fiber discharge.

---

* Wever, E. G., *Theory of Hearing*, John Wiley & Sons (1949).

the beginning of the spiral). The degree of localization is believed to broaden from 2000 cycles downward, until at about 500 cycles-sec^{-1} the whole cochlea is effectively a detector.

It is interesting to see that the volley principle does not of itself rule out the importance of the mechanical resonance properties so well demonstrated by Békésy. Wever has been able to combine the data regarding the variation in the mechanical stimulation along the basilar membrane with the essential known facts regarding the number and behavior of the nerve fibers affected. In this way he has obtained a theoretical sensitivity for the ear as a function of frequency.* The curve representing this function is in fair agreement with the experimental curve known as the "threshold of audibility" contour, to be discussed in Section 9–17.

9–16 Hearing data for the normal ear. Under this heading we shall consider the results of the large number of statistical studies that have been made on hearing properties. It should be remembered that in this field we are dealing fundamentally with the subjective aspect of acoustics. Instruments cannot readily be placed within the hearing mechanism, and even if that were possible, data so obtained might bear little relation to the sensation called "hearing." Hearing tests must be performed with care and the results must be interpreted cautiously. Since the hearing characteristics of different individuals differ widely, general conclusions are valid only for the average ear.

9–17 Threshold of audibility. The normal ear is remarkably sensitive to sound waves of very low intensity, the low intensity limit being of the order of 10^{-9} erg-cm^{-2}-sec^{-1}, the usual reference for intensity level. The

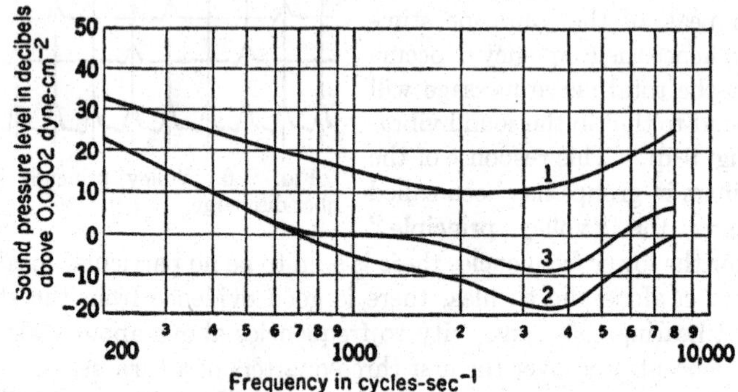

FIG. 9–7. Thresholds of audibility. (*After* Beranek)

* Wever, *op. cit.*, p. 296.

threshold of audibility is usually expressed in terms of the lowest rms acoustic pressure to which the ear will respond, and it varies markedly with the frequency. In Fig. 9–7 are shown three curves plotting the threshold of audibility against frequency. Curve 1 was taken with one ear only, the stimulation originating with an earphone and the sound pressure being measured at the ear. Curves 2 and 3 were taken with the listener in a sound field, making use of both ears. The pressure in the field was measured before the subject entered it. For curve 2 several scattered sources were used and for curve 3 a single source was placed in front of the listener. Despite the differences among the curves, all three show the maximum ear sensitivity to be in the neighborhood of 3000 cycles-sec^{-1}. (The wavy appearance of curve 3 is due to the diffraction effects of the head.) All three curves represent averages of data taken for a large number of subjects.

9–18 Loudness and loudness level. In our discussion of objective sound phenomena, we have been careful to use the word *intensity* instead of *loudness* because, as was pointed out, the two concepts are not identical. Loudness is a purely subjective quantity, not directly measurable with instruments. Loudness increases with intensity, but there is no obvious linear relationship. Each of several sounds may be classified by a listener as twice, five times, or perhaps ten times as loud as another. A scale of loudness based entirely on this type of average listener reaction has been devised. The fundamental unit is the *sone*, defined as the loudness of a pure 1000-cycle note whose pressure level is 40 db (a smaller unit, the *millisone*, equal to .001 sone, is often used). A sound five times as loud as the 1000-cycle, 40-db note would differ from the latter by a loudness of four sones.

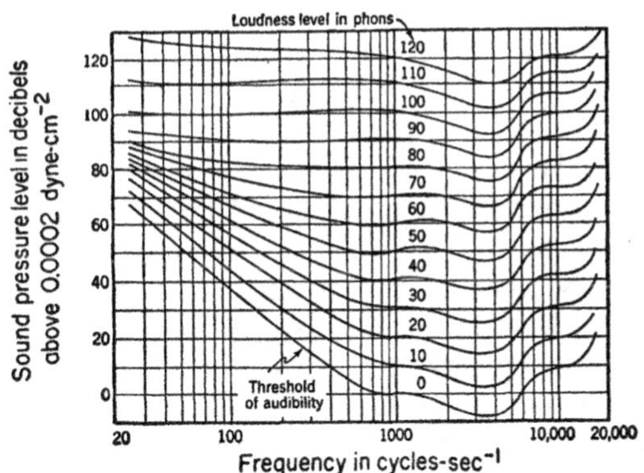

FIG. 9–8. Loudness level contours.

The *loudness level* of a tone of any frequency is defined numerically as the sound pressure level (in db) of a 1000-cycle-sec^{-1} tone which sounds equally loud to the listener. Loudness level is expressed in *phons*. The meaning of loudness level will be made clear by an inspection of Fig. 9–8. The curves are contours of equal loudness for a variety of sound pressures ranging from the threshold of audibility to what is sometimes called the upper threshold of feeling (where the sound becomes painful to the listener). The sound pressures in the free field were measured before the entry of the subject. It will be noted that the numbers corresponding to the loudness level in phons and the sound pressure level in decibels coincide in every case at 1000 cycles-sec^{-1}, in agreement with the definition of loudness level. At lower and at higher frequencies, however, the contours bend upward, the rise for the threshold of audibility at the lowest frequencies being as much as 70 db. Interpreting the general shapes of the contours, one may say that except for very high-level sounds the ear requires much greater sound pressures at low and at high frequencies to match the loudness at the middle frequencies. From curve 4 it will be noted that at 30 cycles-sec^{-1} a sound of loudness level 30 phons will have a sound pressure level (referred to the usual zero reference level) 50 db higher than at 1000 cycles-sec^{-1}. The pressure ratio corresponding to 50 db is 316, so it is obvious that the ear's sensitivity at 30 cycles-sec^{-1} is far below that at 1000 cycles-sec^{-1}.

Data for the plotting of equal loudness contours are relatively easy to obtain. A calibrated source of variable frequency, producing a known intensity at the point of pressure measurement, is adjusted by the auditor until to his ear a 100-cycle note, for instance, and a 1000-cycle note are identically loud. The pressure level produced by the 1000-cycle note is then the loudness level in phons for both notes. Since the pressure level of the 100-cycle note is known, two points for the loudness contour have been determined. The process is repeated for the determination of other contour points.

The relationship between loudness level and the quantitative subjective sensation called loudness is interesting and important but is difficult to determine with accuracy. One experimental procedure is as follows. A calibrated frequency source is first adjusted so as to produce a 1000-cycle note at a pressure level of 40 db. By definition, the loudness is then 1000 millisones. The intensity of the note is next altered until it seems to a listener to be, say, one-half as loud. The loudness is then 500 millisones and the corresponding pressure level can be read from the calibrated source. Similarly, the pressure level corresponding to 2000 millisones, say, is determined by adjusting the intensity until the note seems to be of twice its original loudness. Similar measurements at a sufficient number of different intensities enable loudness to be determined as a function of pressure level. The latter is in this case also *loudness* level, since the frequency we have selected is 1000 cycles. The relationship between loudness and loudness level at other frequencies can be determined by making use of the contours of Fig. 9-8.

Figure 9-9 is a plot of loudness, determined as above, against loudness level as a parameter.* It is interesting to study the shape of this curve

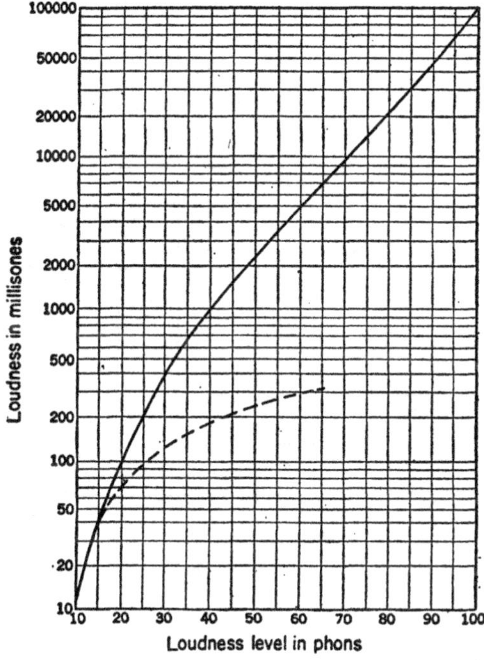

FIG. 9-9. Relationship between loudness and loudness level.

in the light of the assumed law of sensory response referred to in Section 2-20. If, as was suggested, the response is proportional to the logarithm of the stimulus, we might expect the actual loudness numbers to be linear with the decibel scale of loudness level (this scale being logarithmic). This is approximately so for loudness levels below about 15 phons but for higher levels there is considerable deviation from linearity. The broken curve of Fig. 9-9 indicates the hypothetical shape if loudness were strictly linear with loudness level. (Such a curve will not be straight on this graph because loudness has been plotted on a logarithmic scale.) At

* Fletcher and Munson, *Jour. Acous. Soc. Amer.* **9**, 1 (1937).

higher sound levels the loudness increases much faster than the response law would predict.

9–19 Differential intensity level sensitivity of the ear. In introducing the decibel scale of intensity comparison earlier, it was mentioned that a difference of about one decibel is necessary before a change in intensity is discernible by the ear. Actually, this minimum detectable change in intensity level varies greatly with the intensity itself, the frequency, and, to some extent, the complexity of the sound. The curves of Fig. 9–10 show

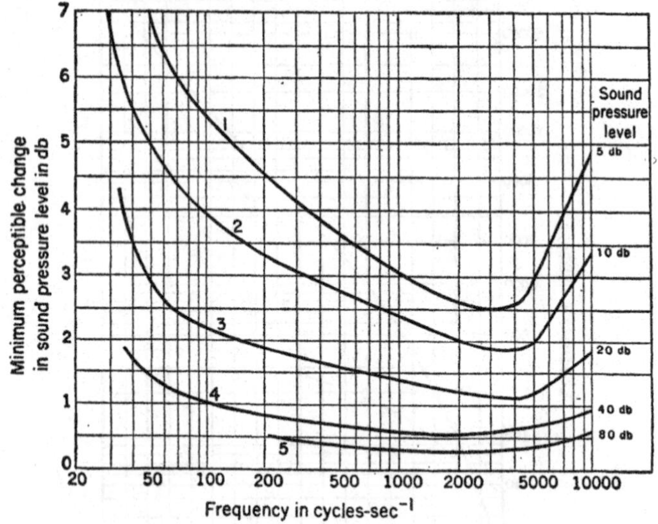

Fig. 9–10. Curves showing sensitivity of the ear to changes in sound pressure. (*After* Fletcher)

the minimum detectable change in intensity level as a function of intensity and of frequency.

Smaller changes in intensity can be detected at high sound levels than at low levels. Also, the differential sensitivity of the ear is much less at high and at low frequencies than it is at the middle frequencies. At a sound pressure level of 80 db and for a frequency of 2000 cycles-sec^{-1} (the most favorable conditions), the ear is aware of changes as small as 0.25 db; on the other hand, near the threshold of audibility changes of the order of 5 to 10 db produce barely audible effects for the very low and the very high frequencies.

9–20 Pitch vs frequency. The words pitch and frequency are commonly used interchangeably, but in acoustics *frequency* is always the objective vibration rate, while *pitch* is reserved for that subjective sensation by which the listener classifies a note as high or low. Conversion scales have been devised whereby one may relate pitch to frequency numerically, just as one relates loudness to intensity. Such scales are based largely on psychological tests.* The relationship is intimately connected with theories of hearing. We shall confine our attention here to several special aspects of this part of the hearing process.

9–21 Differential frequency sensitivity of the ear. In Fig. 9–11 are shown graphs of the ratio $\Delta f/f$ plotted against frequency, where Δf is the minimum detectable frequency shift.† Curves are given for four different

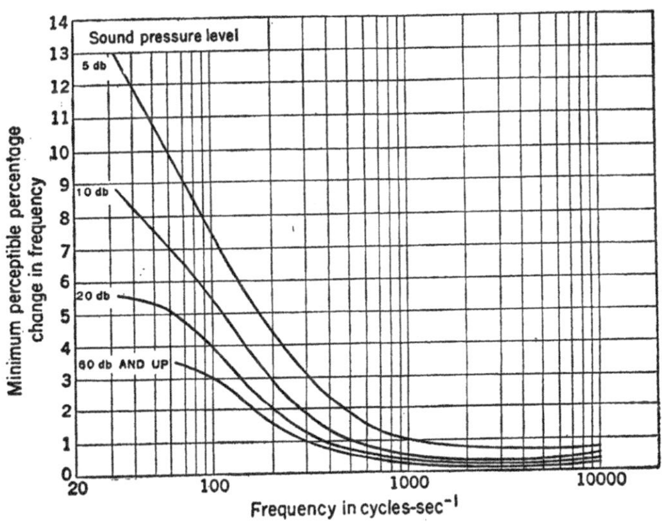

Fig. 9–11. Curves showing sensitivity of the ear to changes in frequency. (*After* Shower and Biddulph)

sound pressure levels. Clearly, the differential frequency sensitivity of the ear, treated as a fraction of the frequency, is greatest for the frequencies above about 1000 cycles-sec^{-1} and becomes relatively poor at frequencies below that. In general, sensitivity decreases with the sound intensity. These data are for pure tones only. The ability of the ear to detect an off-pitch note in the low register of the piano is not necessarily as poor as these data might indicate. The low piano notes are rich in harmonics,

* Stevens and Volkman, *Am. J. Psychol.* **53,** 329 (1940).

† Shower and Biddulph, *Jour. Acous. Soc. Amer.* **3,** 275 (1931).

especially if struck vigorously, and the presence of the higher frequencies makes possible greater frequency discrimination.

It will be noted (Fig. 9–11) that at a 10-db level a 30-cycle note must be changed in frequency by 9% (i.e., to 32.7 cycles-sec^{-1}) before a change in pitch is detectable. Hence two pure notes of frequencies 30 and 32 cycles-sec^{-1} respectively will have nearly identical pitch. This is one example of the importance of using the words *pitch* and *frequency* carefully and properly.

On the basis of the graphs of Fig. 9–11 one may compute the total number of frequencies distinguishable by the average ear within any given frequency interval. The number will obviously be greatest for the frequency range above about 1000 cycles-sec^{-1}. About midway between the threshold of audibility and the threshold of feeling, for instance, there are something like 10,000 distinguishable frequencies in the range from 1000 to 2000 cycles-sec^{-1}. The number in a 1000-cycle interval is smaller at the high frequencies and very much smaller at the low frequencies.

9–22 Shift in pitch (or apparent frequency) at high intensities. When the ear is exposed to a pure note of constant frequency and of great intensity (a sound pressure level of 60 db or more), there is an apparent change in the frequency (or in the pitch sensation). This change is usually a decrease, the amount of the change being a function of the frequency of the source

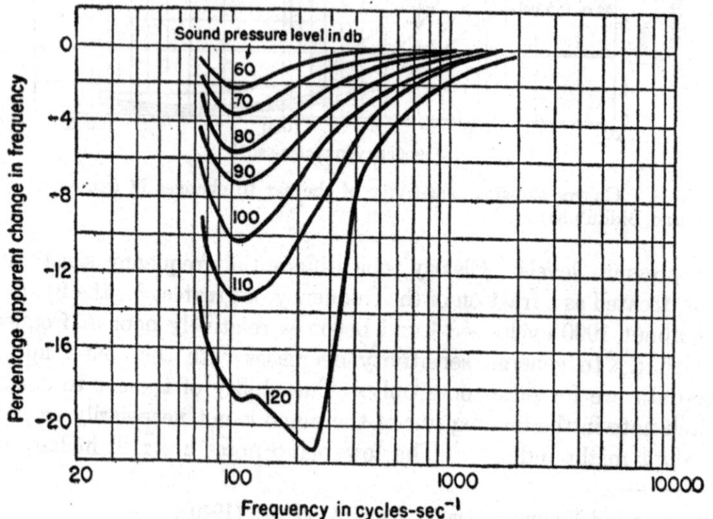

FIG. 9–12. Apparent shift in frequency with high sound pressures. (*After* Snow)

and the intensity at the ear. The graphs of Fig. 9–12 are from data by Snow.* Apparent frequency changes are given in percent, all changes being negative. Data by Fletcher indicate a slight rise in pitch for frequencies above 2000 cycles-sec^{-1}. The 10% change indicated in some cases is not unusual. Many listeners can detect a drop in apparent frequency when certain loud notes are sounded on the organ, an instrument of great acoustic power. This may be due to actual fluctuations at the source because of large air amplitudes in the pipe or may be an effect occurring at the ear.

9–23 Masking. The phenomenon of the masking of a useful sound by the presence of background noise (*noise* being undesired accompanying

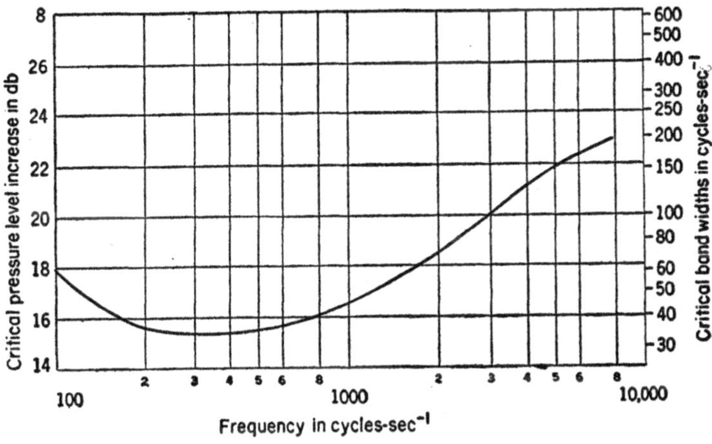

FIG. 9–13. Masking effects. (*After* French and Steinburg)

sound, usually consisting of a heterogeneous frequency mixture) is a familiar one. From a practical point of view it is necessary to know by exactly how much the level of useful sound must be raised in order to be just detectable or, in the case of speech, just intelligible above background noise. It turns out that only the frequencies adjacent to the frequency of a useful pure tone affect the masking. The band width of such pertinent frequencies varies with the desired frequency. The curve in Fig. 9–13 shows the variation in band width plotted against the frequency of the desired pure tone. The *spectrum level* for a random background noise in the neighborhood of any one frequency may be defined as the effective pressure, due to the noise, in a portion of the band one cycle in width. Fletcher has shown

* Snow, *Jour. Acous. Soc. Amer.* **8**, 14 (1936).

that to be barely audible, the level of a pure tone must be raised above the spectrum level a definite number of decibels directly related to the band widths just mentioned. The numbers corresponding to this critical level increase are shown in the figure in relation to various band widths. Such data are of great value in acoustical engineering, for instance in the problem of designing public address systems for noisy factories.

9–24 Sum and difference frequencies. As was emphasized in Chapter 1, the phenomenon of beats does not indicate a true difference frequency. The so-called beat frequency is merely a slow periodic variation in the amplitude of the sum effect of two higher frequencies. Mathematical analysis will fail to reveal the presence of a third frequency component, if the primary frequencies represent pure simple harmonic vibrations. It is nevertheless possible to *hear* difference frequencies under certain conditions and even to detect them objectively with instruments of the microphone type.

One clue to the existence of these combination frequencies lies in the general requirement that for their production the two primary tones must be of high intensity. Vibrations of large amplitude at the source will call into play both nonlinearities of source properties and of the behavior of the air immediately surrounding the source. That is, we may no longer consider the vibrations in the medium to be of the small amplitude type. Any nonlinearity in a transfer process results in intermodulation, which simply means that when two SHM's co-exist, the amplitude and wave form of one of the frequencies is affected by the presence of the other. It can be readily shown that any modulation process is equivalent to the production, among other things, of sum and difference frequencies. Once such frequencies appear near the source, they will propagate and, if of sufficient intensity, will affect the ear. The greater the amplitude of the two primary vibrations, the greater will be the amplitudes of the sum and difference frequencies (this is the reason why these effects are observed in general only with sounds of high intensity).

It is sometimes possible to detect difference frequencies when the primary sounds are of only moderate intensity. Under these conditions it is probable that nonlinearities somewhere in the hearing process are responsible. If, for instance, the eardrum were stiffer for inwardly directed forces than for those directed outward, its motion during the sound cycle would be asymmetrical, as indicated in Fig. 9–14a, and the symmetrical pattern associated with beat formation would be distorted to the shape shown in Fig. 9–14b. Such distortion amounts to what is called "partial rectification" in an electrical circuit. Figure 9-14c shows how this rectification has brought into existence the true difference frequency.

Studies of the electrical characteristics of the nerve action associated with the hearing process reveal certain marked nonlinearities in behavior. It therefore seems likely that these characteristics are more important in

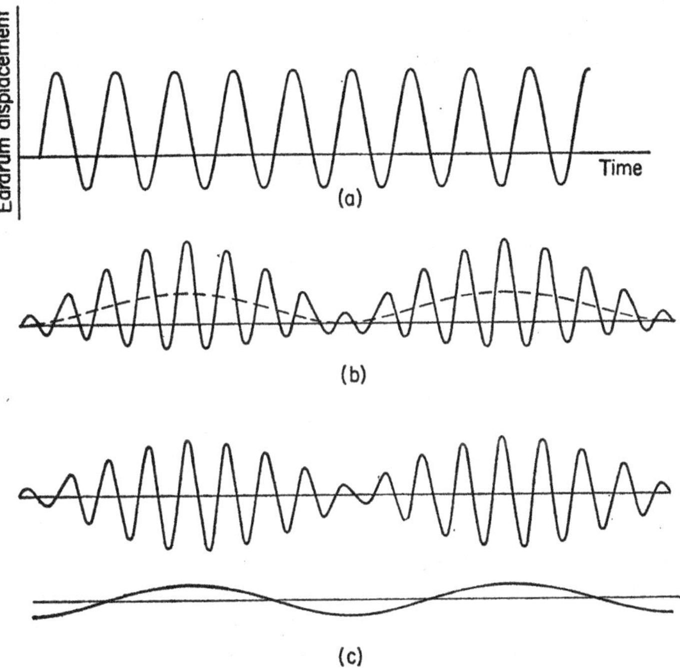

FIG. 9-14. Possible rectifying action of the ear.

accounting for the combination frequencies not present in the original acoustic wave than are the nonlinearities in the mechanical behavior of the ear mechanism, at least for sounds of moderate intensity.

For sound waves of average intensity and when many frequencies are received simultaneously, the sum and difference frequencies are so weak as to be of negligible importance for ordinary hearing purposes.

9-25 The response of the ear to a harmonic series. As we have pointed out, many of the notes played on musical instruments are rich in harmonics, some of which may be more prominent than the fundamental. Presented with an array of frequencies constituting a harmonic series, the ear will still assign a characteristic pitch to the combination, this pitch being that associated with the fundamental frequency of the series. So definite is this pitch sensation that it is possible, by means of filters, to eliminate the fundamental frequency entirely without any observable effect upon the pitch! The ear apparently will supply the fundamental, provided the correct harmonics are present. It is this rather surprising property of the ear that enables a small loudspeaker, inadequately baffled, to give the impression of good radiation in the low frequency region. Because the speaker is a fairly efficient radiator for the frequencies of the harmonics, the listener believes he is actually hearing the low frequencies, when instead he is hearing only multiples of these frequencies and his ear is supplying the fundamental. It is possible, by deliberate distortion of the harmonics associated with low musical notes, to make a very small radio set, totally inadequate in the low frequency range, sound somewhat like a larger, acoustically superior console set. Such synthetic bass is, to the critical ear, inferior in sound to true bass reproduction, where the harmonic content is closer to that of the original sound.

It seems likely that this whole phenomenon is capable of explanation in terms of the production of difference frequencies somewhere in the hearing process, as discussed in the previous section. The difference in frequency between the various terms in a harmonic series is, of course, identical with the frequency corresponding to the fundamental of the series.

9–26 The importance of the transient period to sound quality. Phase effects. Many of the vibrations characteristic of speech and music are transient in nature and not susceptible to the conventional Fourier analysis. Such short duration motions have certain build-up and decay times whose particular values contribute in an important way to the over-all quality as perceived by the ear. Fletcher has recently demonstrated a remarkable synthesizer for artificially reproducing with startling realism the quality typical of various musical instruments. Both the steady-state harmonic content and the proper build-up and decay time constants of the transient components are included in the simulation. The transient features are particularly important in the case of the drum, since the motion of the drumhead is highly damped and the membrane is excited by a single blow. Such a synthesizer, largely electrical in design, has considerable flexibility of adjustment and would seem to lend itself, as Fletcher suggests, to the development of new interesting sound qualities unobtainable with standard musical instruments.

Ordinarily the phase relationships between the various frequency components in a complex sound are of negligible importance. The ear seems to be a fairly effective Fourier analyzer, responsive primarily to the frequency structure in the wave and with little regard for over-all wave shape. The wave shape, of course, is radically affected by the relative phases. There are some exceptions to these statements. In the case of steep-fronted pulses, phase is apparently of some importance. In the field of steady-state sounds, experiments by Firestone and his associates * indicate some change in loudness level and observed quality with the shift in relative phase of a single frequency and its second harmonic, both being sounded together. Some of these effects are attributed to nonlinearities in the hearing process and are apparently of minor importance in the case of a general mixture of sound of the musical variety. For instance, in building electronic amplifiers for acoustical purposes no particular care need be taken to eliminate phase shifts in the circuits. Such shifts, even though they are different for different frequencies, do not appreciably alter the effect of the final reproduced sound.

9–27 Binaural effects. The ability of a person equipped with two good ears to locate the direction of a small source of sound is well known. There is no complete agreement as to the mechanism of this direction sensitivity but it seems likely that for frequencies below about 1500 cycles-sec^{-1} it is largely due to the difference of phase in arrival at each of the two ears. Rayleigh recognized that phase is related to binaural hearing. Experiments by G. W. Stewart † and by Hartley and Fry ‡ showed the effect of shifting the relative phases of two sounds of the same frequency fed independently to each ear. With no phase difference, the apparent position of the source was directly in front of the individual. With a gradual shift in phase the sound source could be made to "move" to the right or to the left, the direction being towards the ear receiving the leading phase. When listening to sound sources in free space one would expect phase effects to play a part only for those frequencies where the wavelength is some appreciable fraction of the distance by which the two ears are separated. At higher frequencies, the wavelength is much smaller than this distance; hence only *intensity* differences could conceivably give a sense of direction. By varying the relative intensity at the separate ears Stewart did produce directional effects, although the degree of angular shift was smaller than that produced by simple phase shift. The actual interpretation of phase or intensity differences probably occurs in the

---

* Chapin and Firestone, *Jour. Acous. Soc. Amer.* **5**, 173 (1934).

† *Phys. Rev.* **15**, 425 (1920).

‡ *Phys. Rev.* **18**, 431 (1921). Also see Kock, *Jour. Acous. Soc. Amer.* **22**, 804 (1950) for a consideration of "conscious" binaural localization.

nervous system or in the brain, rather than in the mechanisms of the ear.

Regardless of the explanation, the aural perception of direction is certainly real. Some attempt at further realism in the reproduction of music has been made along these lines. It is possible to pick up sound from the various sections of an orchestra, send the electrical equivalent of the sound from each section through separate electrical channels, and finally deliver the energy separately to loudspeakers arranged in the same relative positions as the sections of the original orchestra. This is a complicated procedure and has been attempted only on a limited experimental scale. It does result in greater musical realism.

9–28 Hearing defects. The anatomical and neurological details of the causes of deafness, partial or complete, belong to the subject of physiology and medicine and will not be considered here. The superficial facts of hearing impairment are, however, clear enough. The threshold of hearing curve varies widely in shape and general level even among so-called normal ears. When the level is raised unduly, the hearing is said to be impaired. It is a relatively simple matter to prepare audiograms which represent the threshold of hearing as a function of frequency. The threshold may be higher over some particular part of the spectrum or there may be a general, more or less uniform rise from the lowest frequency to the highest. A decreased sensitivity for the upper frequencies is especially unfortunate, of course, since this part of the spectrum is so essential to the intelligibility of speech.

An amplifier of the hearing aid type is the usual tool of the partially deaf but a simple rise in the level of all received sound is not always an adequate solution to the problem. As the level of the received sound is artificially raised in this way, the upper threshold of feeling is not necessarily raised by a similar amount, with the result that pain may set in not far above the threshold of audibility. This effect may seriously reduce the ability to differentiate useful sounds from background noise. Other related difficulties often complicate the problem.

9–29 Musical intervals. Scales. A full consideration of the subject of musical scales and their effect upon the ear would take us well outside the field of physical acoustics. However, since the acoustician is often dealing with sound of a musical nature, he should have at least a speaking acquaintance with the system of intervals used in western music.

The basic *diatonic scale* extends over a frequency range of two to one; above and below such a range the pattern is repeated. Each total interval is called an octave. Despite the fact that the frequency doubles with each successive octave, each interval of an octave seems, to the ear, to be a frequency *difference* interval. This effect compares with the ear's roughly logarithmic perception of intensity. Within the interval of one octave the diatonic scale introduces smaller intervals according to the scheme of Table 9-1. The numbers in the first row represent the frequency ratios of each note to the lowest note, C, called the tonic. In the second row is given the frequency ratio of two adjacent notes. From the latter it is seen that the intervals within the range from the lower C to the C an octave higher are also determined more on a frequency *ratio* basis than on a frequency difference, the ratios varying between 9/8 and 16/15. The larger ratios, 9/8 and 10/9, describe a whole tone interval; the smaller one, 16/15, is a half-tone interval. Intervals smaller than this are rarely used in western music. (The Arabs and the Persians divide their note cycle into a larger number of intervals; quarter-tone intervals are frequently used, one of the features of their music which makes it seem very strange to the western ear.)

TABLE 9-1

Musical note	C		D		E		F		G		A		B		C′
Frequency as related to the tonic	1		$\frac{9}{8}$		$\frac{5}{4}$		$\frac{4}{3}$		$\frac{3}{2}$		$\frac{5}{3}$		$\frac{15}{8}$		2
Frequency ratio for successive intervals		$\frac{9}{8}$		$\frac{10}{9}$		$\frac{16}{15}$		$\frac{9}{8}$		$\frac{10}{9}$		$\frac{9}{8}$		$\frac{16}{15}$	

If music were always written using the same frequency for the tonic (middle C actually has, on the modern piano, a frequency of 261.6 cycles-sec^{-1}, based on a standard A of 440 cycles-sec^{-1}), the exact relationships of the diatonic scale could always be maintained. If, however, it is desired to choose a variety of notes as the tonic, difficulties develop. Table 9-2 shows the frequencies present if we choose to use the key of D instead of the key of C, the intervals in each case being based on the diatonic scale. (If we establish an A of 440 cycles, the diatonic scale requires C, the tonic, to be 264 cycles.)

TABLE 9-2

Musical note	C	D	E	F	G	A	B	C′	D′
Key of C	264	297	330	350	396	440	495	528	
Key of D		297	*334	*371	396	*445	495	*557	594

The four new notes marked with asterisks would have to be added for the scale of D. If all possible musical keys were to be provided for, *seventy-two* notes would be needed for each octave. To avoid this tremendous complication, what is known as the *equally tempered scale* has been devised (used by J. S. Bach and probably earlier musicians). In this scheme there are 12 half-tone intervals in every range of an octave, adjacent notes a half-tone apart bearing the constant ratio of the twelfth root of 2, i.e., 1.05946. Simple as this scheme is, it results in no one scale corresponding to a given tonic being exactly like the diatonic. Since the ratios of the diatonic scale were originally selected to suit the preferences of the ear (being made up of three sets of major triads, each of which constitutes a harmonious combination), this means that music which makes use of the equally tempered scale is not quite so pleasant to the ear. The difference, however, is apparently slight to any but the most critical.

9–30 Consonance and dissonance. What constitutes a pleasant combination of frequencies and what is an unpleasant combination has long been a subject for discussion among both musicians and physicists. The disagreeable sound of certain combinations was attributed by Helmholtz to beat effects, either between the fundamental frequencies themselves or between some of their harmonics. He believed that for the middle of the audible spectrum the difference frequency or beat rate which produced aural irritation covered the range of about 30 to 130 cycles-sec^{-1}. Two adjacent half-tones in the neighborhood of 440 cycles-sec^{-1} differ by about 25 cycles and when sounded together are on the verge of being disagreeable to the ear. Two adjacent whole tones in this region differ by about 45 cycles and the combination of two such notes is usually considered dissonant. Even though the fundamental frequencies of two sounds with rich harmonic structures are far enough apart to produce consonance, there may be a particular harmonic of one tone which is close to a harmonic of the other, and the result is an over-all effect of dissonance.

There is a strong individual subjective element in the matter of consonance and dissonance that often determines the final impression. There is also no doubt that musical fashions change. Much of the music written in recent years is highly and continuously dissonant to ears accustomed to Haydn and Mozart, yet adherents of this newer musical style welcome each crashing "discord" with great satisfaction. Such sharp cleavages in musical taste make it difficult to draw any very certain conclusions in the matter of consonance and dissonance.

PROBLEMS

1. The lowest note within the range of a bass voice has a fundamental frequency of about 80 cycles-sec^{-1}. (a) In view of the dimensions of the vocal cavities, is it likely that air resonance will occur for this frequency? (b) Would the throat and mouth, considered as a horn, constitute an efficient radiating system for such a frequency? (c) In view of the above considerations, suggest an explanation of the observed pitch, corresponding to a frequency of 80 cycles-sec^{-1}. (See Section 9–22.)

2. Assume the existence of an ideal electroacoustic transducer whose acoustic radiation is proportional to the electrical power delivered to it, at all possible audible frequencies. Making use of the loudness level contour (Fig. 9–8) corresponding to 40 db, (a) plot the power gain (ratio of output to input power) vs the frequency f, in db, for an amplifier designed to make sounds of equal intensity appear equally loud. Assume the gain in db at 1000 cycles-sec^{-1} to be zero. (b) Assuming the power output of the amplifier to be proportional to the square of the output voltage, plot the required *voltage* gain of the amplifier vs the frequency. Use as ordinates the ratio of output voltage to input voltage. (c) Would the use of such an amplifier make the acoustic output sound perfectly "natural"? Explain.

3. Compare the "dynamic range" of the ear (useful range of sound intensities in db) at 40, 1000, and 10,000 cycles-sec^{-1}, using Fig. 9–8.

4. The graph representing the harmonic content of the vowel sound "oh" (Fig. 9-1b), indicates that the most prominent frequency present is 625 cycles-sec^{-1}. On the basis of the discussion in Section 9–22, explain why, to both a singer and the hearer, the characteristic frequency appears to be 125 cycles-sec^{-1}.

5. Considering the ear mechanism as a mechanical system, what is the probable nature of its transient response (that is, its response to an acoustic "pulse")?

6. In the "volley" theory of the hearing sensation, would a variation in the recovery time of an individual nerve fiber seriously affect the ear's ability to perceive the frequency property in a sound wave?

7. Consider the graphs of Figs. 9–8 and 9–9. If the sound pressure level rises from 60 db to 80 db at a frequency of 100 cycles-sec^{-1}, what will be the change in loudness in millisones?

8. (a) For the average human ear, what is the minimum perceptible change in rms acoustic pressure (in dynes-cm^{-2}) if the frequency is 100 cycles-sec^{-1} and if the sound pressure level is 10 db? Make use of the graphs of Fig. 9–10. (b) Answer the same question if the frequency is 4000 cycles-sec^{-1}. (c) What will be the corresponding minimum perceptible changes for (a) and (b) if the sound pressure level is 40 db?

9. (a) Discuss the significance of the graphs of Fig. 9–11 in connection with accurately tuning a piano at very low and at very high frequencies. (b) By how much would a 100 cycle-sec^{-1} note have to be off tune at the several sound levels shown to be detectable by the average ear? Express your answer as a percentage and as a frequency difference in cycles-sec^{-1}.

10. Is the effect discussed in Section 9–20 and illustrated in the graphs of Fig. 9–12 likely to be of significance in (a) speech, (b) the singing voice, (c) the music from a symphonic orchestra, (d) the music from a large pipe organ?

11. Assume that the ear is a nonlinear mechanism and that the relationship between the driving force and the displacement is $F = x^{\frac{1}{2}}$. If the driving force is

$$F = (F_m)_1 \cos \omega_1 t + (F_m)_2 \cos \omega_2 t,$$

show that the response of the system, x, will include both sum and difference frequencies.

CHAPTER 10

SOUND MEASUREMENTS. EXPERIMENTAL ACOUSTICS

10–1 Precise acoustic measurement. The theoretical and mathematical aspects of modern acoustics are grounded in the thorough studies of the physicists and mathematicians of the nineteenth century and earlier. Articles published in recent years have helped to fill many gaps in the earlier formulations, but almost all such contributions have been in the matter of detail rather than of fundamental theory. It is in experimental acoustics that striking progress has been made, largely because of the parallel developments in electrical equipment, particularly of the electronic variety. Modern electronic measuring devices have almost entirely replaced the earlier mechanical type. Their greatly increased sensitivity, ease of manipulation, and general flexibility of design have contributed much to the precision of acoustic measurements. It is now possible to test many of the conclusions of the earlier theoreticians and, what is perhaps more important, to test the performance of acoustic equipment when theoretical predictions are lacking due to mathematical difficulties.

Because of the widespread use of electrical equipment in acoustic measurement, we shall make frequent reference to specific circuits and electronic laboratory equipment. When questions arise as to details of the circuits and the equipment, one of the numerous books on circuitry and instrument design should be consulted. In the experimental field, acoustics and electronics are most intimately connected; a good understanding of electronic circuits is invaluable to laboratory work in acoustics.

10–2 Free-space measurements. Anechoic rooms. When fundamental data are required on the radiation properties of an acoustical device such as a loudspeaker, measurements are usually taken under free-space conditions, without the presence of complicating stationary wave patterns. The simplest procedure is to take all the equipment outdoors, but even there, reflections from the ground may introduce an important lack of symmetry. Many such tests are made with both sound source and receiver mounted high above the ground, on a tower. The use of directional microphones (Section 10–9) will further help to discriminate against energy reflected from the ground.

The advantages of an indoor method of free-space measurement are obvious. The weather does not always cooperate with the scientist, for one thing, and moreover the laboratory is a safer place to set up complicated electrical equipment. Indoor free-space measurements require an

A modern anechoic room. (*Courtesy* Bell Telephone Laboratories)

anechoic room. As the name implies, this is a chamber with completely nonreflecting walls (walls that are complete absorbers). In addition to complete elimination of stationary wave patterns it is necessary to thoroughly insulate the chamber against all disturbing sounds from the outside. Until lately sound test rooms have fallen far short of meeting these stringent requirements but recent study * has resulted in design data for a most

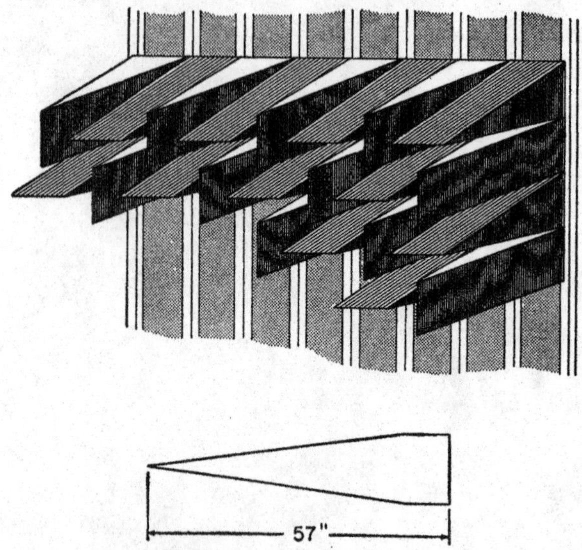

Fig. 10–1. Wedge structure on wall surfaces of anechoic chamber. (*After* Beranek and Sleeper) Below, typical Fiberglas wedge.

efficient type of anechoic chamber. The wall structure is interesting and will be described briefly.

The entire surface of all six walls of a room of rectangular shape is lined with wedge-shaped units, the sharp edge of the wedge pointing directly into the room. Figure 10–1 shows how the wedges look from the inside of the room. The effective length of each wedge is from 4 to 5 feet and the direction of the sharp edges is alternately vertical and horizontal. Between each wedge is a wedge-shaped air cavity. Any portion of a sound wave entering these cavities at an angle will be reflected back and forth several times. The wedges are constructed of Fiberglas,† a material of loose

* Beranek and Sleeper, *Jour. Acous. Soc. Amer.* **18,** 140 (1946).

† A trade name.

structure and with such highly absorbent properties that there is great attenuation of the wave with each reflection. In addition, there will be losses due to increased viscous effects towards the small end of the air wedge. It is usual to suspend a tightly stretched steel net near the center of the chamber in lieu of a floor; upon this net the observer may set up the acoustic equipment.

A graph showing the reflecting power of this type of wall surface is given in Fig. 10-2. The ordinates represent the ratio, in percent, of the acoustic pressure in the reflected wave to the acoustic pressure in the incident wave, at the wall. Since for plane waves the intensity is proportional to the square of the pressure, the ratio of the corresponding intensities, in percent, will in each case be the *square* of the ordinate. From a frequency of 1500 cycles down to about 90 cycles, the pressure ratio is less than 5%. The intensity ratio is therefore less than 0.25%. In terms of absorption, this means that 99.75% of the incident energy is absorbed. This is very close to the conditions in free space. In an anechoic chamber 38 × 50 × 38 feet, the inverse square law characteristic of free space was found to hold to within a variation of 0.5 db in pressure level for distances up to 20 feet from a small source and for any frequency above 65 cycles-sec^{-1}. At 50 cycles-sec^{-1} the absorption is less effective, as the graph of Fig. 10-2 indicates, but even there the deviations from the inverse square law amount to only 3 or 4 db.

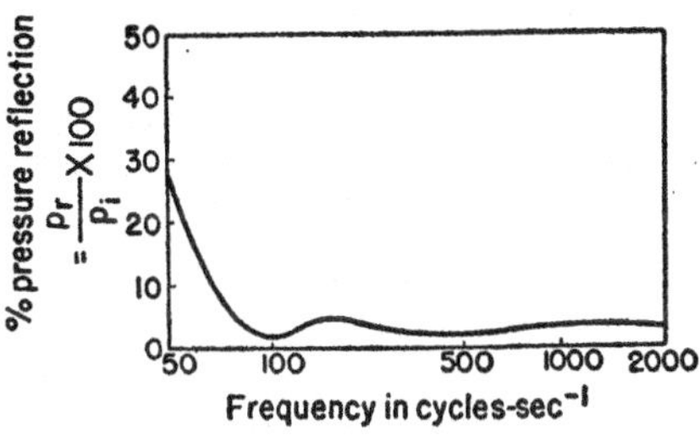

FIG. 10-2. Reflection properties of wall surface like that of Fig. 10-1.

10-3 Reverberant chambers. In many problems related to the practical reproduction of speech and music it is desirable to take measurements in rooms having partially reflecting walls of the ordinary type. Since every room is different in size, shape, and the reflecting properties of its wall surfaces, all reported measurements should, to have any physical significance, be accompanied by a full description of the room and of the location of the acoustical apparatus. We shall return to this discussion in the next chapter, in connection with the indoor performance of loudspeakers.

10-4 Standard sound sources. The thermophone. Two procedures for the calibration of microphones are in common use. One is a free-space method, the other limits the location of the microphone and the source to the confines of a small cavity. In the latter method it is sometimes convenient to have a source which will produce a known acoustic pressure. Several such sources whose behavior can be predicted with some degree of precision have been developed.

The thermophone, in its modern form, is due to Arnold and Crandall,* although experiments were conducted as far back as 1898 by F. Braun. As its name suggests, the thermophone makes use of a strip of foil or a fine wire, heated by passing through it superimposed alternating and direct current. The periodic fluctuations in the temperature of the foil or wire produce a heat diffusion wave in the gas in the immediate neighborhood of the conductor. This diffusion wave gives rise to an acoustic wave that propagates away from the heated surface. The purpose of the d.c. bias is to ensure in the acoustic wave a frequency which is the same as that of the alternating heating current. Without such a steady current the frequency would be doubled.

The thermophone is generally used within a cavity whose dimensions are small compared with the wavelength of the sound wave. Under these conditions the acoustic pressure within such a volume is fundamentally dependent upon the temperature fluctuation at the surface of the heated conductor, which in turn is a function of the values of the d.c. and a.c. components of the currents. The main body of the cavity, including the walls and the gas within the cavity (up to within a few mm of the heated surface), may be considered to remain at constant temperature. An expression can be derived for the acoustic pressure within the cavity. In the case of the heated foil, this pressure is a quantitative function of the currents used, the frequency of the alternating component, the physical properties of the gas used, and the mean temperatures of the foil and of the surrounding gas in the cavity.† In view of the rather complicated form of this equation it is rather surprising that the pressure value which it yields agrees so well with experiment. The pressure in the cavity may be experimentally determined by using the reciprocity principle (Section 10–13). Agreement with the theoretical behavior of the thermophone is to within less than 1.0 db, considered a small error in acoustical measurements.

10–5 The pistonphone. It is possible to construct a true acoustic piston which will behave in a manner similar to that of the ideal piston mentioned

* Arnold and Crandall, *Phys. Rev.* **10**, 22–38 (1917).

† See Beranek, *Acoustic Measurements*, pp. 165–168.

so frequently in this book. One such design is illustrated in Fig. 10-3.* The piston, driven sinusoidally in the manner of a loudspeaker unit, projects into a small cavity in which is mounted the microphone or other device to be tested. If the dimensions of the cavity are small compared with the wavelength, the excess pressure p within the cavity at any one instant may be simply computed on the basis of the elasticity of the enclosed gas:

$$p = \frac{xS_p\mathcal{B}}{V_0}, \qquad (10\text{-}1)$$

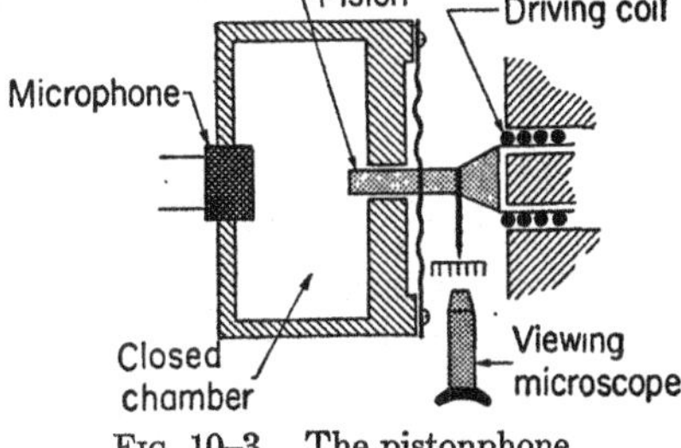

FIG. 10-3. The pistonphone.

where S_p is the area of the piston face, V_0 is the mean volume within the cavity, $\mathcal{B}$ is the adiabatic elastic bulk modulus of the gas for the mean pressure being used, and x is the displacement of the piston. (If the rms displacement is used, the pressure will be the rms value.) The amplitude of motion of the piston may be observed directly with a microscope, as shown in the figure.

Equation (10-1), in practice, must be corrected for heat conduction at the walls. The pistonphone may be used only at low frequencies, primarily because of limitations associated with the inertia of the piston and also because of the requirements regarding cavity dimensions.

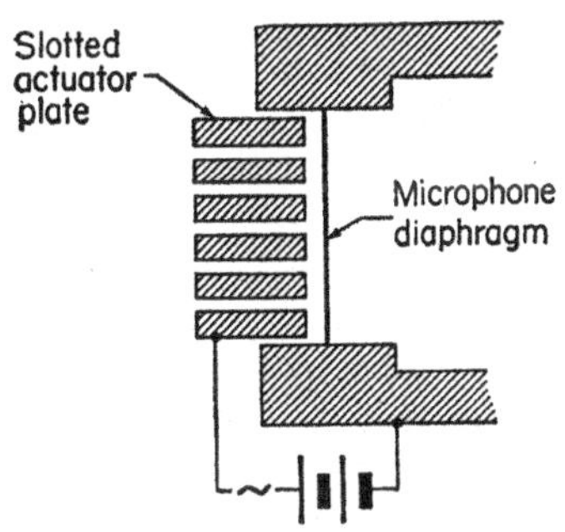

FIG. 10-4. The electrostatic actuator.

10-6 The electrostatic actuator. At high frequencies both the thermophone and the pistonphone are inadequate because of the presence of phase differences within the cavity. Under these conditions a microphone may be calibrated with an electrostatic actuator. Strictly speaking, the "actuator" is not a source of acoustical waves at all, but a calibrated mechanism whereby direct forces may be applied to a diaphragm like that of the microphone. Since the standard sound sources, the thermophone and the pistonphone, are of primary utility in the calibration of sound receivers, we shall briefly describe the electrostatic actuator at this point.

---

* After Glover and Baumzweiger, *Jour. Acous. Soc. of Amer.* **10,** 200-202 (1939).

As shown in Fig. 10–4, a thick metallic plate having a number of slots constitutes one plate of a capacitor, the other plate being the diaphragm of the microphone to be calibrated. If a sinusoidal variation in potential is applied between the two plates of the capacitor, along with a biasing d.c. potential (the bias being used for the same reason as in the case of the thermophone) the electrostatic force per unit area of the diaphragm will be

$$p_{\mathrm{rms}} = \frac{8.85\, E_0\, E_{\mathrm{rms}}}{d^2}\, 10^{-7} \text{ dyne-cm}^{-2}, \tag{10–2}$$

where E_0 is the bias voltage, E_{rms} is the rms value of the alternating component, and d is the effective spacing between the plates, corrected for the presence of the slots. This equation neglects the effect of the air loading, which the presence of the slots in the fixed plate is designed to minimize. This method of microphone calibration is particularly convenient, since the value of p does not in any way depend on the conditions within a gas. The actuator may be used well into the ultrasonic frequency region.

10–7 Measurements in a field of sound. The Rayleigh disk. At the receiving end, there are a few detectors of sound waves whose exact response to a particular intensity or pressure level can be predicted with good accuracy. Among such absolute sound measurement devices is the Rayleigh disk, suggested by Lord Rayleigh in 1882. In Fig. 10–5 are shown the flow lines around a thin flat rigid disk set at an angle to the general direction of motion of a fluid. It will be noted that there are two symmetrical points, a and b, which represent regions where the component flow parallel to the plane of the disk changes direction. At these points the fluid (gas or liquid) is relatively stagnant. Opposite each point, on the other side of the disk, is a region where the flow is relatively rapid. According to the theorem of Bernoulli, there will be a difference in pressure on the two sides of the disk in the neighborhood of points a and b, the sense of the net forces being such that a torque on the disk results. For the situation as shown, the torque is that of a couple in the clockwise direction. If the direction of motion of the fluid is reversed, the flow lines remain unchanged in shape. If such a disk is exposed to an advancing longitudinal wave there should be a net average torque, despite the reversals in the direction of motion of the air particles.

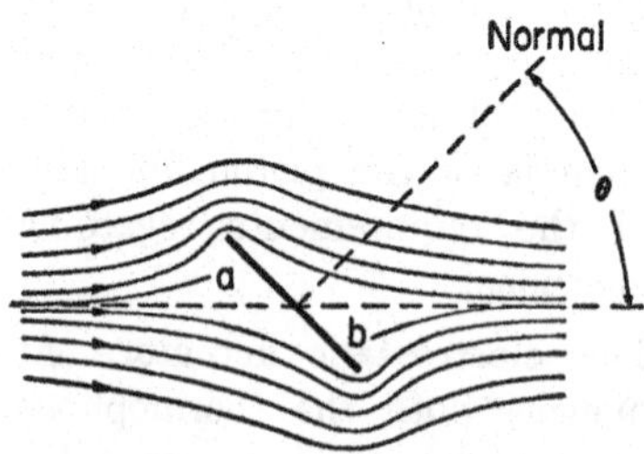

FIG. 10–5. Flow lines around a Rayleigh disk.

For the steady stream velocity of an incompressible fluid around a thin rigid circular disk (strictly, a thin ellipsoid), the torque L can be shown to be

$$L = \tfrac{4}{3}\,\rho_0 r^3 u^2 \sin 2\theta, \tag{10-3}$$

where ρ_0 is the fluid density, r the radius of the disk, u the stream velocity, and θ the angle between the normal drawn to the plane of the disk and the direction of motion of the undisturbed stream. For the periodic motions associated with a sound wave the average torque will be determined by the average squared velocity. It should then be correct to use for u the rms particle velocity in the wave, $\dot{\xi}_{\text{rms}}$.

From the form of Eq. (10-3), the torque is zero if the disk is placed either parallel or perpendicular to the direction of the stream velocity. (The latter position is actually that of stable equilibrium.) Because the torque is a maximum when $\theta = 45°$, the disk is suspended by an elastic fiber with the plane of the disk making a 45° angle with the direction of propagation of the wave. A torsion head adjustment may be used to maintain the disk in this position in the presence of the torque due to the sound wave. Knowing the elastic constant of the fiber, the value of $\dot{\xi}$ may be computed by use of Eq. (10-3).

In actual practice a much more complicated equation than (10-3) must be used.* Small second order displacements of the disk arise because of lack of complete rigidity in the suspension and, since the size of the disk is always finite in relation to the wavelength, diffraction effects must be taken into account. (The radius of the disk must in any case be fairly small compared with the wavelength λ if the velocity $\dot{\xi}$ is to be assumed uniform in the immediate vicinity of the disk.) If the radius of the disk is sufficiently small compared with λ, a slight modification in form of the steady flow equation (10-3) may be taken as an accurate expression for the average torque, L_{av}:

$$L_{\text{av}} = \frac{4}{3}\,\rho_0 r^3 (\dot{\xi}_{\text{rms}})^2 \sin 2\theta \left(\frac{m_1}{m_1 + m_0}\right). \tag{10-4}$$

In this equation, m_1 is the mass of the disk and m_0 is its so-called "hydrodynamical mass," equal to $\frac{8}{3}\rho_0 r^3$. (This is the same quantity that enters into the index of refraction of an acoustic lens of the obstacle type, discussed in Section 6-7.) Equation (10-4) may be used either for a progressive sound wave or at a position of a velocity antinode in a stationary wave pattern.

---

* See Beranek, *Acoustic Measurements*, pp. 149-152.

Unfortunately, the Rayleigh disk may be used for precision measurements only in regions where particle velocities are high. If used without a resonator, the pressure level must be at least 50 db in air and 85 db in water. By the use of a tuned resonator, sensitivities comparable to that of the ear may be reached, but in this case the disk ceases to be a directly calibrated precision measuring instrument. With a resonator, however, the Rayleigh disk was a convenient comparison detector before the advent of electrical microphones.

10–8 Other absolute detection methods. Indirect observation of the amplitudes of motion of air molecules in the presence of a sound wave is possible if the region is filled with finely divided smoke particles. After careful microscopic measurement of the path lengths of the smoke particles (whose motion was the result of the vibration of the much smaller air molecules), Andrade and Parker * concluded that the amplitude of the smoke particles and the amplitude of the air particles were nearly identical (within 2%) up to a frequency of 5000 cycles-sec^{-1}. This method of measurement of the particle displacement ξ in a wave is hardly a practical technique for routine acoustical measurements.

The existence of the phenomenon of radiation pressure is the basis of another "absolute" measurement technique. In acoustic waves such pressure is the result of second-order variations in the pressure in front of the surface upon which the wave impinges. It will be recalled that for small amplitude waves we may assume that the relation between P and V is a linear one. Actually, the graph is hyperbolic, not a straight line. Therefore if we consider the situation in front of a rigid reflecting surface, as the particle layers surge towards and away from the boundary, we must recognize that the *increases* in pressure above the undisturbed value are slightly greater than the decreases. The time average of the acoustic pressure is then not zero, but a small positive value.

An interesting simple treatment of acoustic radiation pressure, due to Larmor,† is worth consideration. Assume that a steady train of incident waves in which the energy density is e_i impinges normally upon a rigid, stationary, perfectly reflecting wall. The average energy density in the reflected wave will be the same as in the incident wave, so that the total energy density in front of the wall, e_{total}, will be $2e_i$. Now imagine the wall to be slowly advancing to meet the waves with a steady velocity, u. In one second the wall will intercept a column of length $c + u$, where c is

* Andrade and Parker, *Proc. Roy. Soc. London* **A159**, 507–526 (1937).

† Larmor, *Enc. Brit.*, 11th ed. **22**, 786 (1911).

the wave velocity. This energy will be returned to the medium, but it will occupy a column of reduced length, that is, the length will be $c - u$. Hence the energy density e_r in the reflected wave will be greater than that in the incident wave according to the ratio

$$\frac{e_r}{e_i} = \frac{c+u}{c-u} = 1 + \frac{2u}{c-u}. \tag{10-5}$$

The increase in the total energy density in the region, if u is much less than c, will be

$$e_r - e_i = e_i\left(1 + \frac{2u}{c}\right) - e_i = \frac{2ue_i}{c}. \tag{10-6}$$

In a region in the form of a column in front of the wall, of length c and of unit cross section, the total increase in energy will be $c(e_r - e_i) = 2ue_i$. The work done by the force necessary to move the wall must account for this energy. The force on the end of a column of unit cross section may be called a pressure, P. Using the wall velocity, u, we may therefore equate the work performed per second to the energy increase in the medium per second:

$$Pu = 2ue_i.$$

Since the wall velocity u cancels out, the result does not depend on its value, which we may imagine to be vanishingly small. Under these conditions, $2e_i$ is the total energy density in front of the wall, so that

$$P = e_{\text{total}}, \tag{10-7}$$

an astonishingly simple result.

A more rigorous derivation of the pressure due to impinging radiation shows Eq. (10-7) to be correct also in cases of imperfect reflection. In the special case of a perfectly absorbing wall, the incident wave only is involved, so that the value of e_{total} is just one-half what it is for a perfectly reflecting wall. The radiation pressure is hence also one-half as great.

The existence of radiation pressure is more interesting than useful for purposes of sound measurement. Rarely, even in the case of very high intensity sounds, does the energy density exceed the value of one erg-cm^{-3}. According to Eq. (10-7) the corresponding range of radiation pressure is from one to two dynes-cm^{-2}, depending on the absorption coefficient of the surface. Sounds of average intensity would result in much lower pressures, possibly as low as 10^{-13} dyne-cm^{-2} near the threshold of audibility. In the field of ultrasonics the effect is more useful, since waves of high intensity are commonly encountered both in air and in water.

10-9 Detectors requiring calibration. Microphones. By far the most useful type of acoustical detector is some form of electrical microphone in combination with an electronic amplifier. Such systems yield an electrical response that varies with the pressure in the wave and in most cases with the frequency also. Their absolute sensitivity calibration may sometimes be estimated, but it is almost universal practice either to compare their response with that of some absolute laboratory standard, such as the Rayleigh disk, or else to deduce their calibration from measurements based on the reciprocity principle, as will be discussed presently. Unlike the detectors previously mentioned, a good microphone and amplifier system will approach the sensitivity of the human ear. Such an arrangement is rugged and easy to adjust, in contrast to many of the earlier devices.

Microphones in general may be classified as either sensitive to acoustic *pressure* or to *pressure gradient.* (The latter are often called *velocity microphones.*) Microphones of all types almost universally have some sort of diaphragm exposed to the wave. The resulting motion of this diaphragm actuates the mechanism peculiar to the particular type of microphone. If one side only of the diaphragm is exposed, it is called a *pressure microphone.* Such a microphone is relatively insensitive to the direction from which the wave is coming, since the force on a surface in a fluid under pressure is independent of the orientation of the surface. Strictly speaking, it is the pressure aspect in the wave that actuates practically all microphones. However, if *both* sides of the diaphragm are exposed, the net effective force per unit area will be the *difference* between the pressures on the two sides. For a diaphragm of given thickness, the pressure difference will be small but it will be a definite function of the intensity and of the frequency and will actuate the diaphragm accordingly. If the diaphragm is mounted parallel to the wave front, the pressure difference will be a maximum. If the plane of the diaphragm lies perpendicular to the wave front, the pressure difference will be zero. For this reason, such a *pressure gradient* microphone will be directional, often a very useful practical property. For a given diaphragm, it is the pressure gradient (or pressure change per unit distance in space) that determines the microphone response, hence the name. Since the particle velocity $\dot{\xi}$ is also a vector quantity in the same direction as the pressure gradient, the name *velocity microphone* is also appropriate. (The response to pressure gradient and to velocity is somewhat different, however, as we shall see presently.)

10–10 Microphones. A thorough discussion of the design features of all the numerous types of microphones would occupy more space than is allowable in a book of this size. We shall have to be content with a brief description of a few of the more important kinds.

1. *The carbon microphone.* About the only remaining example of this early variety of microphone is the telephone transmitter, familiar in the home and also still used in certain military applications where circuit simplicity and mechanical ruggedness are important requirements. The active element in the microphone is a loosely confined collection of carbon granules, in contact with which is an electrode attached to the actuating diaphragm. A simplified diagram of the mechanism and the essential associated circuit is given in Fig. 10–6. The vibration of the diaphragm varies the resistance of the collection of carbon granules, and this variation modulates the current in an electrical circuit. It will be noted that a d.c. source of potential is required. The battery supplying this potential is the source of power in the circuit, the sound vibrations merely serving as the triggering agent. Assuming that for small displacements the variation in the resistance of the microphone is proportional to the displacement of the diaphragm, the current in the circuit due to a sinusoidal acoustic pressure can be shown to constitute a harmonic series, with the lowest frequency corresponding to that in the wave. The presence of the higher frequency terms is not too detrimental for purposes of speech. The size and stiffness of the diaphragm (the latter is a "plate," not a "membrane") are such as to cause it to have its main resonance at about the middle of the important range of frequencies for speech (about 2000 cycles-sec^{-1}). The type of response to be expected at a constant pressure level is shown in Fig. 10–7.

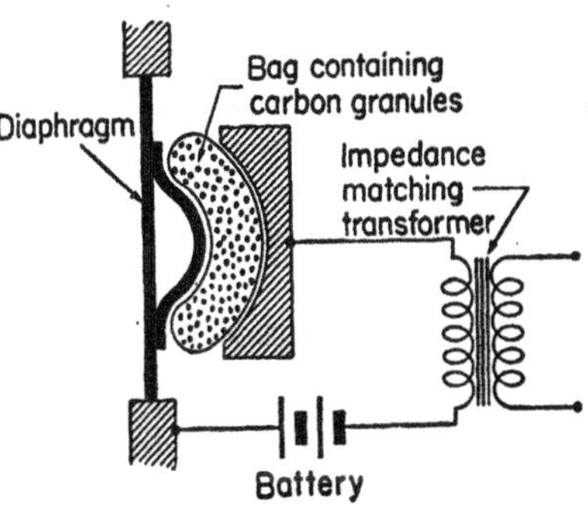

FIG. 10–6. Essential parts and circuit for modern military carbon microphone.

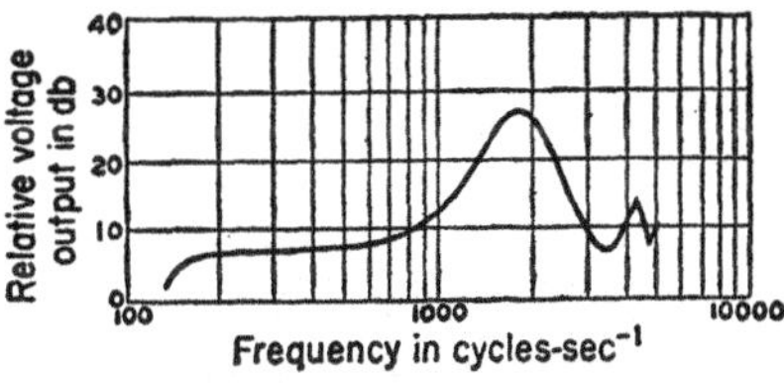

FIG. 10–7. Typical response characteristic of carbon microphone.

A double-button carbon microphone makes use of two carbon packs, one on each side of the diaphragm. As the diaphragm moves, the resistance of one pack increases, while that of the other decreases. The microphone is connected into a push-pull type of circuit, very similar to the electronic amplifier circuit of the push-pull variety. As in the case of the latter circuit, the even harmonics generated by the microphone cancel out (see any book on electronic amplifiers). Improved forms of the two-button carbon microphone compare well in uniformity of frequency response with the best microphones of other types. The chief drawbacks of all carbon microphones are the presence of background noise of the hiss type and the fact that the calibration of a carbon microphone may not be relied upon for any considerable length of time.

2. *The capacitor microphone.* The capacitor microphone is one of the earliest precision acoustical instruments. The diaphragm is a stretched thin metallic membrane which forms one plate of an air dielectric capacitor,

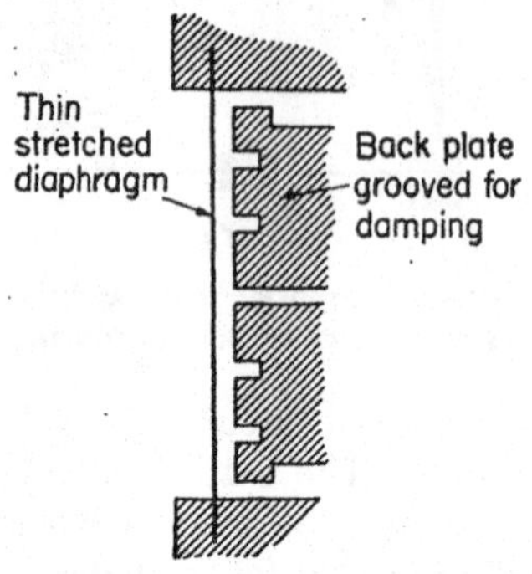

FIG. 10–8. The capacitor microphone.

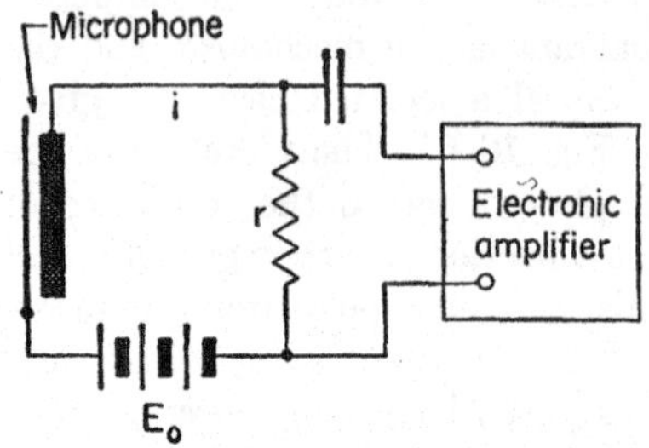

FIG. 10–9. Essential circuit for a capacitor microphone.

the other plate being relatively thick and rigid (Fig. 10–8). With waves of small amplitude, the capacity C may be made to vary sinusoidally according to the equation

$$C = C_0 + (C_m)_s \cos \omega t, \qquad (10\text{–}8)$$

where C_0 is the quiescent capacity and $(C_m)_s$ is the maximum value of the increment of capacity due to the effect of a particular impinging wave. The microphone is connected in series with a resistor and a polarizing d.c. potential, as indicated in Fig. 10–9. The current i in the microphone-resistor loop is determined by the differential equation

$$ri + \frac{1}{C}\int i\,dt = E_0, \qquad (10\text{–}9)$$

where E_0 and r are the values of the d.c. potential and the series resistor, respectively, and C is the instantaneous value of the capacity in the presence of the wave. If C from Eq. (10–8) is introduced into Eq. (10–9), the solution for small amplitude diaphragm motions for the steady state a.c. component i_{ac} yields,

$$i_{ac} = \frac{E_0(C_m)_s}{C_0\sqrt{r^2 + (1/\omega C_0)^2}} \cos(\omega t + \alpha). \qquad (10\text{–}10)$$

The microphone is thus seen to supply an open circuit emf of

$$e = E_0 \frac{(C_m)_s}{C_0} \cos \omega t. \tag{10-11}$$

For small amplitude motions, e will be proportional to the diaphragm amplitude x_m, since the capacitance $(C_m)_s$ will be nearly linear with x_m. According to Eq. (1-33), the amplitude x_m of a particle (in this case an "effective" particle) executing forced vibrations will be proportional to $1/\omega Z$, where Z is the mechanical impedance of the particle system. If the microphone diaphragm is stretched so that its fundamental resonance is at the upper end of the audible spectrum, at lower frequencies it will offer to the sound wave an impedance Z which is largely reactive and of the capacitive type, of the form K/ω. Therefore with an acoustic pressure of constant peak value the amplitude of motion will be constant, quite independent of frequency. A typical response curve for a capacitor microphone is shown in Fig. 10-10. The resonance peak near 8000 cycles-sec^{-1} is clearly in evidence.

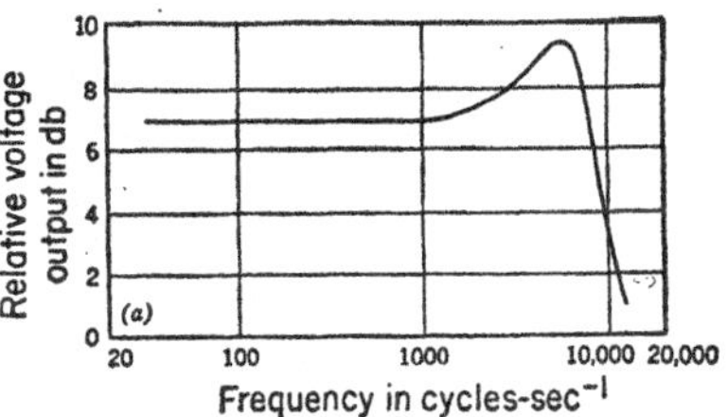

FIG. 10-10. Typical response characteristic of a capacitor microphone.

The chief drawback to the use of the capacitor microphone in applied acoustics is the high electrical internal impedance of a capacitive nature. This necessitates placing an electronic amplifier very close to the microphone, since with long lines of high distributed capacitance, the electrical output will otherwise be very low. The necessity for a relatively high polarizing potential is also a detriment. In spite of these disadvantages, the capacitor microphone is widely used for precise acoustical measurements. The response is uniform, the calibration may be relied upon, and in view of the small physical size of some of the recent designs, diffraction difficulties can be held to a minimum.

3. *Electrodynamic microphones.* Microphones of the electrodynamic type are rugged and may be designed to have a wide frequency response. Two general varieties are in use, the moving coil and the ribbon type. The electrical output in both cases is the result of the motion of a conductor in a magnetic field, the conductor being in the first type a helical coil and in the other a flat metallic ribbon or strip (Fig. 10-11). Since the potential developed is due to electromagnetic induction, it is, in either case, proportional to the *velocity* of the conductor.

For the velocity of the moving system to be independent of frequency, assuming constant sound pressure, the mechanical impedance must also be independent of frequency. This means, according to Eq. (1–34), that Z must be primarily *resistive*, rather than reactive. The usual design achieves this end by heavily loading the diaphragm with an acoustic network in the body of the microphone. The important resistive element is often supplied by a porous silk membrane placed close to the rear of the diaphragm. As a result, the motion of the latter is more than critically

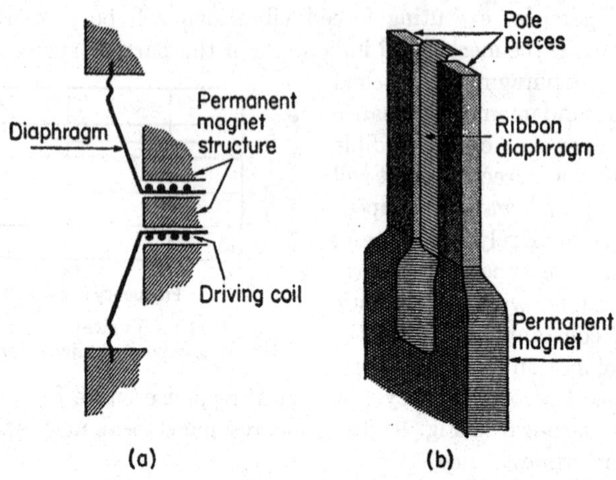

FIG. 10–11. (a) Moving coil electrodynamic microphone. (b) Ribbon microphone.

damped and the response can be designed to be practically free of peaks and valleys.

Both the moving coil and ribbon types of microphone have relatively low internal impedance (that of the ribbon type is lower). The use of long lines is perfectly feasible with the use of impedance shifting transformers at the microphone. No polarizing potential is necessary, of course.

The ribbon type of construction is especially adapted to a microphone of the *pressure gradient* type. All that is necessary is to expose both sides of the ribbon to the wave; since the ribbon is narrow, the phase of the acoustic pressure is virtually constant across its width. In this microphone the moving diaphragm (the ribbon) is also the electrical conductor, the induced potential appearing across its ends.

Inspection of Eq. (2–19e) shows that for plane waves having constant maximum pressure (p_m), the pressure gradient ($\partial p/\partial x$) will be proportional to $1/\lambda$ or to the frequency f. If the ribbon is designed to have a mechanical impedance which is essentially a mass reactance ($X = \omega m$), its velocity response $\dot{x}$ to the actuating force associated with the pressure gradient will be independent of frequency. Therefore, for a fixed microphone orientation, its electrical output ($\propto \dot{x}$) will be a direct measure of either the sound pressure or of the particle velocity in the wave, since for plane waves these two qualities are proportional to each other. To give to the ribbon a mechanical reactance of the mass type, the resonance frequency is adjusted to be at the *lower* end of the audible spectrum.

Many practical microphones of the ribbon type may be converted so as to act as nondirectional *pressure* detectors. To accomplish this, a simple mechanical arrangement is provided whereby the wave is denied access to the back of the ribbon.

4. *Crystal microphones.* Many nonmetallic crystals become electrically polarized with deformation of the crystal shape and a variety of microphone designs are based on this piezoelectric effect. If metal foil is cemented to certain surfaces, the potentials developed can be applied to the input circuits of an electronic amplifier, as in the case of other electrical microphones. The type of deformation and its relation to the direction of the crystal axes and planes determine the magnitude of the resulting potential. Deformations of the bending, shear, and compression type have been utilized, the change of shape being brought about by means of linkages arranged between the crystal and a diaphragm exposed to the sound wave.* Such a diaphragm type of crystal microphone is shown in section in Fig. 10–12.

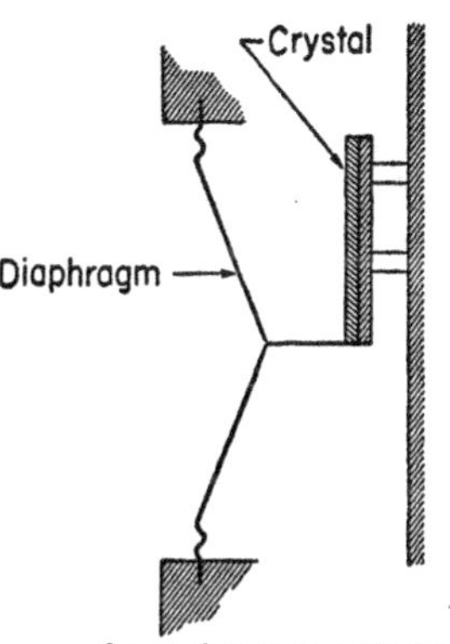

FIG. 10–12. Crystal microphone of the diaphragm type.

The alternating voltage appearing on the foil cemented to the crystal surfaces, when the deformation is periodic in nature, is proportional (for small deformations) to the *amplitude* of the deformation. Hence, as for microphones of the capacitor type and for the same reasons, it is desirable to arrange the over-all fundamental mechanical resonance of the system to be at some high audible frequency. This can readily be done.

In order to eliminate the undesirable resonances peculiar to a diaphragm, a directly actuated crystal microphone may be constructed. In this case the sound waves impinge directly on the crystal surfaces. The output

* See Olson, *Elements of Acoustical Engineering*, p. 181, D. Van Nostrand Co. (1940).

voltage of this microphone is usually much smaller than for one actuated by a diaphragm.

The potentials developed at the terminals of a crystal microphone, especially of the diaphragm type, are considerably greater than for many electrodynamic microphones and hence less amplification is required. The most commonly used crystal, Rochelle salt, is a hydrate, and the crystal must be carefully housed to prevent dehydration. In addition, temperatures above 115°F will permanently injure the crystal. Nevertheless, by proper design, the crystal microphone can be made into a rugged precision instrument, especially if some of the newer, more stable crystals are used. One such laboratory tool reveals an absolutely flat frequency characteristic (for constant sound pressure) from 20 to 20,000 cycles-sec^{-1}.

10–11 Relative sensitivities of different types of microphones. It is interesting to compare the sensitivities of the three principal types of microphone just described. For this purpose it has become common to specify the electrical response of a microphone to a free-field rms acoustic pressure of 1 dyne-cm^{-2}. With acoustic pressures of this magnitude, the output voltage of any microphone is very small, being of the order of 10^{-3} or 10^{-4} volt. It is convenient to use the decibel scale to specify this voltage sensitivity. The output voltage E, compared with 1.0 volt, using the above standard acoustic pressure, may be expressed in decibels as

$$\text{Output voltage in db} = 20 \log_{10} \frac{E}{1.0}.$$

On this basis the appproximate output of the three principal types of microphone may be stated.

Type of microphone	*Approximate output voltage in db above 1 volt, for a pressure of 1 dyne-cm^{-2}*
Capacitor	−50 db
Electrodynamic (after output impedance is brought to 500 ohms with a transformer)	−80 db
Crystal (diaphragm type)	−55 db

10–12 The calibration of microphones. For many purposes in the laboratory a microphone is simply a comparison device, used, for instance, to compare the acoustic pressure at one point in space with that at another. For such uses, especially if one frequency only is to be used, no absolute calibration is necessary. There are many problems, however, in which it is necessary to know the acoustic pressure corresponding to a given voltage appearing at the output of the amplifier. The so-called "gain" (ratio of output to input voltage) of an electronic amplifier is readily determined. But the exact relation between the potential delivered by the particular microphone used to the acoustic pressure at the diaphragm or, perhaps more important, to the acoustic pressure at the point *before* the introduction of the microphone, is quite another matter. For certain simple geometry, such as that of the ribbon microphone, the pressure-output voltage relationship may be computed with fair accuracy. In any case, careful experimental calibration is called for.

For many years the calculated response of the Rayleigh disk was the standard of comparison for free-field measurements. As has been pointed out, the Rayleigh disk is not a sensitive detector, and in recent years it has been observed that it is subject to many errors of a second-order variety. Consequently, ever since the published work of Dubois * and of MacLean † on the "reciprocity method" of calibration, this newer technique has been widely used, instead of comparison with a primary standard. We shall describe this method briefly.

10-13 The reciprocity method for calibrating microphones. The principle of reciprocity, well established in electrical circuit theory, is basic to the calibration technique bearing the same name. It may be shown that for any passive four-terminal network, no change of observed current will take place if the generator and ammeter, shown in Fig. 10-13a, are interchanged as indicated in Fig. 10-13b. To test this principle, a circuit which can be "reversed" in this sense is necessary (any ordinary circuit will have this property). It is possible to construct many kinds of sound sources or sound detectors that possess electroacoustic reversibility, i.e., the same device can be used with equal facility as a source of sound or as a receiver. We call such a device a reversible *transducer*, a transducer being any agent capable of receiving power from one system and transferring it to a different system.

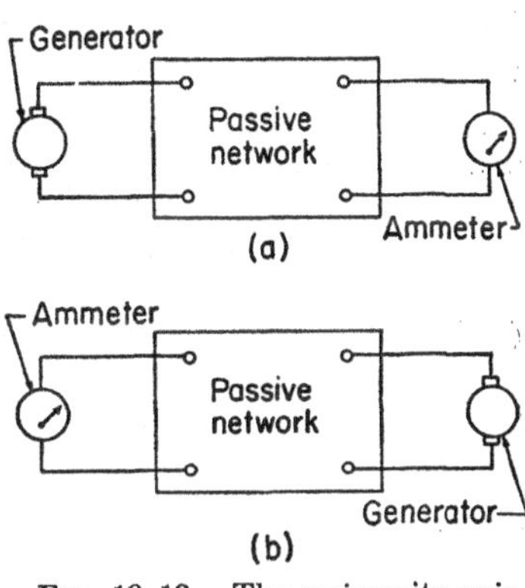

FIG. 10-13. The reciprocity principle.

* Dubois, *Revue d'acoustique* **2,** 253-287 (1932).

† MacLean, *Jour. Acous. Soc. of Amer.* **12,** 140-146 (1940).

As we have seen, it is possible to write the electrical analogs of many of the quantities important in acoustical and mechanical phenomena. Similarly, analogous "circuit" equations may be written for the behavior of acoustical and mechanical systems. When a system is part acoustical, part mechanical, and part electrical, unifying equations for the whole may also be written, although we have given no example of this procedure. It is therefore possible to represent a reversible transducer as coupled or "connected" to the surrounding medium and acting either as a generator, delivering power to the medium when acting as a detector, or as an impedance to which power is being delivered.

The following experimental technique is based on the above point of view. To calibrate a microphone it is necessary to have also a small reversible transducer and a small source of virtually spherical waves. Measurements are made in free space or in an anechoic chamber. The microphone T_x and the transducer T_r (used as a detector) are placed successively at the same distance d from the source T. The ratio of the two open-circuit voltages, E_x/E_r, is noted. Then T_r is driven with a constant known alternating current I_r, T_x is placed the same distance d from T_r, and the open circuit voltage E'_x is noted. Data sufficient for the calibration of the microphone have now been obtained. The circuit representing the transducer and its coupling to the medium (the free field) is drawn and its equation set up.* The final expression for the calibration constant M_x, representing the ratio of the open circuit voltage at the microphone to the sound pressure present before the microphone was introduced, is

$$M_x = \sqrt{\frac{E'_x}{I_r}\left(\frac{E_x}{E_r}\right)\frac{2\,d\lambda}{\rho_0 c}} \times 10^{-7} \text{ volt-dyne}^{-1}\text{-cm}^2. \qquad (10\text{–}12)$$

(It is also possible to compute the calibration constant of the transducer.)

It should be noted that no previous calibration of either T or T_r is implied or necessary. While there are some restrictions on the method as to frequency (the frequency may not be too high), the technique is remarkably straightforward. The distance d is in some doubt largely because of the finite size of the transducer. The errors in calibration, however, are in general small, and the computations take account of diffraction effects, always present in free-space measurements.

The reciprocity technique may be extended to the "closed chamber" type of calibration mentioned earlier in this chapter in connection with the thermophone and the pistonphone. As in free space applications, no primary standard source of pressure is required for the use of the reciprocity

* See Beranek, *Acoustic Measurements*, pp. 116–122, for details of this analysis.

principle. The closed chamber calibration of a microphone must always be corrected if the instrument is to be used for free-space measurements, since diffraction effects then enter to alter the pressure pattern.

10-14 Measurement of frequency in a wave. The frequency of a pure sinusoidal wave may be measured indirectly by determination of the wavelength λ and the propagation velocity c, using the relation $f = c/\lambda$. The wave velocity in air can be determined, as described in Chapter 6, by the use of any convenient source of known frequency. Knowing the velocity c, the wavelength λ in the disturbance can be measured by setting up a stationary wave pattern in front of a reflecting surface. The wavelength will be twice the distance between nodes (the latter points having been located with a microphone of small dimensions).

The frequency in the wave may be compared directly with that of a calibrated laboratory oscillator of the fixed frequency type or, more conveniently, one whose frequency is continuously variable. Fixed primary standards may be of two general types. The older form consists of a temperature-controlled tuning fork gently driven by means of electrical feedback circuits; a very pure sinusoidal voltage may be extracted from the motion of the prongs by induction. By the use of special alloys, the variation of fork frequency with temperature can be minimized. If the temperature is controlled to within 0.1°C, the frequency can be maintained constant to within one part in 100,000.

A frequency standard of somewhat higher precision makes use of an electronic oscillator whose frequency is controlled by the use of a quartz crystal. Such crystals are cut to resonate in the neighborhood of 50 to 100 kc, a frequency which is, of course, far above the audible range. Precise lower frequencies within the audible band may be obtained by means of "frequency dividing" circuits coupled to the oscillator. The latter are essentially multivibrator circuits, which themselves generate oscillations of the relaxation type, very rich in harmonics which are all multiples of the fundamental. If the master crystal-controlled oscillator has a frequency near that of one of the harmonics generated by the multivibrator, the latter will "lock in" with the master oscillator and so will the complete harmonic series. In this way frequencies lower than that of the master oscillator will partake of a similar high precision. Constancy to within one part in 10^7 may be attained.

Generators of alternating voltages of continuously variable frequency are often used in acoustic measurement. Such generators are considered to be secondary standards whose calibration must be checked with primary standards of the kind just described. Their frequencies may rarely be assumed known to within 1%. Two general types of electronic circuits are used, that of the beat frequency oscillator and that of the R–C oscillator. The first employs two radio frequency oscillators whose outputs are coupled into a common circuit. If the separate frequencies are adjusted to differ by a small amount, the beat variation can be rectified and a difference frequency extracted. By varying the frequency of one of the oscillators a variable audio frequency can be obtained. The R–C oscillator is small, compact, and easy to adjust, although somewhat less stable than the beat frequency oscillator (at least at the higher frequencies). This interesting circuit employs no inductances, but instead utilizes the resonant properties of a resistive-capacitive network, together with electronic feedback circuits. It is particularly useful at very low frequencies, since it can be made to operate with good wave form at frequencies of less than 1 cycle-sec^{-1}.

For comparing an unknown frequency with any laboratory standard, the most convenient instrument is the oscilloscope. The voltage output of the standard frequency generator is applied to one pair of deflecting plates while the amplified voltage from a microphone located in the sound field is applied to the other pair of plates. As explained in Chapter 1, the spot due to the electron beam will then execute Lissajous' figures. The interpretation of these figures furnishes a ready means of frequency identification. The simplest procedure, if the standard frequency source is continuously variable, is to turn the dial of the instrument until the pattern on the screen becomes a steady circle, ellipse, or straight line. The two frequencies are then the same, and the unknown frequency is simply read from the calibrated dial of the standard frequency generator.

Other types of instruments occasionally used for frequency comparison will not be described here, since the oscilloscope method is almost universally preferred.

10–15 Complex wave analysis. Sounds rarely consist of a single isolated frequency. Steady state sounds from a musical instrument, however, may consist of a fairly small number of distinct frequencies with no energy in the intervening frequency regions, and this type of complex sound can be broken down into an equivalent Fourier series. To determine the particular members of the series that are present, all that is required is an oscillographic record of the wave shape. This record can be analyzed by graphical methods or in one of the many mechanical or electronic analyzers referred to in Chapter 1. The precision and detail of this analysis depend upon the patience of the computer in the one case and upon the complexity and precision of the instrument in the other.

A convenient and rapid analyzing instrument for determining the individual frequencies and their amplitudes is the heterodyne type of analyzer, a simplified functional diagram of which is shown in Fig. 10-14. The output from a variable frequency oscillator (tuned to a frequency in the neighborhood of 50 kc) is combined with the signal voltage and the sum frequency is sent through a narrow band electrical filter. The latter may be designed for a band width of only a few cycles-sec^{-1}. To the output of the filter is connected an amplifier and a recording meter. A signal

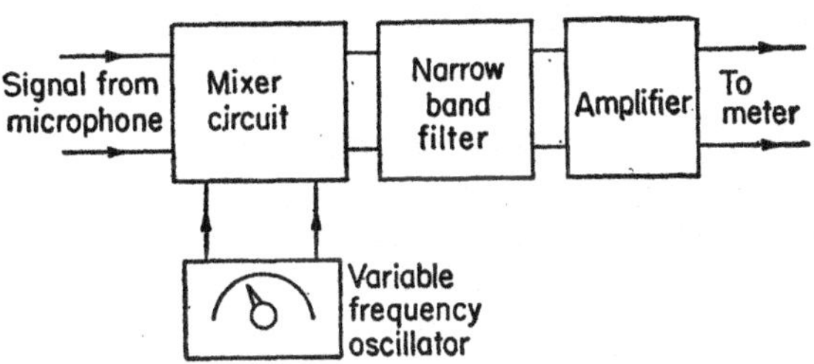

FIG. 10-14. Functional diagram of the heterodyne type of complex wave analyzer.

will reach the meter only when the frequency of the local oscillator, plus that present in the acoustic wave, lies within the narrow pass band. The setting of the local oscillator will obviously differ for every such frequency component in the wave, and consequently the dial can be calibrated in terms of the acoustic input frequency. The operation of the circuit is very similar in principle to the heterodyne circuit used in practically all modern radio receivers. The final recording in the analyzer may also be interpreted in terms of the relative amplitudes of the several components. This type of instrument is useful, too, for the study of noise, to be discussed in the next section.

Tuned mechanical reeds can be used to determine the frequencies and amplitudes of a complex sound wave, and a modern arrangement of this sort has been described by Hickman.* An optical beam reflected from a mirror attached to each reed was used to determine the amplitude of a particular frequency. Hickman used 144 reeds for the range of 50 to 3109 cycles-sec^{-1}.

The possibility of constructing an acoustic diffraction grating for the analysis of sound was mentioned in Chapter 4. Such a grating is bulky but quite convenient and rapid to manipulate. Meyer and Thienhaus † used a vibrating ribbon as a source analogous to the optical slit. The

* Hickman, *Jour. Acous. Soc. Amer.* **6**, 108–111 (1934).

† See Meyer, *Jour. Acous. Soc. Amer.* **7**, 88–93 (1935).

grating may be used for either a "bright line" spectrum (such as that characteristic of a violin note) or for the continuous spectrum typical of "noise."

The analysis of transient sounds presents a special problem. The acoustic grating and analyzers of the heterodyne and other types can, however, be adapted to the short times of observation involved in transient study. Oscillograph pictures and records can also be taken of short phenomena, to be analyzed at leisure.

The calibration of analyzers is a complicated procedure and current technical literature should be consulted for details.

10–16 Noise. The continuous acoustic spectrum. *Noise* is here defined as undesired sound. Sometimes the sound may be musical in nature but undesirable because it interferes with conversation or some other useful occupation. Such sound at times consists of a relatively small number of frequency components, but more often it is a heterogeneous mixture of frequencies and amplitudes, both of which change rapidly with time.

Because of its great practical importance as a masking agent of useful sound, the general nature of noise, both in the acoustical and in the electrical form, has been carefully studied from a mathematical and statistical point of view. Besides its own intrinsic interest, this study is important if one is to design instruments for correctly measuring noise level. The final indicating device is some sort of rectifying meter, different types of which respond differently to noise voltages of the random type. (See Beranek, *Acoustic Measurements*, for a full discussion.)

For noise assessment, much practical use is made of the sound level meter (noise meter), which includes a microphone whose angular pick-up range should be 90°. The microphone is connected to an electronic amplifier in whose circuit are included three standard correction networks. It will be remembered from the appearance of the loudness level contours of Fig. 9–8 that the sensitivity of the ear varies greatly with frequency. In addition, the shape of the loudness level contours is a function of the magnitude of the level. The correction networks are designed to give to the amplifier a gain characteristic which is the inverse of the ear characteristic at three definite arbitrary levels, i.e., 40 phons, 70 phons, and 100 phons. At the output of the amplifier is usually connected a rectifier and a d.c. meter whose reading gives the sound level. The characteristics of the whole circuit must be such as to indicate *the square root of the weighted sound pressures of the different single frequency components in the complex wave.*

The readings of the American type of sound level meter, described above, are only true loudness level readings at the three standard levels. Since the loudness contours vary in shape continuously as the level is changed, the use of only three correction networks means that the readings are approximate at other levels. It is important to note that the contours of Fig. 9-8 were derived using only one frequency at a time. There is much evidence that the ear does not judge sounds of wide frequency range in the same manner as it does single frequency tones. The sound level meter in its present practical form is a compromise.

It is usual to introduce a time constant somewhere in the electrical circuit of the sound level meter so that the fairly rapid fluctuations of the instantaneous sound level are averaged out at the meter. Depending on the purpose to which the instrument is to be put, this time constant may be varied considerably. For steady sounds, the calibration must in any case agree with the requirements stated above.

Approximate sound levels for certain typical environments are given in Table 10-1. It is to be noted that even in a "quiet" office the level is as much as 30-40 db above the threshold of audibility. The average person is so used to sound levels of this order of magnitude that a really quiet environment, such as exists within an anechoic chamber, is apt to disturb rather than to relax.

When the distribution of energy in the different parts of the audible spectrum must be known, a selective instrument is needed. In this case, band pass filters may be included to cover the frequency range in steps, the frequency width of each filter being a matter of choice.

TABLE 10-1 *

Typical sound environment	*Sound level in db*
Threshold of pain	120-130
Riveting machine 30-40 ft away	100
Subway with train passing	90
Average city street	70
Average restaurant	60
Average conversation 3 ft away	60-70
Outdoor minimum in city	30-40
Quiet office	30-40
Outdoor minimum in country	10
Threshold of audibility	0

A spectrum of particular theoretical interest is the so-called "white noise" spectrum. Such a distribution is uniform over the whole frequency

* Adapted from Colby, *Sound Waves and Acoustics.* Henry Holt & Co. (1938).

range. It may be conveniently produced electrically by amplifying the noise originating at the input of an electronic amplifier. Corrective networks may be necessary to make the output truly "white." Such a generator, connected to a loudspeaker, constitutes a useful laboratory test instrument.

10–17 The measurement of acoustic impedance. As we have seen in Chapters 5 and 8, the concept of acoustic impedance is most useful in problems concerned with the radiation and absorption of sound energy. While in the path of plane and spherical waves in free space the value of z_s may be predicted for certain simple types of sources, there are many cases where this cannot be done. In particular, the important impedance z_n, characteristic of partially reflecting surfaces, must be determined experimentally. We shall describe briefly a few of the more important methods of experimental determination.

1. *Direct measurement of p and $\dot{\xi}$ at a surface.* Since specific acoustic impedance is defined as the ratio of the pressure to the particle velocity, it may be computed both in magnitude and in phase once p and $\dot{\xi}$ are known. The pressure can be measured with probes small enough to avoid serious distortion of the sound field. A velocity type of microphone (the pressure gradient type) could be used to measure $\dot{\xi}$, but at the present time none of small enough dimensions is available. A more fruitful method of approach for porous surfaces is to force a known volume of air through the surface per unit time (this is the "volume current" $\dot{X}$) and to measure p at the surface with a probe tube. The analogous impedance z_a can then be computed by the relation $z_a = p/\dot{X}$. A small vibrating diaphragm and the surface of the sample may be placed at opposite ends of a very short cylindrical tube. The volume current is calculated from the known distribution of velocities over the surface of the circular diaphragm.

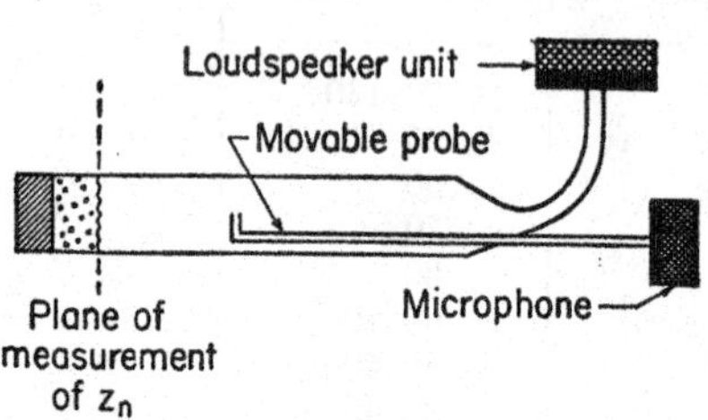

FIG. 10–15. Laboratory arrangement for the measurement of z_n.

2. *Acoustic transmission line methods.* Of the several methods falling under this classification, that which involves the exploration of a stationary wave pattern is the most widely used. The fundamental arrangement is indicated in Fig. 10–15. The sample at the surface of which the acoustic impedance is to be determined is placed at one end of a cylindrical tube, while into the other end is piped the energy from a loudspeaker unit. A movable probe coupled to a de-

tecting microphone may be slid along the axis of the tube. The probe commonly used for acoustical sampling is simply a small-bore tube whose outside dimensions are sufficiently small so as not to disturb the pressure distribution in the region being studied. In general, its effect is to attenuate the wave by the time it reaches the microphone diaphragm. This attenuation, as a function of frequency, can be calculated theoretically and also checked experimentally; hence the electrical output of the microphone can be corrected to indicate the acoustic pressure at the entrance to the probe.

In Section 8–7 it was shown that with the *partial* reflection of plane waves at the end of a pipe, a pattern will be produced within the pipe consisting of two superimposed sets of stationary waves whose antinodes do not coincide. The pressure amplitudes at these antinodes are

$$(p_m)_{\text{total}} = (p_m)_i + (p_m)_r \tag{10–13}$$

and

$$(p_m)'_{\text{total}} = (p_m)_i - (p_m)_r, \tag{10–14}$$

where the subscripts i and r refer respectively to the incident and to the reflected waves. By sliding the probe along the tube axis, the positions of greatest and least acoustic pressure can be located and the total pressures on the left-hand side of Eqs. (10–13) and (10–14) can be measured. The maximum pressures in both the incident and the reflected waves can thus be computed from these same equations. By Eqs. (8–23) and (8–24) the ratio $(p_m)_r/(p_m)_i$ is directly connected with the normal specific acoustic impedance z_n at the boundary of the reflecting material. If the material is such that z_n is essentially real, and this is quite often the case, the angle θ in Eq. (8–23) is zero, and we may write Eq. (8-24) as

$$z_n = \rho_0 c \left(\frac{1 + M}{1 - M}\right). \tag{10–15}$$

In this way we can compute the impedance z_n from the experimental pressure data.

Actually, the nature of the stationary wave pattern is significantly affected by attenuation effects at the walls of the tube. In addition, z_n may well be complex. Under these conditions, the equations given in Chapter 8 are not sufficient to determine the real and imaginary parts of z_n, but additional equations may be deduced to make the computation possible. It is necessary also to measure the distances from the reflecting surface at which the nodes and antinodes occur, as well as to know the attenuation properties of the tube.*

---

* See Beranek, *Acoustic Measurements*, pp. 321–329.

It should be mentioned that when z_n has been measured at the surface of an absorbing material, the absorption coefficient for normal incidence, α_n, has also been measured indirectly. Once the pressure ratio M has been determined, the value of α_n can be computed from the relation

$$\alpha_n = 1 - M^2.$$

Since θ does not enter, the determination of α_n is somewhat simpler than the determination of z_n.

To determine the Sabine absorption coefficient, which assumes sound waves arriving from all directions, a quite different technique is usually employed. In Chapter 12 this measurement will be discussed in connection with the reverberation properties of rooms.

3. *Bridge methods.* There are a number of methods of measuring acoustic impedance which are closely analogous to the bridge methods of electrical measurement. Many of these arrangements involve considerable electrical as well as acoustical detail, and their operation would be difficult to make clear within our allotted space. One of the earlier types of bridges, devised by G. W. Stewart,* is almost wholly acoustic in character and its theory is fairly simple. We shall therefore describe it briefly, as illustrative of the measurement possibilities along this line.

The construction of the bridge is shown in Fig. 10–16. Two long, straight, cylindrical tubes are coupled, at one end, to loudspeaker units, the

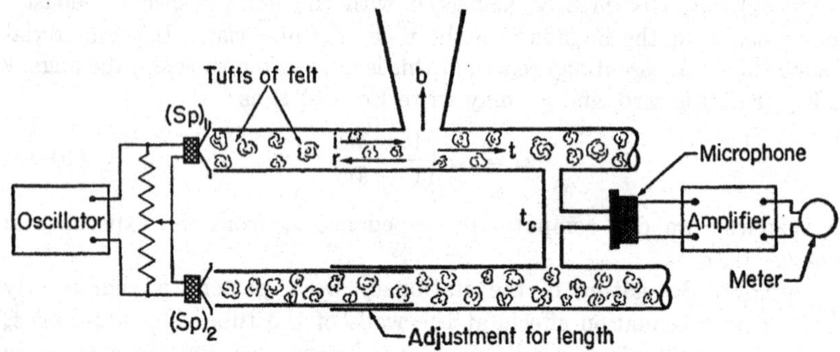

FIG. 10–16. Acoustic impedance bridge. (*After* Stewart)

two units being identical in design. The loudspeakers are driven from a common oscillator by means of a dividing network, so that the relative acoustic energy delivered to the tubes can be continuously adjusted. At some distance from the loudspeaker ends of the tubes, an acoustical con-

* Stewart, *Phys. Rev.* **28**, 1038 (1926).

nection is made by means of the tube t_c. Another pipe, leading to a microphone, is attached at the center of this branch tube. The microphone has its own amplifier and measuring meter. By means of a telescoping arrangement, the length of one tube can be varied as shown.

This bridge is particularly well suited to the measurement of the acoustic impedance at the end of a horn or acoustical conduit. The device to be used is attached at one side of the upper straight tube as shown. Assuming the loudspeakers to be operating at a fixed frequency, we may express the principles of continuity of pressure and of volume current $\dot{X}$ (equal to $S\dot{\xi}$) at the junction in this way:

$$\left.\begin{aligned} p_i + p_r &= p_b = p_t \\ \text{and} \qquad \dot{X}_i - \dot{X}_r &= \dot{X}_b + \dot{X}_t, \end{aligned}\right\} \tag{10-16}$$

where the subscript i refers to the wave incident at the junction, originating at speaker Sp_1, the subscript r to the reflected wave, the subscript b to the wave progressing into the horn or conduit, and the subscript t to the wave traveling towards the far end of the straight tube. To eliminate complicating reflections at the ends of the long tube, the latter (and its parallel neighbor) is partially filled with tufts of felt.

In addition to the above relationships, we may write

$$\begin{aligned} \dot{X}_i &= \frac{p_i S}{\rho_0 c}, \qquad & \dot{X}_r &= \frac{p_r S}{\rho_0 c}, \\ \dot{X}_t &= \frac{p_t S}{\rho_0 c}, \qquad & \dot{X}_b &= \frac{p_b}{(z_a)_b}. \end{aligned} \tag{10-17}$$

The quantity $(z_a)_b$ represents the analogous impedance at the entrance to the side branch, and is the quantity to be determined. The quantity S is the cross-sectional area of the bridge conduit.

By combining Eqs. (10-16) and (10-17), we obtain the ratio

$$\frac{p_t}{p_i} = \frac{2(z_a)_b}{2(z_a)_b + (\rho_0 c/S)}. \tag{10-18}$$

In this equation we must recognize that the pressures are the instantaneous values in the complex form, and that p_t and p_i are not in general in phase. We express this phase relationship by stating that

$$p_t = p_i \frac{(p_m)_t}{(p_m)_i} \epsilon^{j\theta},$$

where θ is the phase angle, and $(p_m)_t$ and $(p_m)_i$ represent peak values of the acoustic pressures. Hence Eq. (10-18) may be written

$$\frac{(p_m)_t}{(p_m)_i} \epsilon^{j\theta} = \frac{2(z_a)_b}{2(z_a)_b + (\rho_0 c/S)}. \tag{10-19}$$

Equation (10–19) is the fundamental relationship for this type of bridge. If we can measure the ratio $(p_m)_t/(p_m)_i$ and the phase angle θ, we can, by reducing Eq. (10–19) by the methods of complex algebra, compute the real and imaginary parts of $(z_a)_b$.

The experimental procedure is as follows. With the branch b removed and the hole closed, the voltages applied to the speakers are varied until, due to interference, the detecting system attached to the cross tube reads zero. (A small adjustment of the length of the lower tube may be necessary to equalize the phases.) Then the branch b is introduced and the balance again made, both for pressure amplitude and for phase. The ratio $(p_m)_t/(p_m)_i$ may be computed by

$$\frac{(p_m)_t}{(p_m)_i} = \frac{E_1/E_2}{E_1'/E_2'}, \tag{10–20}$$

where E_1/E_2 is the ratio of the two voltages applied to the loudspeakers with the branch out, and E_1'/E_2' is the ratio with the branch connected. To obtain the phase angle, note is made of the change of length, d, of the lower tube necessary to equalize the phases after the branch is inserted. The phase angle (for which compensation is made in the lower tube) is then

$$\theta = 2\pi \frac{d}{\lambda}. \tag{10–21}$$

With this information, $(z_a)_b$ can be computed in the manner indicated above.

Measurements made on this bridge are subject to serious errors under certain conditions. The analyzing of the chief sources of such errors is left for the reader.

10–18 Conclusion. No attempt has been made in this chapter to give a complete detailed survey of the many special techniques of acoustical measurement. The apparatus of measurement is increasing steadily in variety and is constantly being improved. To be well informed, it is necessary to keep abreast of current technical literature.

PROBLEMS

1. Measurements of the absorbing properties of the wedge type of wall structure used in modern anechoic rooms (Fig. 10–1) indicate a reduced effectiveness at the lower frequencies (see the graph of Fig. 10–2). A major factor in this behavior is the phenomenon of diffraction. Explain.

2. From the graph of Fig. 10–2, find the absorption coefficient for the wedge type of wall surface at a frequency of (a) 100 cycles-sec^{-1}, (b) 1000 cycles-sec^{-1}.

3. (a) In the case of the pistonphone (Section 10–5), why should all dimensions

be small compared with the wavelength? (b) Derive Eq. (10–1).

4. (a) If there were no slots in the plate of the electrostatic actuator (Section 10–6), what error would enter into the calibration of the microphone? (b) Would the calibrated sensitivity of the microphone then be higher or lower than the actual sensitivity when operating in a field of sound?

5. (a) For the Rayleigh disk, why is the position parallel to the direction of stream flow unstable? (b) Why is the position perpendicular to the stream flow stable? Make this clear with a vector picture.

6. The double-button carbon microphone is briefly described in Section 10–10. (a) Draw a simple circuit for such a microphone, making clear the push-pull action. (b) Explain just how the push-pull action eliminates *even harmonic* distortion due to the nonlinear behavior of each button, using an argument based on the graphical representation of the current variation for either button.

7. The spacing between the two metallic surfaces in the capacitor microphone should be small compared with the diameters. Why? (There is more than one reason.)

8. (a) From Eq. (10–11), why is a polarizing potential needed with the capacitor microphone? (b) For the voltage across the resistor in Fig. 10–9 to be independent of the frequency in the sound wave, the value of the resistance must be very large. Why? (c) Would the presence of distributed capacitance in a cable between the microphone and the resistor affect the *sensitivity* of the microphone at any one frequency? (d) Would this same capacitance affect the *uniformity of response to waves of different frequency?* (e) Answer parts (c) and (d) for the case where the cable capacitance operates between a tap on the resistor and one end.

9. In view of the differential sensitivity of the ear for sounds of variable intensity, is the peak in the curve of Fig. 10–10 significant? (See Chapter 9.)

10. Practical microphones of the pressure type exhibit somewhat greater sensitivity at high than at low frequencies, due to diffraction effects. (The effect is greatest with normal incidence.) Show that this behavior is reasonable in the light of the radiation properties of an acoustic piston, making use of the general reciprocity principle.

11. (a) Discuss all the possible sources of error in the measurement of z_a by means of an acoustic bridge of the type designed by G. W. Stewart. (b) If the two speakers had somewhat different resonant frequencies, would the precision be affected? (c) Would one expect the precision to be greater at low frequencies or at high frequencies? Explain.

CHAPTER 11

REPRODUCTION OF SOUND

11-1 Introduction. With the almost simultaneous invention of the telephone and the phonograph there began a historic development in applied physics which was to draw strongly upon three important fields, i.e., those concerned with the principles of electricity, of mechanics, and of acoustics. The appearance of radio telephony in the 1920's furnished still further stimulus to experimentation and improvement, as did also the revived interest in the phonograph in its newer, electrical form.

It is interesting to note that the acoustical aspects of the reproduction of speech and music were at first sadly neglected. For many years the telephone transmitter and receiver remained practically in their original form as electromechanical and electroacoustic devices. The early radio loudspeakers were simply glorified telephone receivers, and almost any short horn that had some kind of flare was deemed adequate as a coupling agent.

This lack of interest in the acoustic link in the chain seems at first difficult to understand, in view of the long and honorable history of the subject of acoustics. The explanation of the delay is probably twofold. For one thing, the novelty and excitement of the appearance of the phonograph and the radio set were such that for a long time no serious objections were raised to the acoustic inadequacies. As the novelty wore off, listeners became more critical, and engineers began to think about the acoustical design, or rather, *lack* of design. The second reason for slow development along this line is that it took some time before the principles of the important "wave filter" of G. A. Campbell were applied to electromechanical systems. With the papers of A. G. Webster, W. P. Mason, A. C. Bartlett, and others, the concept of mechanical and acoustical impedance rapidly took shape, and the analogy between the behavior of electrical circuits and that of mechanical and acoustical systems was recognized as a tool of great usefulness. Eventually this knowledge began to affect the design of transducers of all sorts, such as the mechanism of the phonograph pickup, the telephone receiver and transmitter, and the loudspeaker constituting the coupling device between the electrical circuit and the sound medium.

Important advances have recently been made in the design of electroacoustic transducers, but it is still correct to call these the weakest link in the chain of reproduction of speech and music. This is not due to lack of attention and study in recent years, but to the intrinsic difficulties in effectively transforming electrical energy into acoustical energy *over a wide range of frequencies*. In this chapter, we shall try to point out some of the chief difficulties, and to describe a number of practical solutions.

11-2 The general problem. The problems connected with the reproduction of sound over a wide range of frequencies really begin at the source itself. Whether the final acoustic output is from a phonograph, a public address system, or a radio loudspeaker, any initial pressure variation is picked up, in modern systems, by an electrical microphone and is then amplified electronically. Subsequently, the signal may have a varied history. In the phonograph it may be supplied to the cutter head of the recording mechanism or to the magnets of a tape recorder. In either system, the recorded information must eventually be extracted from the record or the tape in electrical form and again amplified. Eventually electrical power is delivered to an electroacoustic transducer (actually an electro-mechanico-acoustic transducer) and sound waves are set up in the air which it is hoped are faithful replicas of the original waves in front of the microphone. In the radio set, there is the additional complication that between the first and last amplifier appear the radio transmitter and the radio receiver, with all their complicated circuits and behavior.

Since this is a book on acoustics, we would naturally choose to concentrate on the acoustic aspects of this complicated chain of transformations, and this chapter is devoted primarily to the final transformation back to sound waves. But a loudspeaker is primarily a mechanical system, free to vibrate, and the forces which make it vibrate are electrical or magnetic, or perhaps electromagnetic in nature. We are therefore led inevitably to a consideration of the mechanical and electrical links in the chain. Indeed, as we shall see, some features of the electronic amplifier that delivers its energy to the final transducer are closely linked with the behavior of the loudspeaker, so that we shall have to make some general comments on that score in connection with our discussion of the transducer itself.

If the system described above were to handle vibrations (currents) of a single frequency only, the problem would be greatly simplified, especially from the acoustic point of view. For most applications, the electrical energy supplied to the final transducer is small, so that the matter of efficiency is of little interest. (We shall comment on efficiency later.) It is not difficult to build circuits and vibrating mechanical systems whose response to amplitude variations is essentially linear at any one frequency. Therefore a single sinusoidal electrical voltage at the input of the system can be amplified with little difficulty, and the electrical energy may be eventually transformed back into sound energy with little distortion of wave shape. Almost any type of loudspeaker, if not driven to too high an amplitude of motion, will give rise to a variable sound pressure which at any one frequency is a reasonably close replica of the electrical variation applied to it.

Speech and music, however, contain a wide range of frequencies. As we have seen in Chapter 1, the existence of a resonant frequency is one of the fundamental features of the motion of a particle free to vibrate. In the neighborhood of that frequency the response to a given outside periodic force may be large; at other frequencies the response will, in general, be much smaller. The radiating plate or diaphragm is more complicated in its behavior than is a particle, as we saw in Chapter 7; there may be *many* resonances. Suppose, then, that a voltage having a complex wave form that faithfully follows the shape of the acoustical wave is applied to the input of an electronic amplifier, the output of the latter being connected to an electroacoustic transducer. Both amplifier and transducer must be capable of handling all the frequency components that are present, so as to maintain their original relative prominences, and, one would expect, the original relative phases. Such a system must therefore satisfy the triple requirement of linearity with respect to amplitude variations, uniform over-all response over a frequency range of the order of 500:1, and a net over-all phase shift which is ideally zero degrees.

Any failure to meet the above requirements is called a source of distortion. As we have mentioned earlier, phase shift or phase distortion is not ordinarily important to the ear in sounds having a large number of assorted frequency components, which is fortunate, as zero over-all phase shift at the upper end of the audible spectrum is difficult and expensive to achieve. (It is difficult enough to minimize phase shift in electronic circuits; it is practically impossible to keep such shifts to a low value in the mechanical response of the transducer diaphragm, especially in the neighborhood of the inevitable resonances. See the discussion in Section 1–21.) It is now possible to build circuits and amplifiers which are linear for amplitude variations and have little or no frequency discrimination over a range greater than that to which the ear will respond. When it comes to the behavior of the electroacoustic transducer, however, the problem is much more difficult. In the next section we shall discuss some of the difficulties to be overcome.

11–3 An ideal transducer. Suppose the rectangular area in Fig. 11–1 to represent an acoustic piston whose mass and stiffness (K) are negligibly small. The mode of support involves no appreciable dissipative force.

Only the right-hand face is exposed to the air, so that it is a single source. The area S of its radiating face has a diameter large compared with the wavelength, and a beam of plane waves will therefore be produced.

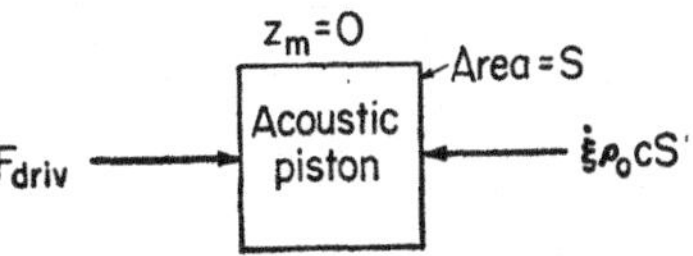

FIG. 11-1. Driving force and air reaction force for an acoustic piston of negligible mechanical impedance.

Under these conditions, the instantaneous driving force, F_{driv}, will always equal the total force due to the air, or

$$F_{\text{driv}} = pS = \dot{\xi}\rho_0 cS, \qquad (11\text{-}1)$$

where p is the acoustic pressure. Since the velocity $\dot{x}$ of the piston is equal to the particle velocity in the air near it, we may also write

$$F_{\text{driv}} = \dot{x}\rho_0 cS. \qquad (11\text{-}2)$$

The average power delivered by the driving force will then go entirely into wave motion and will equal

$$(F_{\text{driv}})_{\text{rms}}\dot{x}_{\text{rms}} = (\dot{x}_{\text{rms}})^2\rho_0 cS = (\dot{\xi}_{\text{rms}})^2\rho_0 cS. \qquad (11\text{-}3)$$

According to Eq. (11-1), the acoustic pressure in the radiated wave is proportional to the driving force. For a given value of the driving force, $\dot{x}$ (or $\dot{\xi}$) is constant and therefore, by Eq. (11-3), the acoustic power is constant. Hence, if the amplifier and driving mechanism are such that the force applied to the piston is proportional to the acoustic pressure in the original wave, independent of the frequency, it should be clear that the pressure and intensity in the radiated wave will follow the corresponding variations in the original disturbance. Such an acoustic piston may then be called an "ideal" radiator for complex sound waves.

The quantity $\rho_0 cS$ in Eq. (11-2) is the effective mechanical impedance of the air-loaded piston, or its "radiation impedance." Entirely apart from the effect of the air load, for any actual vibrating diaphragm there will be a mechanical impedance for the driven mechanism, due to its inertia, its stiffness, and the presence of frictional forces at the supports. If z_m is the mechanical impedance of the piston without the effect of the air, Eq. (11-2) must be written

$$F_{\text{driv}} = \dot{x}(z_m + \rho_0 cS), \qquad (11\text{-}4)$$

where z_m is, in general, complex. We may also write

$$(F_{\text{driv}})_{\text{rms}} = \dot{x}_{\text{rms}}\sqrt{(R + \rho_0 cS)^2 + X^2}, \qquad (11\text{-}5)$$

where R and X are the magnitudes of the real and imaginary parts of z_m.

With a constant driving force, the maximum value of $\dot{x}$ in Eq. (11-4) will not remain constant at all frequencies, due to the variations in z_m, whose reactive component is a function of frequency. Hence the average real power delivered by F_{driv} will vary with the frequency. Moreover, not all of this real power will be radiated as waves, because of the dissipative part of z_m, that is, R. The average radiated acoustic power is $(\dot{x}_{rms})^2\rho_0 cS$, which is less than the total power delivered by the driving force, i.e., $(\dot{x}_{rms})^2(R + \rho_0 cS)$. In addition to these deficiencies, any actual acoustic radiator will never have dimensions large compared with the wavelength except at the higher frequencies. At medium and low frequencies it will therefore become a source of diverging waves, the wave front shape becoming spherical at very low frequencies. For purposes of sound diffusion this is good, but the radiation efficiency drops at those frequencies, as shown by the graph of Fig. 5–11. The specific radiation impedance at the piston becomes complex at the lower frequencies, and its real part falls below $\rho_0 c$. All of these effects cause distortion of the radiated complex wave.

We must not neglect the electrical aspects of transducer design. Since transducers differ in this respect, we shall now consider the particular features of some specific driver mechanisms.

11–4 Early types of transducers. 1. *Magnetically driven iron reed.* To this classification belongs the ordinary telephone receiver, with its circular iron diaphragm clamped at the edge. The plate is made to vibrate by virtue of the varying magnetic force between it and the pole of an electromagnet which carries the signal current. Such a plate is attracted at each peak of the current, and to prevent this virtual doubling of the frequency of the plate vibration as compared with the frequency of the current variation, a constant flux component is introduced into the magnetic circuit by means of a permanent magnet. The signal current will then strengthen or weaken the pull associated with the total flux at a rate identical with that of the signal. An analysis of the behavior of the system shows that a sinusoidal variation in the magnet current will give rise to a periodic force containing frequencies which are harmonics of the current frequency, as well as the current frequency itself. By making the constant flux bias much greater than the differential flux caused by the signal, the harmonics can be suppressed. The driving force can be shown to be proportional to the value of the steady flux.

As an earphone transducer, this design is simple and adequate. The thick iron diaphragm has strong resonances near the middle of the speech band, which aid efficient reproduction of speech but which would cause considerable distortion in the reproduction of music. Heavier units can be designed to handle considerable power, but the defects remain. A magnetically balanced version of the reed mechanism was used in early loudspeaker design (Fig. 11-2), and this design, with a light armature coupled mechanically to a large paper cone, makes possible an acoustic radiator capable of better low frequency radiation and somewhat less marked resonance distortion. The fundamental resonant frequency in this design can be lowered almost to 100 cycles-sec^{-1}, a change in the right direction.

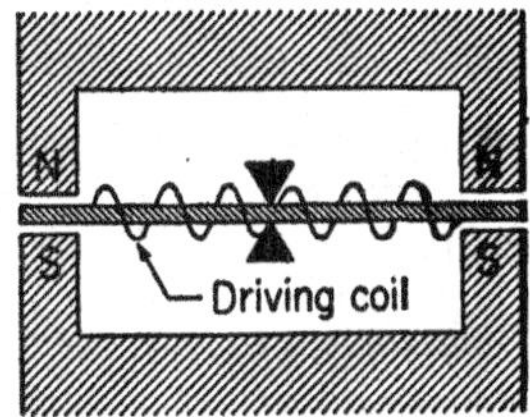

FIG. 11-2. Balanced armature type of loudspeaker mechanism.

2. *The electrostatic loudspeaker*. It is possible to employ the design of the capacitor microphone in an acoustic radiator. As in the microphone, a polarizing d.c. potential, E_0, is used. If a sinusoidal variation in potential, e, due to the signal, is then introduced in series with E_0, there will be a periodic force of attraction between the plates. If one of the plates is a flexible membrane, it will vibrate and radiate sound waves. As with the transducer just described, harmonics are generated which can be kept low in amplitude by the use of a high polarizing potential. The instantaneous driving force per unit area is given by

$$F_{\text{driv}} = \frac{E_0 e}{4\pi d^2},$$

where d is the spacing between the plates. (All units are electrostatic.)

To ensure a sufficiently large driving force, E_0 must be large (of the order of hundreds of volts) and d must be small. The membrane must then be stretched tightly to prevent the two plates from touching under the steady electrostatic force. A push-pull arrangement has been used to eliminate the steady force of attraction, setting the thin membrane between two stationary plates, each charged to the same potential. Even so, the device is fragile and requires its own potential source, so that this type of transducer has never been widely used.

3. *Piezoelectric-driven system*. A piezoelectric crystal is a reversible transducer. An alternating potential applied to sheets of foil or to metal plates cemented or clamped to a pair of crystal surfaces will produce mechanical deformation of the crystal. This motion can be transferred, by the use of a lever, to a separate plate or diaphragm which then acts as the sound radiator. The system is not well adapted to the radiation of large amounts of acoustical power in the range of audible frequencies, since the required amplitudes of motion are in danger of shattering the crystal. In the range of ultrasonic frequencies, however, the crystal transducer has been an exceedingly useful source of high intensity longitudinal waves. In this frequency region the required amplitudes are

smaller. Crystals may be cut so as to resonate at frequencies in the neighborhood of 50 kc-sec^{-1} or higher. The mode of internal vibration is in this case perpendicular to the face of the crystal, so that the crystal surface becomes the radiating surface. We shall have more to say of such generators in the next chapter.

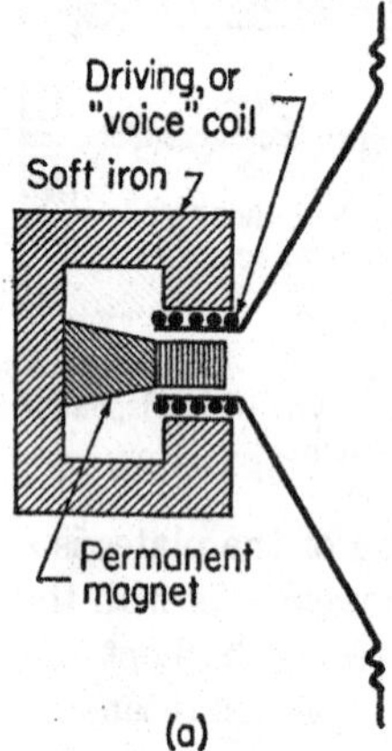

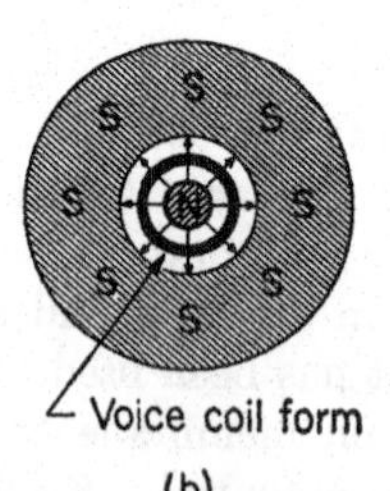

FIG. 11–3. Electrodynamic loudspeaker. (a) Essential parts. (b) End view of magnetic structure.

11–5 Transducer with electromagnetic drive. The electromagnetic type of drive mechanism has become almost universal among modern transducers designed for the audible range of frequencies, and we shall therefore discuss its operation in some detail. The essential parts of the mechanism are shown in Fig. 11–3. The vibrating diaphragm is in the form of a truncated cone, made of paper in the larger sizes or of light metal or plastic in units designed for use with a horn. At the base of the cone is attached a short cylindrical form, on which is wound a relatively small number of turns of wire. The cone is held in position by two elastic rings, as shown, so that the driving coil is normally held symmetrically within a radial magnetic field of constant strength. The source of the field may be either an electromagnet or a slug of permanently magnetic material. Connections to the coil are made by flexible leads.

Since the conductors making up the coil lie everywhere perpendicular to the radial field, the presence of an alternating current in the coil, constituting the signal current, will result in a periodic axial force upon the coil and therefore upon the cone to which it is attached. The axial vibration of the cone will then set up longitudinal sound waves in the air. The vibrating system, even with the cone exposed to the air, is always underdamped, so that there is a fundamental resonant frequency. This frequency, for practical loudspeakers, may range from several hundred down to 30 or 40 cycles-sec^{-1}, depending on the size of the cone and its particular construction. This resonant frequency is an important factor in the acoustic behavior of the loudspeaker, as we shall see presently.

11–6 "Blocked" vs "motional" impedance. If the movable part of the mechanism under discussion is clamped mechanically, the electrical impedance as measured at the coil terminals is called the "blocked" impedance, the complex expression for which we shall label z_b. The *magnitude* of z_b (i.e., the modulus, Z_b) as so measured for a typical loudspeaker, usually varies with frequency in the manner shown by the solid curve of Fig. 11–4. Up to a frequency of several hundred cycles-sec^{-1}, Z_b is practically constant, and is only slightly greater than the ohmic resistance of the coil. As the frequency is raised, Z_b rises gradually, due to the increased importance of the coil inductance. Even though there is iron within the coil, the coil inductance is small, since the iron is near saturation because of the large steady flux passing through it.

If the cone and the attached coil are free to move (as is true when the loudspeaker is in use), the graph of the electrical impedance Z_b vs frequency has a form similar to that shown by the broken curve in Fig. 11–4. The peak at the lower end of the frequency scale occurs at the mechanical resonance frequency of the moving system; the value of Z_b at this frequency may be many times the value in the mid-frequency region. In the complex form, *the difference between the complex electrical impedance with the cone clamped and the complex electrical impedance with the cone free to move is called the "motional" impedance.* For this impedance we shall use the symbol z_{em}. It is the existence of this motional impedance that entirely accounts for the production of sound waves.

11–7 Motional impedance and mechanical impedance. The relationship between motional impedance and the mechanical impedance of the moving system is easy to discover in the case of the electrodynamic transducer. The driving force acting upon the coil is given by

$$F_{\text{driv}} = Bli, \tag{11–6}$$

where B is the flux density in the gap, l is the total length of the conductor, and i is the instantaneous current. If the coil has an axial velocity $\dot{x}$, the emf induced therein due to its motion is

$$e = Bl\dot{x}. \tag{11–7}$$

The electrical impedance due to the motion of the coil is

$$z_{em} = \frac{e}{i} = (Bl)^2 \frac{\dot{x}}{F_{\text{driv}}}, \tag{11–8}$$

from Eqs. (11–6) and (11–7). The mechanical impedance is connected with the driving force through the relationship

$$F_{\text{driv}} = \dot{x} z_{m'}, \tag{11–9}$$

where $z_{m'}$ is the total mechanical impedance of the cone system *plus* the impedance due to the presence of the air (mention of the latter was made in Section 11-3). By combining Eqs. (11-8) and (11-9), we obtain

$$z_{em} = \frac{(Bl)^2}{z_{m'}}. \tag{11-10}$$

In the mks system, this equation may be used as it stands, provided the appropriate units are used for B and $z_{m'}$ (in the case of $z_{m'}$ the unit is *not* the mechanical ohm characteristic of the cgs system). In the cgs system, Eq. (11-10) must be rewritten,

$$z_{em} = \frac{(Bl)^2}{z_{m'}} 10^{-9} \text{ electrical ohm.} \tag{11-11}$$

(The reader should check the factor 10^{-9}.)

If there is no baffle around the speaker cone and if no horn is coupled to it, the total mechanical impedance $z_{m'}$ is mainly due to the impedance z_m of the cone and coil system, with the presence of the air playing a relatively minor part. At the middle and lower frequencies, this system may be treated as approximately equivalent to a particle. At the frequency of mechanical velocity resonance, z_m is a minimum and all real, since the mechanical reactance is zero. Therefore, from Eq. (11-11), z_{em} will be a maximum and real. This is the reason for the peak in the broken curve of Fig. 11-4. It is interesting to note that due to the inverse relationship, at frequencies *lower* than the resonant frequency (where the reactive part of z_m is predominantly due to compliance), the reactive part of the motional impedance is predominantly *inductive*. For the same reason, at frequencies higher than that for resonance, the reactive part of z_{em} is predominantly *capacitive*. At resonance, z_{em} is resistive.

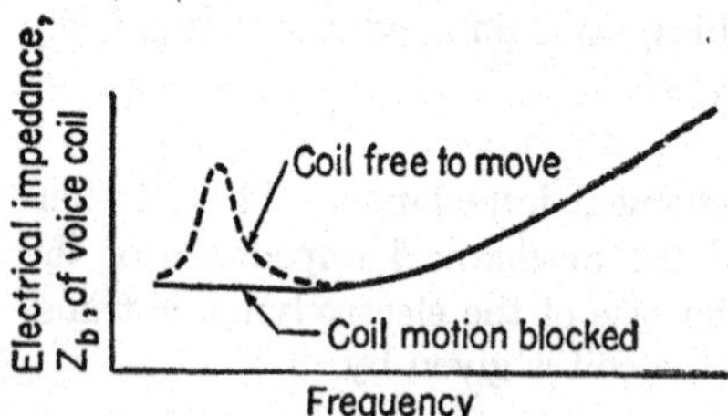

FIG. 11-4. "Free" and "blocked" electrical impedance of electrodynamic loudspeaker.

11-8 Motional impedance and acoustic radiation. If we apply the results of the previous section to the "ideal" transducer considered in Section 11-3, we find that since in this case $z_{m'}$ is due solely to the air and is, for the plane wave radiator, $\rho_0 cS$, the motional impedance is constant at all frequencies and is real. Hence the average electrical power radiated as sound power is given by $I^2_{rms}Z_{em}$, where in this case

$$Z_{em} = \frac{(Bl)^2}{\rho_0 cS} 10^{-9} \text{ ohm.} \tag{11-12}$$

The motional impedance of *practical* loudspeakers cannot be discussed in terms of Eq. (11–12) because it is necessary to add the mechanical impedance of the piston itself to the mechanical impedance due to the presence of the air. The correct expression is that of Eq. (11–10) or (11–11). Expanding the denominator, we may write

$$z_{em} = \frac{(Bl)^2 10^{-9}}{[r + S(z_s)_R] + j\left[\left(\omega m - \frac{K}{\omega}\right) + S(z_s)_X\right]}. \tag{11–13}$$

In this expression, r is the internal damping coefficient of the cone, $(z_s)_R$ and $(z_s)_X$ represent the real and imaginary parts, respectively, of the average specific acoustic impedance over the surface of the cone, and S is the effective cone area.

To estimate the acoustic efficiency of an actual loudspeaker and to see qualitatively how this efficiency may vary with the driving frequency, it is useful to examine Eq. (11–13).

The situation in the neighborhood of resonance is the simplest to analyze. While all but one of the terms in the denominator on the right-hand side of Eq. (11-13) have strong frequency dependence, the constants for actual loudspeakers are such that velocity resonance virtually occurs when the expression in j is zero. We may therefore write that at resonance

$$z_{em} \cong \frac{(Bl)^2 10^{-9}}{r + S(z_s)_R}. \tag{11–14}$$

Substituting in Eq. (11–14) values which are typical of an actual cone speaker,* of diameter 6 inches, mounted in a very large (virtually infinite) baffle, we have

$$z_{em} \cong \frac{[(10{,}000)(250)]^2(10^{-9})}{100 + (130)(1)} \cong 27 \text{ ohms.}$$

Such a speaker may have an "ohmic" resistance (virtually equal to the "blocked" impedance) of several ohms. Hence, at resonance, most of the total electrical impedance is due to z_{em}.

Not all of the power delivered to z_{em}, however, goes into sound radiation. Since for a given voltage E_{rms} across the voice coil, the power Ω delivered is inversely proportional to the electrical impedance, we may write

$$\Omega \propto [r + S(z_s)_R].$$

---

* Numerical data of an experimental nature used in this discussion have been kindly furnished by Dr. A. Wilson Nolle, Department of Physics, University of Texas.

Only the power associated with the term $S(z_s)_R$ represents radiated sound. Hence the efficiency of the speaker at resonance is

$$\text{Efficiency} \cong \frac{S(z_s)_R}{r + S(z_s)_R} 100\%$$

$$\cong \frac{130}{100 + 130} 100 \cong 56\%. \tag{11-15}$$

At the higher frequencies $(z_s)_R$ becomes larger and approaches the value $\rho_0 c$ (see Fig. 5–11). However, at frequencies well above resonance the reactive part of $z_{m'}$ predominates over the resistive portion, due primarily to the magnitude of the term ωm. If we write Eq. (11–13) in the form

$$z_{em} = \frac{(Bl)^2 10^{-9}}{R' + jX'},$$

where R' and X' represent the real and imaginary parts, respectively, of $z_{m'}$, we see, upon rationalizing the denominator, that the real part of z_{em} becomes

$$(z_{em})_{\text{real}} = \frac{R'(Bl)^2 10^{-9}}{(R')^2 + (X')^2}. \tag{11-16}$$

At the upper end of the audible spectrum the magnitude of X' is so great that the value of $(z_{em})_{\text{real}}$ may be reduced to a small fraction of an ohm. This is considerably less than the ohmic resistance of the voice coil. As a result, the over-all efficiency as a sound source is low and may drop to much less than 1% at the high frequencies.

In general, the efficiency of a loudspeaker at resonance represents a maximum. Considering the whole range of audible frequencies, few ordinary transducers have an average efficiency of greater than 10%.

11–9 Behavior of the transducer in a vacuum tube circuit. The electrical impedance of the voice coil is at all times low. To efficiently deliver power to the coil from a vacuum tube amplifier, an impedance matching transformer is necessary between the coil and the high-impedance plate circuit of the tube. The effect of this transformer is to introduce into the tube circuit an electrical impedance Z_e, mainly resistive, of several thousand ohms, rather than the few ohms characteristic of the coil itself. The way in which the reactive components of the total impedance of the voice coil are affected by the transformer action is somewhat complicated and will not be discussed here. In general, the transformed impedance vs frequency follows a graph very similar to that shown in Fig. 11–4 for the coil itself, except, of course, at the higher numerical values of the transformed impedance.

The behavior of the cone is markedly affected by whether the vacuum tube circuit can be considered to supply a constant current to the speaker, or a constant potential. If the internal resistance of the vacuum tube is much *higher* than the transformed speaker impedance, the variations in Z_e with frequency will not greatly affect the current. With a constant current there will be a constant force on the coil. Due to the mechanical resistance, the velocity and amplitude of the cone will be greatly increased in the neighborhood of the resonant frequency (Fig. 11–5a).

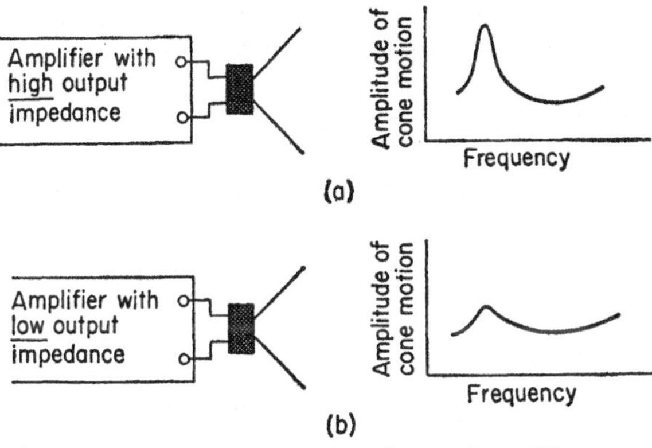

FIG. 11–5. Effect of electrical output impedance of amplifier upon loudspeaker steady state cone motion in the neighborhood of resonance.

If the vacuum tube has a *low* internal resistance, the transformed impedance may be considered the main impedance in the circuit. In this case the applied potential is approximately constant for a signal of fixed amplitude, regardless of frequency, and therefore when Z_e is high (near resonance), the current will be *low*. This will reduce the driving force in the neighborhood of resonance, thus tending to make the response in that frequency region much less pronounced (Fig. 11–5b). This latter situation is generally preferred, partly to reduce the excessive response at the resonance frequency and partly for the following reason.

From the mechanical point of view, the cone system is actually underdamped. Therefore when the signal has stopped, a transient vibration may continue for a fraction of a second, producing sound which was not in the original wave. By connecting the loudspeaker to a vacuum tube of low internal resistance, *electromagnetic damping* due to the motion of the coil in the speaker magnetic field will be relatively rapid, since the induced currents will be large. This effect is concerned with the loudspeaker's transient response and is important in high fidelity sound reproduction.

11–10 Behavior of the cone vs the acoustic piston. It should be clear from the discussion of Section 11–3 that the mechanical impedance of the moving system (the cone and the coil) for the mechanism being considered should be as small as possible in comparison to the radiation impedance. Therefore the cone mass should be small and the compliance as great as possible. Also, in the interests of efficiency, the dissipative forces at the supports should be small. (Some dissipation at the supports is desirable, however, to reduce the time of the transient motion of the cone.)

A cone of paper has small mass and, because of its shape, has an amazing degree of rigidity under the action of axial forces. The rigidity is desirable in order that the cone behave as nearly as possible like an ideal acoustic piston; the motion of its surface will then most closely reproduce the motion of the small coil attached to its apex. At frequencies below about 500 cycles-sec^{-1} the cone is apparently quite rigid. At higher frequencies, however, the cone begins to behave like a thin plate; flexure sets in, both along and at right angles to the elements of the cone, and transverse motion of the paper results. At certain frequencies, stationary wave patterns are set up over the surface of the cone. These transverse motions affect the radiation of longitudinal sound waves in a very complicated manner, since the vibrations of certain regions on the cone will obviously be out of phase with the motions of other regions (Fig. 11–6). The many peaks and valleys in the sound pressure observed in front of a loudspeaker of this type, throughout the higher range of frequencies, are largely due to this so-called "cone breakup." By pressing into the paper circular corrugations concentric with the axis of the cone, many of the transverse wave motions just referred to are discouraged.

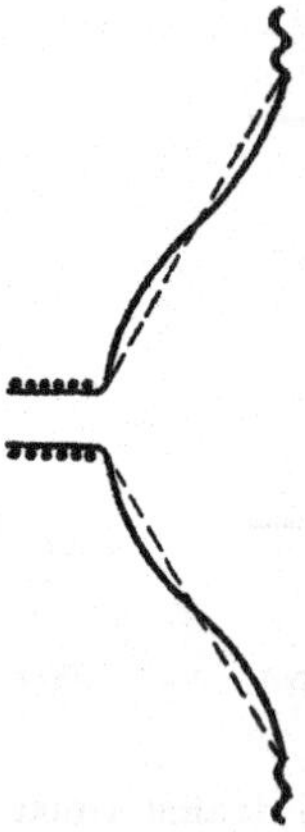

FIG. 11–6. Cone breakup.

From another point of view, these somewhat elastic regions can be considered elements in a mechanical low-pass filter, the inertial elements being the relatively rigid regions between the corrugations (see Chapter 12). Thus the higher frequency motions never reach the outer sections of the cone. This has the beneficial effect of reducing the total mass which is in motion at the higher frequencies. At the upper frequencies, the mechanical impedance of the cone system is mainly a reactance of the inductive type, and it is desirable to keep the mass small; otherwise the velocity response ($\dot{x} = F_{\text{driv}}/\omega m$) will fall off rapidly and so will the intensity of the radiated sound. One might also expect that the reduction in the effective area of the vibrating surface would cause the radiated sound to fall off markedly at the higher frequencies. Actually, however, the rest of the cone, relatively passive in the presence of vibrations at the apex, supplies some of the coupling action of a horn (as discussed in Chapter 5) and so prevents too serious a decrease in efficiency.

11-11 Acoustic coupling problems. In Chapter 3 it was pointed out that when acting as a double source, an acoustic piston is a very poor acoustic radiator. To ensure single source action, one simple solution is to use a baffle. The size of the baffle depends on the desired low frequency limit for efficient radiation, and the effect is best investigated experimentally. It is found that the transition between single and double source behavior takes place rather critically when the baffle dimension (assuming a circular shape) is a little less than $\lambda/2$. For example, a plate 5 ft in diameter will ensure single source action (each side of the cone acting independently) above a frequency of about 100 cycles-sec^{-1}. Below this frequency the radiation will fall off rapidly because of interference between the front wave and the back wave. When the distances from the microphone to the front and to the back surfaces of the piston (or cone) differ by about one wavelength, a sharp dip in the acoustic pressure is observed (Fig. 11-7). This critical cancellation is to be expected, since the front and back waves start out just 180° out of phase.

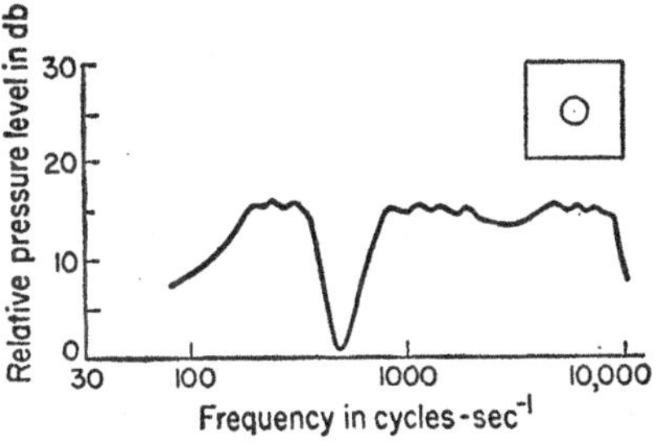

FIG. 11-7. Relative acoustic pressure level 10 ft in front of an 8-inch loudspeaker mounted in the center of a square flat baffle 3 ft on a side. Dip at 500 cycles-sec^{-1} is due to destructive interference of front and back waves. (*After* Olsen)

In the frequency region where the baffle is effectively infinite in area, a constant velocity imparted by the driving mechanism to an acoustic piston does not ensure uniform radiation of energy at all frequencies. As the graph of Fig. 5-11 indicates, even with an *infinite* baffle the transmission coefficient τ, a measure of relative radiated energy, falls off with frequency. Assuming constant piston velocity, τ is approximately proportional to the square of the frequency at the lower end of the spectrum, unless the piston has an impossibly large diameter. This would seem to be an insurmountable difficulty in the design of a radiator of wide frequency range.

The fundamental mechanical resonant frequency of the cone and coil system of the average electromagnetic transducer designed for direct sound radiation lies in the lower part of the audible spectrum. The best design places the resonance at the lowest frequency which it is desired to radiate, or preferably slightly below this limit. The reason is simple. At frequencies above resonance, the reactive component of z_m (large compared with the resistive component, except near resonance) soon becomes essentially inertial in nature and its magnitude is equal to ωm (neglecting the smaller reactance due to the compliance). As a result, the velocity response of the cone is not uniform with frequency, assuming a constant driving force. As the frequency is lowered, $\dot{x}$ varies inversely with f. Thus the increased velocity at the lower frequencies may be made to approximately compensate for the reduction, at those frequencies, in the real part of the radiation impedance. This artifice, however, may lead to nonlinear distortion if the amplitude of the cone motion becomes too great at frequencies near cone resonance.

Below resonance both effects are in the same direction, since the mechanical impedance becomes that due to compliance and therefore *increases* with a drop in frequency. As a result, little radiation occurs below the resonant frequency. It is obviously desirable, then, to place the resonant frequency as low as possible. This is not difficult in cones with a diameter of at least one foot. In the smaller sizes it is hard to reduce the stiffness of the cone suspension sufficiently to achieve a low resonant frequency and still retain a rugged construction. However, the smaller speakers are used largely in compact radio sets where the baffling is very inadequate anyway, and an increased efficiency associated with a higher resonant frequency is a desirable goal.

At the higher frequencies, where according to the graph of Fig. 5–11 the specific acoustic impedance at the piston becomes constant at the value $\rho_0 c$, the increase in the mechanical reactance with frequency, due to the choice of the resonant frequency, becomes a detriment and leads to reduced radiation. As shown in the discussion of Section 11–10, the behavior of the cone at the higher frequencies departs radically from that of a rigid piston, and other factors often control the radiation efficiency. The use of a separate unit for the high frequencies is becoming increasingly popular as a solution to the difficulties inherent in that region (Section 11–17).

11–12 Back of cone completely enclosed. To effectively ensure the single source type of radiation down to a frequency limit of 50 cycles-sec^{-1}, a flat baffle at least 10 feet in diameter is required. A structure of such dimensions is hardly feasible (or tolerable!) for the average living room.

To prevent the back wave from interfering with the front wave, one might try the rather obvious solution of completely enclosing the back of the loudspeaker in a sound tight box, thus entirely eliminating one component of the double source. Unfortunately, this procedure has an undesirable consequence. As we have seen in discussing the Helmholtz resonator, the air trapped in such an enclosure will behave like a simple spring, at wavelengths large compared with the box dimensions. The effect upon the motion of the loudspeaker cone will be to add an elastic stiffness to the forces already acting upon the moving system, with a consequent definite rise in the fundamental mechanical resonant frequency. The effect is quite marked, as the graph of Fig. 11–8 indicates. The peak in the response for the loudspeaker when enclosed is about 200 cycles-sec^{-1}, whereas in free air it is at about 150 cycles-sec^{-1}. As will be remembered from Section 5–9, the elastic force due to an enclosed volume of air, upon a given area of the container, is inversely proportional to the volume. If the natural cone resonance is placed very low, say at 30 cycles-sec^{-1}, and if the volume of the enclosure is large (8 to 10 ft^3 is none too large), the resonant frequency of the combination can be kept close to the lower limit of usable frequencies. This type of enclosure is sometimes called an "infinite baffle," although its behavior, due to the stiffness of the enclosed air, is really quite different from that of a true infinite plane baffle.

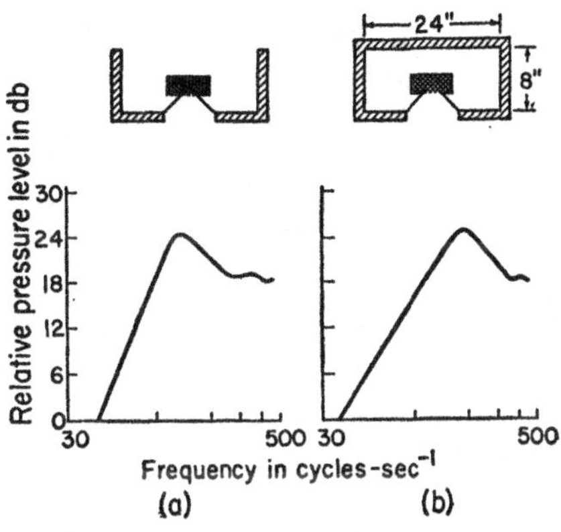

FIG. 11–8. Effect upon response characteristics of complete back enclosure of loudspeaker. Natural resonance frequency without enclosure is 150 cycles-sec^{-1}. Enclosure in (b) is 2 ft × 2 ft × 8 inches. (*After* Olsen)

The presence of reflecting surfaces within the enclosure will give rise to stationary waves at certain frequencies where the wavelength is smaller than the dimensions of the box. It is usual to line the box with sound absorbing material to cut down the effect of such resonances upon the motion of the cone. (Such material has little effect upon the stiffness loading of the cone at the very low frequencies.)

11–13 Loudspeaker cabinet with open back. In most radio sets designed for the home, the loudspeaker is enclosed in the same cabinet that contains the electrical equipment. The cabinet may be small or large, but it is almost invariably open at the back. This is partly for cooling purposes and partly to prevent the stiffness effect upon the speaker cone, as discussed above. Unfortunately, this acoustical system is rather poor. The air in the open box may resonate in a number of ways, the lowest frequency of resonance corresponding to a mode of vibration partly of the Helmholtz type and partly like that for a pipe open at one end only. The exact frequency of this resonance is difficult to predict, but experimentally it is found to be in the vicinity of 100–150 cycles-sec^{-1} for the average console radio, which is much too high in the audible spectrum. Radiation in this frequency region is unduly enhanced because of the efficiency of radiation from the open side of the box, which acts like a large piston. Below this resonant frequency, the acoustic radiation is very poor in the important frequency band between 100 and 50 cycles-sec^{-1}. For small table radio sets the resonance is even higher, so that very little energy is observed below frequencies of 200 or 300 cycles-sec^{-1}.

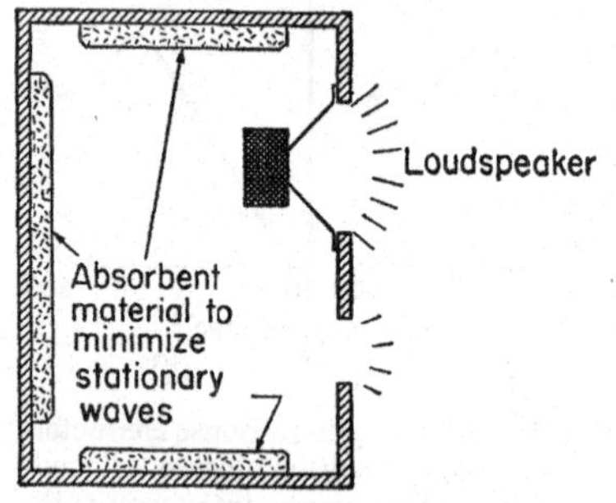

FIG. 11–9. Loudspeaker enclosure acting as a phase inverter.

11–14 The acoustic phase inverter.* This device, in its effect upon the low-frequency radiation of the transducer, is a considerable improvement over the ordinary open-backed cabinet. The phase inverter enclosure is a simple box which encloses the back of the loudspeaker but provides an opening which serves to couple the motion of the air within the box to the outside air. The position of the opening is not critical, but it is usually placed not far from the speaker cone (Fig. 11–9).

The acoustic behavior of the phase inverter can be explained in several ways. From the standpoint of resonance, the system may be considered as two closely coupled tuned circuits of a mechanical type. Energy is fed into the loudspeaker, which constitutes a tuned primary whose separate resonant frequency is that of the cone-plus-voice coil system (with the added effect of the acoustic loading at the front of the cone). This primary is closely coupled, through the back of the cone, to a secondary comprising an acoustic oscillator of the Helmholtz type, the enclosed air supplying the stiffness and a plug of air at the opening furnishing the mass. The

* Dickey, Caulton, and Perry, *Radio Engineering* **8**, No. 2, p. 104 (1936).

frequency of the Helmholtz resonator is determined by the dimensions and other factors discussed in Chapter 5. It is usual to tune the "secondary" to the same separate frequency as that of the loudspeaker "primary." Students of electricity will remember that under these conditions a complete electrical transformer whose windings are closely coupled will have an over-all *double* resonance, one corresponding to a frequency higher than the separate frequency of the primary or secondary and the other lower by the same amount. The two resonances will appear quite symmetrical when measured in the laboratory (Fig. 11-10). Exactly the same double resonance occurs in the case of the speaker and the air cavity to which it is coupled. It turns out that a resonance of the Helmholtz type can occur at a frequency considerably lower than that for an open-backed cabinet of the same size. Within limits, the resonant frequency can be controlled by varying the size of the opening; a small opening gives a lower resonant frequency than a large opening (Section 5-10). The system is damped (due partly to radiation) and the double resonance gives a bandlike boost to the response (Fig. 11-11).

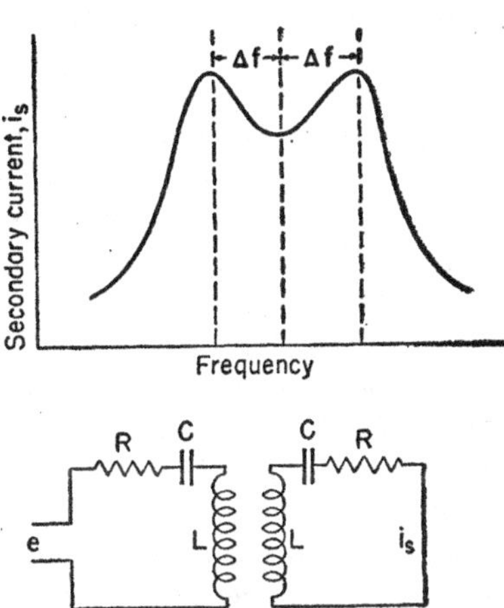

FIG. 11-10. Behavior of two resonant circuits with moderately close coupling. With very close coupling the peaks are noticeably different in height.

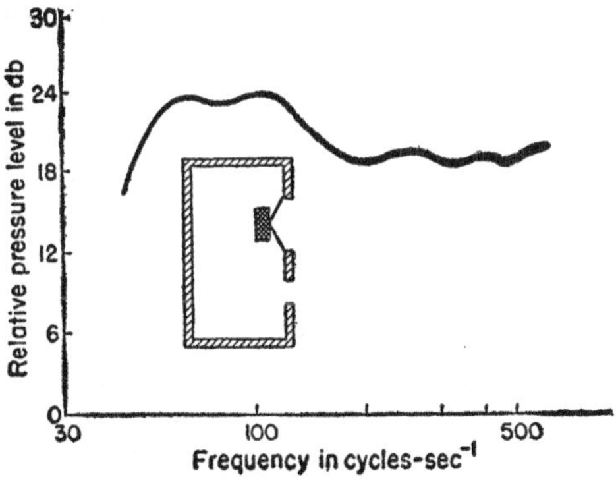

FIG. 11-11. Typical low frequency response characteristics of loudspeaker mounted in enclosure of the phase-inverter type.

The phase inversion feature of the behavior of the enclosure is even more important than its particular resonance properties, since it accounts for the radiation efficiency in the low-frequency region. Following is an approximate analysis of this behavior. Imagine a short neck to be attached to the opening near the speaker, as in our analysis of Chapter 5. Consider a steady state simple harmonic vibration of the cone, of a frequency low enough so that any dimension of the box is much smaller than the wavelength. Under these conditions, if the cone is displaced inward, the pressure will build up simultaneously throughout the interior of the box. There will then be a net *outward* pressure on the plug of air in the tube. If we neglect all dissipative forces (in the neck, and due to radiation from the face of the plug), we may consider the plug to be acted upon only by the periodic force due to the rise and fall of the pressure within the cavity. This net force, at the moment under discussion, is *outward*, and the acceleration of the plug is therefore also outward. It will be remembered that in pure simple harmonic motion there is a 180° relationship between the acceleration and the displacement. Therefore at the moment being considered, the displacement will be *inward*. Since the speaker cone is also displaced inward, the motions of the cone and of the plug of air are in phase, and their acoustic radiations will be additive. The opposite is true if both sides of the cone are exposed to the open air. It is from this effective phase inversion that the enclosure takes its name.

Just as for all the enclosures we have discussed, the primary improvement in uniformity of radiation properties brought about by the acoustic phase inverter is at the low frequencies. A partial lining of the interior of the enclosure prevents undesirable resonances at the higher frequencies. The phase inversion feature operates only over about two octaves at the low end of the audible band. Due to the dissipative factors in the neighborhood of the opening, the phase inversion is not the ideal one of 180° assumed above, since the frictional forces on the plug must be considered, as well as the elastic force. The phase shift is, however, sufficient to considerably enhance the over-all radiation in the neighborhood of resonance.

11–15 The half wavelength pipe. Another arrangement designed to increase the low frequency radiation of a relatively small acoustic piston depends upon the production of stationary waves in a pipe. In the usual design, the pipe is coupled to the back of the cone and its length is equal to $\lambda/2$ for a frequency in the neighborhood of 100 cycles-sec^{-1}. At this frequency the motion of the air at the open end of the pipe is 180° out of phase with the motion at the back of the cone. Therefore the acoustic radiation from the open end of the pipe is in phase with that from the front of the cone, and there is enhanced radiation at that frequency. The inside of the pipe is heavily lined with absorbing material, and this fact, together with the presence of radiation from the open end of the pipe, brings about a rather high degree of damping. Hence the resonance is not sharp. The absorbing material serves also to damp out most of the higher resonances; at the middle and higher frequencies the pipe becomes virtually nonconducting, and practically all of the radiation comes from the front of the cone. At low frequencies, for which the pipe length becomes $\lambda/4$, a different type of resonance occurs. The open end is a velocity antinode, as for the half-wave resonance, but in this case the wave reflected from the open end arrives back at the cone out of phase with the motion of the cone, so that the amplitude of motion of the latter is greatly reduced at that frequency. In terms of acoustic radiation impedance, this effect may be described as due to the large impedance offered to the back of the cone by the air in the pipe. A complete analysis of the pipe behavior at the quarter wavelength resonance shows a good fraction of the acoustic impedance at the back of the cone to be real and connected with the radiation of real power from the open end of the pipe. Since the pipe walls are lined and introduce additional damping forces, the quarter wavelength response is fairly broad. The over-all effect of both half and quarter wavelength resonance is to enhance the low frequency radiation over a frequency interval of more than an octave, with no very pronounced peak.

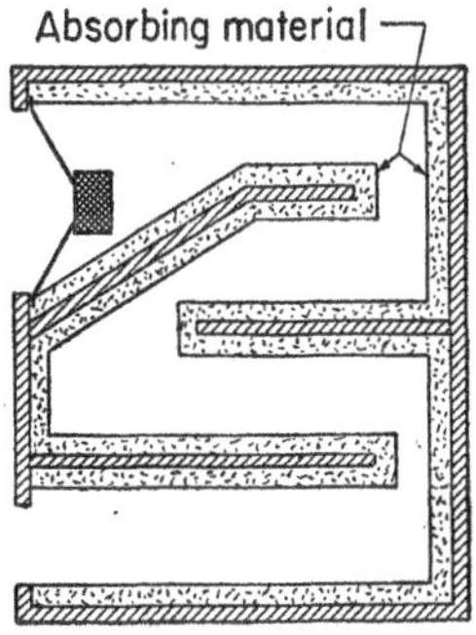

Fig. 11-12. Half wavelength pipe, or "acoustic labyrinth," coupled to back of cone loudspeaker.

A well-designed loudspeaker has a natural resonance frequency at least as low as 50 cycles-sec^{-1}. The wavelength corresponding to this frequency is about 20 feet, and so the pipe is made 5 feet long to properly damp the cone motion at the quarter wavelength resonance. Its half wavelength resonance will occur for the frequency 100 cycles-sec^{-1}. Good acoustic radiation will take place over the approximate range of 50 to 100 cycles-sec^{-1}. The pipe may be straight or may be folded for compactness (Fig. 11-12).

11-16 The use of horns. The essential behavior of a horn was discussed in Chapter 5. For a long exponential horn with a mouth whose diameter is somewhat larger than the wavelength, the specific acoustic impedance at the small end of the horn will be approximately $\rho_0 c$ at all frequencies, nearly down to the cutoff frequency. There is therefore a better real acoustic "load" on the driver diaphragm at the lower frequencies when a horn is used than when the loudspeaker unit is a simple direct radiator surrounded by a flat baffle. It is not so necessary to depend upon cone resonance at these lower frequencies. The use of a horn always increases the efficiency of any practical transducer at the lower frequencies, since the open end of the horn, the effective radiating area, can be enormously larger than any practical equivalent of an acoustic piston. The horn must be bulky to be effective at the low frequencies, and hence is not suited to home use.

11–17 High frequency radiation problems. Multiple loudspeakers. Most of our discussion thus far has centered around the problem of obtaining good low frequency acoustic radiation. As pointed out in Section 11–10, no light paper cone with a diameter as large as 6 to 12 inches can behave as an acoustic piston above a frequency of several hundred cycles because of cone breakup. For cones of small diameter, the breakup occurs at the higher frequencies, but as was seen in Section 5–20, it is desirable to have the cone as large as possible to obtain good acoustic loading at the low frequencies. The obvious solution to the problem is to use at least *two* separate acoustic radiators, one designed for the low frequency range and the other for the middle and high frequency ranges, each unit being electrically supplied with the proper fraction of the whole spectrum of frequencies. The use of a small diaphragm for the higher frequency radiator has more than one advantage. Besides remaining more rigid at higher frequencies, with a consequent smoother acoustic pressure response, a small diaphragm will set up a more diffuse diffraction pattern (at the sacrifice, of course, of some efficiency). A large diaphragm, however efficient a radiator it may be at the higher frequencies, will unfortunately radiate energy in that frequency region as a beam, and not with the spherical divergence desirable for a distributed listening audience. With a reduction in piston diameter the diffraction pattern, it will be remembered, will spread out. To diffuse the energy still more effectively at the very high frequencies, where even the small diaphragm of the high frequency unit (perhaps several inches in diameter) will radiate within a small solid angle, a cluster of small horns may be coupled to the diaphragm (Fig. 11–13).

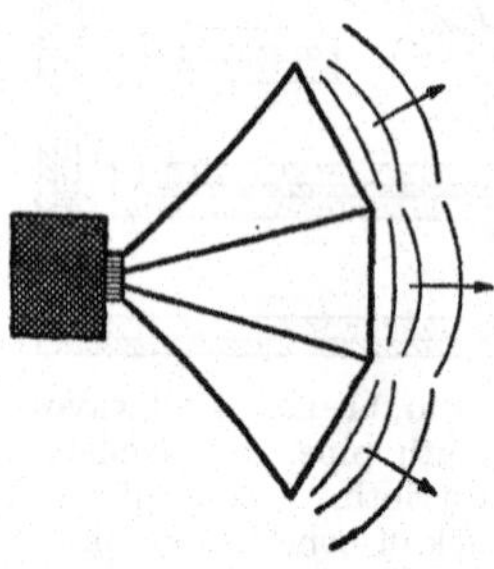

Fig. 11–13. Horn cluster designed to increase divergence of radiation from high frequency unit in multiple speaker systems.

Due to the special problems associated with the design of units which will radiate efficiently above 5000 cycles-sec^{-1}, there is some argument for dividing the complete band of audible frequencies among *three* loudspeaker units. A triple unit radiator of this type is illustrated in Fig. 11–14. The radiator for the middle frequencies, the range from 600 to 4000 cycles-sec^{-1}, is not visible in the photograph. The middle frequencies emerge from a

FIG. 11–14. Three-channel loudspeaker. (*Courtesy* Jensen Mfg. Co.)

short horn whose mouth opening is at the center of the large paper cone (the low frequency radiating surface). The surface of the cone is a continuation of the horn contour and helps to couple the middle frequency unit to the air.

In multiple units of the dual or triple type, electrical networks are necessary to sort out the several groups of frequencies. This is a standard electrical filter problem whose solution is relatively simple. One phase of the acoustical problem not yet completely solved arises from the fact that the several units are not coincident in space, so that interference effects may be observed in the frequency region of crossover (where two units are radiating simultaneously). The coaxial geometry minimizes this difficulty.

11–18 Effect of room resonances. As will now be realized, there are complications enough in the design of an aperiodic acoustical radiator if the transducer is assumed to be radiating into free space. The conclusions reached in the above sections are valid only if the energy radiated from the loudspeaker diaphragm travels indefinitely outward and never encounters

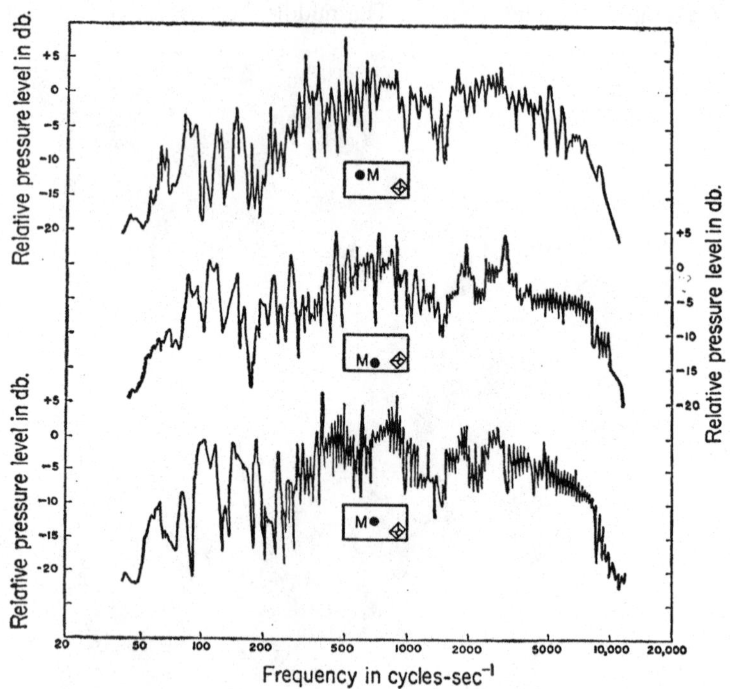

FIG. 11–15. Variation in the measured response of a single loudspeaker located in a "live" room, depending on the position of the microphone, *M*. (*Based on* Technical Monograph #1, Jensen Mfg. Co.)

any reflecting surface. However, electroacoustic equipment is more often than not used indoors and in small rooms. If the rms pressure is measured along the axis of a cone radiator, either outdoors (true free space) or in a good anechoic room, one obtains the type of response discussed so far in this chapter. If the same measurements are made in an ordinary living room with partially reflecting walls, the response curve (acoustic pressure vs frequency) may be entirely different in all but the most essential rough features. The curve taken indoors will have many additional resonances which are due to stationary waves set up within the room. The location of the microphone, that is, whether it is at a pressure node or an antinode for each of the many wave patterns, will greatly affect the magnitude of the measured acoustic pressure (Fig. 11–15).

This does not mean that the general radiation properties of the transducer, as determined on the basis of free-space measurements, are not important, but it does mean that the room acoustics significantly alter the behavior as far as the ear is concerned. To the ear, the low frequency response is somewhat more pronounced within an enclosure like a room than it is outdoors, since the energy is confined and reflected back and forth

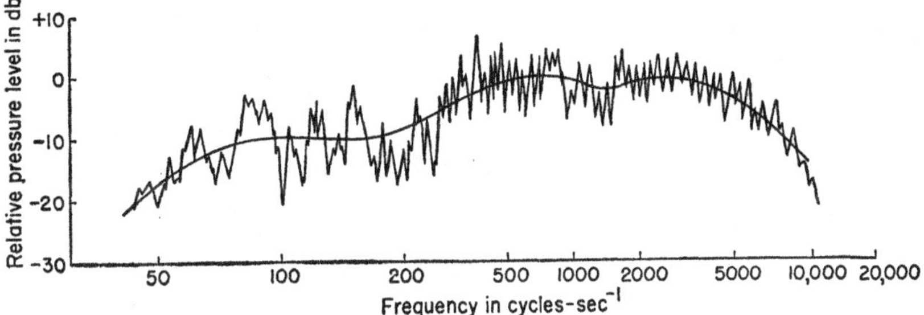

Fig. 11–16. Typical measured response of loudspeaker in "live" room. Smooth line represents the effect to the average ear.

within a limited volume, rather than spread into regions beyond the listener. (This spread is most pronounced at the *low* frequencies, because of diffraction.) Despite room peculiarities, however, it is still true that in the confines of a room a good wide-range transducer will sound better than one of narrow range.

11–19 Electrical equalization circuits. It is not within the scope of this book to discuss the important electronic circuits which are a part of every complete sound reproducing system. Mention must be made, however, of the possibility of correction for deficiencies in the acoustic part of the system by the deliberate introduction of "controlled distortion" into the electrical circuits. The graph of frequency vs sound pressure in front of any actual loudspeaker is a highly irregular curve of the type indicated by the solid line of Fig. 11–16; the general trend of the graph is represented by the smoother curve. (As a matter of fact, the average uncritical listener will picture the curve as somewhat like this anyway, since, as we have noted elsewhere, the ear is insensitive to rather large variations in sound pressure, particularly in the presence of a mixture of frequencies. To the human ear, the transducer will always sound better than the objective measurement with a laboratory microphone would indicate.)

By the introduction of so-called equalizing circuits into the electronic amplifier that feeds the energy to the loudspeaker, it is possible to give to the entire system a response virtually independent of frequency, even though the curve for the transducer has a decided "droop" at both the high and the low frequency ends of the spectrum (Fig. 11–17a). All that is necessary is to electronically amplify both the high and the low frequencies more than the middle frequencies. An amplifier with such a discrimination circuit will show a gain which varies with frequency in the manner indicated by curve (b) of Fig. 11–17. To the ear, the over-all effect will be somewhat like that given by the product of curve (a) and curve (b), that is, as shown by curve (c). In a sense, one distortion has cancelled another. The correction principle is essentially the same as that used to overcome the radiation deficiencies of an acoustic piston; we employ ahead of the acoustic radiation, it will be remembered, a type of mechanical velocity response which is high at those frequencies where the acoustic radiation is low.

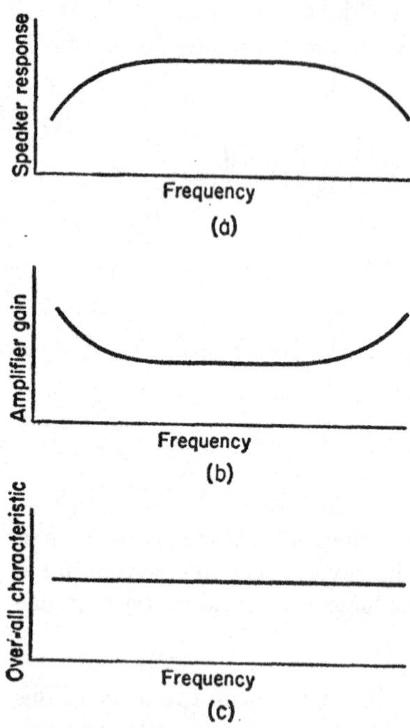

FIG. 11–17. Use of electrical compensation to correct for acoustical shortcomings.

It is not good practice to employ any more electrical equalization than is absolutely necessary, since there is some danger of overloading (with resultant distortion) both the electronic circuits and the mechanical transducer system. Moreover, only gross defects in uniformity of response on the part of the transducer and its acoustic radiation can be corrected in this manner. No reasonably simple circuit has been devised to remove the numerous peaks and valleys due to cone breakup. These latter eccentricities in cone behavior can only be minimized by careful design of the mechanical features of the transducer itself and hence it is still desirable to design the transducer to be as aperiodic as possible.

11–20 Transducers for disk phonograph records. There are several methods whereby the vibrations associated with sound waves can be transferred to a permanent record. Three different media and techniques have been widely used: ordinary flat disk pressing, based on the principles of the early Edison phonograph, the film sound track technique, and the recently revived use of magnetic tape. Each system has its special problems, none of which is fundamentally acoustic in nature although, since the vibrations are due originally to sound waves, acoustics inevitably enters into the larger problem.

In the system enjoying the widest popular use, that employing a flat grooved disk, it is necessary to use an electromechanical transducer (commonly known as the phonograph pickup) between the record and the electrical amplifier that drives the loudspeaker. It is this device that transforms the motion of the needle, resting in the undulating record groove, into an equivalent alternating potential. Such a potential can be amplified electronically and eventually applied to the terminals of a loudspeaker. The design of this transducer involves no acoustic principles. The moving mechanical system, however, is similar in some respects to that used in certain microphones and loudspeakers. This system has a mechanical impedance that varies with the frequency, a variation which plays an important part in the behavior of the phonograph transducer. Since we have paid some attention to the effect of the mechanical impedance of the loudspeaker, it will be interesting to describe briefly the phonograph pickup mechanism, with particular attention to the mechanical impedance of its moving parts.

Many different kinds of phonograph pickups have been proposed and manufactured. Practically all electromechanical transducers are reversible, and can be used either as motors or generators. The most popular types are three in number, the *variable reluctance* pickup, the *electrodynamic* pickup, and the *piezoelectric* pickup. The variable reluctance pickup is really a telephone receiver of the reed type used backwards as a generator. The vibrating needle is attached to a small iron armature whose motion varies the length of the air gap in a magnetic circuit. A coil placed somewhere in the circuit thus receives a variable magnetic flux and experiences an induced emf of the same frequency as that of the needle vibration. The electrodynamic pickup is similar to an electrodynamic microphone, except that the motion of the coil is due to the motion of the needle rather than to sound pressure. In the piezoelectric or crystal pickup, the needle motion deforms a crystal having piezoelectric properties. (The deformation is usually a torsion.)

11-21 Differences in transfer behavior. The three types of transducers discussed above do not all behave alike in their electrical responses to a given motion of the needle. Both the variable reluctance and the electrodynamic pickups give an output voltage which is proportional to the maximum *velocity* of the needle (actually an angular velocity, since the needle-holding mechanism is free to rotate around a pivot). On the other hand, the piezoelectric transducer gives an output voltage proportional to the maximum *displacement* of the needle. These differences in behavior are important because of the manner in which the grooves are cut into the phonograph record. If one examines the record groove with a microscope, he will find that for a pure sinusoidal musical note the position of the center of the groove varies from side to side of the mean position in the manner of an ordinary sine curve. Since the angular frequency of the record on the turntable is constant, the crests of the sine curve will be close for the high frequencies and farther apart for the low frequencies. (The relationship between this spacing and the frequency is obviously not constant for different radii of the disk.)

Over most of the audible frequency range the modulation amplitude is so determined that for a given intensity of the original sound the *maximum velocity* imparted to the pickup needle is approximately the same, regardless of the frequency. Therefore for frequencies for which this is so, the variable reluctance and electrodynamic pickups will faithfully reproduce the original intensity distribution among the various frequencies. The crystal pickup, however, will give a smaller and smaller electrical output with rises in frequency since, with a constant velocity characteristic, the *amplitude* of the undulations decreases with frequency rise (from simple harmonic motion, $x = \dot{x}/2\pi f$). Below about 400 cycles-sec^{-1} the groove motion has a constant *amplitude* characteristic to prevent cutting into the next groove. Hence below this frequency the crystal pickup will respond properly, while the other two types will give reduced output with drops in frequency. Both kinds of response are easily corrected with electrical equalizing circuits, so that the final sound closely simulates the original.

11–22 Mechanical impedance of moving parts. Tracking. For the motion of the needle to faithfully follow the record groove, it must obviously remain *in* the groove. Whether it will or not depends on the *mechanical impedance* of the moving system in relation to the driving force. We may idealize the system and replace it with an equivalent particle resting in the groove under the action of an elastic restraining force, the latter being actually supplied by an elastic torque at the pivot, farther up (see Fig. 11–18 for the equivalence). The equivalent particle has mass and there is also a dissipative force due to the pivot construction, although this force is usually small. The force driving this equivalent particle during the motion of the record is due to the side wall of the record groove; the undulations of the groove move the particle back and forth horizontally in the plane of the paper.

A simple consideration of Newton's second law will show that any displacement of the groove position may cause the particle to "ride up" the

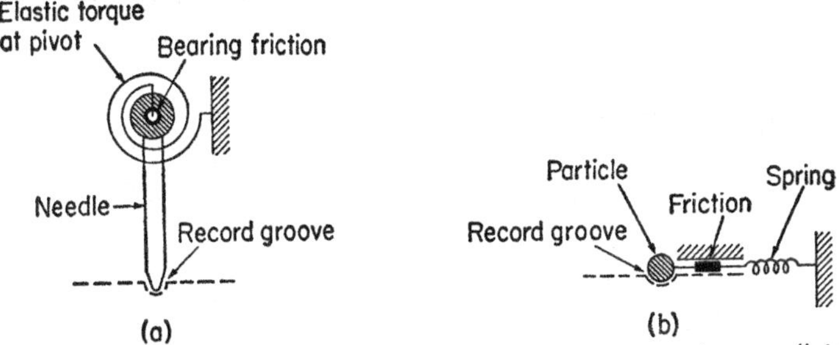

FIG. 11-18. Essential mechanical system in a phonograph transducer or "pickup." (a) Simplified representation of rotating needle system. (b) Equivalent particle system.

wall of the groove, which is inclined at approximately 45°. To prevent this effect, a vertical force is necessary, and the weight of the pickup arm usually supplies this force. In the interests of long needle and record life it is highly desirable to keep this vertical force low, and therefore to keep low the driving force supplied by the sides of the record groove. To accomplish this it is necessary to keep the mechanical impedance of the moving system in the transducer at a minimum.

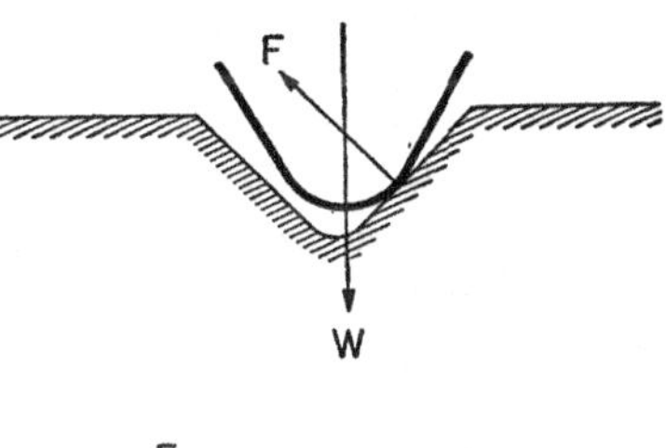

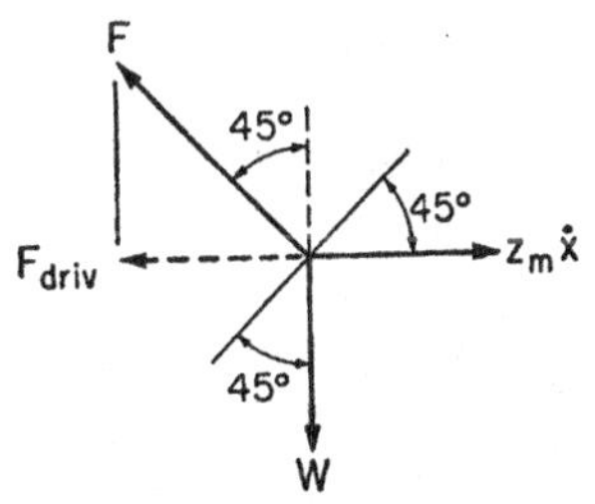

FIG. 11-19. Vector relationship for needle about to be forced out of groove.

Assuming a steady state periodic motion of the needle, the vector force relationship for the equivalent particle is shown in Fig. 11-19, at the critical situation where the needle point is about to slide up the groove wall. Here W is the force due to the weight of the arm, F is the force supplied by the smooth groove wall, and $z_m\dot{x}$ is the magnitude of the "reversed effective" force due to the mechanical impedance z_m of the particle system, whose velocity is $\dot{x}$. (This last vector arises from $F_{\text{driv}} = z_m\dot{x}$, where F_{driv} is the horizontal component of F and constitutes the driving force.) Due to the 45° slope of the wall, it may be seen by resolving W and $z_m\dot{x}$ along the direction of the wall slope that W must be at least as great as the vector $z_m\dot{x}$ for the needle to remain in the groove.

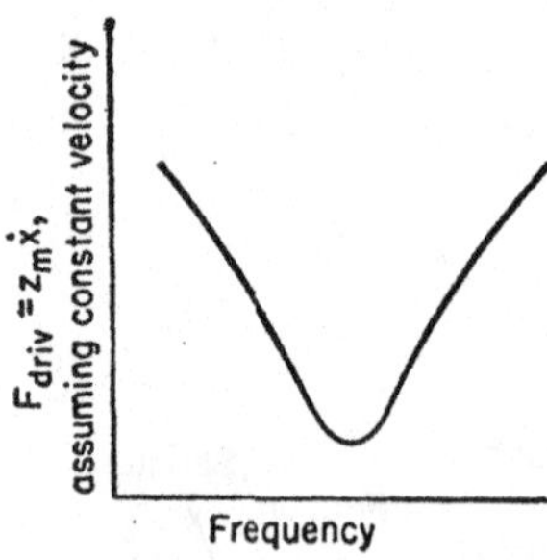

FIG. 11–20. Driving force supplied by record groove wall as a function of frequency. Constant needle point velocity is assumed.

It is usual to place the mechanical resonance of the moving system at about the middle of the frequency spectrum. The mechanical impedance z_m will be a minimum at that frequency and will rise at lower and at higher frequencies. If we assume constant velocity properties for the groove undulation, a plot of $z_m\dot{x}$ (and therefore the driving force) vs frequency will look something like the graph of Fig. 11–20. Since the vertical force W must at all frequencies be at least as great as $z_m\dot{x}$, it must obviously be designed to take care of the greatest recorded value of $\dot{x}$ (about 5 cm-sec^{-1} for 78 rpm recordings) at the *ends* of the frequency spectrum, where z_m is a maximum. If z_m is known at those points, the required value of W is determined, once the characteristics of the recording technique are known.

By the introduction of numerous refinements in the mechanical portions of the phonograph pickup it has been possible to reduce the maximum value of z_m (at very low and at very high frequencies) to an extremely low value, so that the minimum "tracking force" (that is, W) is often as low as 5000 dynes (close to 5 gm in weight). The result is a much longer record and needle life than formerly, when units of much higher mechanical impedance were employed. It should be pointed out that the motion of the needle end is controlled in amplitude by the presence of the groove wall. No unnatural electrical response takes place at the frequency of mechanical resonance. The variations that occur affect the wall-needle point *force* only.

11–23 Conclusion. With this illustration of one more application of the principle of electrical analogy, we shall close this chapter on the reproduction of sound. There are many other aspects and details of this branch of applied acoustics which have not been covered in this brief summary. For further information the reader is referred to the many articles appearing in the current journals, as well as to the specialized books on the engineering aspects of sound.

PROBLEMS

1. Consider a loudspeaker mechanism of the electromagnetic type, in which the electrical losses in the voice coil due to heating effects are negligible. The cone, of diameter 10 inches, behaves as a rigid acoustic piston radiating plane waves. The internal mechanical damping constant is 1.5×10^4 dynes-cm^{-1}-sec. (a) What is the efficiency of the loudspeaker as a radiator of sound waves? (b) What would be the efficiency for a cone of one-half this diameter, still assuming plane wave radiation and the same value for the internal damping constant?

2. The moving parts of a loudspeaker mechanism have a mechanical impedance which is at all times large compared with the impedance due to the air load. The resonant frequency is 60 cycles-sec^{-1}. (a) Plot an approximate curve for frequencies above resonance to show the variation of the velocity $\dot{x}$ with the frequency, assuming a constant driving force. (b) Assuming the cone to act as an acoustic piston of constant diameter, radiating plane waves, plot an approximate graph to show the relationship between total radiated power and the frequency.

3. A loudspeaker, whose cone has a diameter of 12 inches, has an internal mechanical impedance which is mainly resistive and which is always large compared with the impedance due to the air load. The cone is surrounded by a very large flat baffle. (a) Making use of the graph of Fig. 5–11, plot an approximate graph of the total radiated power vs the frequency, assuming a constant driving force. (b) Is this loudspeaker arrangement ideal? If there are any disadvantages, discuss them.

4. It is desired to radiate maximum acoustic power with a cone-type loudspeaker at a *single* fixed frequency of 1000 cycles-sec^{-1}. Discuss the design features of the loudspeaker from the standpoint of (a) the size of the cone, (b) the resonant frequency, and (c) the value of the real and imaginary components of the mechanical impedance z_m.

5. Compute the rms driving force for an electrostatic loudspeaker (see Section 11–4) of diameter 10 inches where E_0 is 1000 volts, d is 5 mm, and the maximum value of the signal voltage e is 50 volts.

6. A paper cone may sometimes generate sound waves which are *subharmonics* of the applied sinusoidal voltage, that is, the frequencies are *less* than that of the signal. This is due to flexure along the elements of the cone, under the action of the axial force at the apex. Paper cones molded so as to have a contour somewhat like that of an exponential horn do not exhibit this property. Explain how this shape eliminates the danger of such subharmonics.

7. Making use of Eq. (11–13), derive an expression for the motional impedance z_{em} in the form $a + jb$.

8. From the form of the equation obtained in problem 7, discuss in detail the effect upon z_{em} of varying the frequency of the voltage applied to an ideal acoustic piston mounted in an infinite baffle.

9. Assume an acoustic piston to be radiating plane waves. Its effective mass is 2 gm, the elastic factor K is 3×10^8 dynes-cm^{-1}, and the damping coefficient r is 200 dynes-cm^{-1}-sec. (a) Find its resonant frequency, neglecting the effect of the air. (b) Find, at the resonant frequency, the value of the motional impedance z_{em} if the diameter of the cone is 10 inches. The voice coil has a diameter of one inch and consists of 60 turns. It lies in a magnetic field where the flux density B has a value of 8000 gauss. (c) How, in general, will the resonant frequency be affected by the presence of the air?

10. Compute the over-all efficiency, at the resonant frequency, of the loudspeaker described in problem 9. The voice coil has a blocked electrical resistance of 3 ohms.

11. The diameter of the effective radiating area of the cone of a certain

loudspeaker is reduced to one-half when the frequency of the applied signal is raised from 1000 to 2000 cycles-sec^{-1}. Assuming the radiation of plane waves in both cases, what will be the relative acoustic power output in the two cases with (a) constant cone velocity, and (b) constant cone amplitude?

CHAPTER 12

MISCELLANEOUS APPLIED ACOUSTICS

12–1 The acoustic properties of rooms. One of the earliest branches of applied acoustics to receive serious theoretical and experimental attention was that concerned with the acoustic properties of rooms. Because of the reflecting power of the average wall surface, at least part of the acoustic energy reaching the walls is returned to the room. Hence the whole nature of the energy distribution around a source located within a room is quite different from that around the same source in free space.

The most obvious result of the reflections at the walls will be what is usually called *reverberation.* When the vibrations of a sound source within the room are stopped, so that the influx of energy is cut off, the acoustic energy does not instantly become zero throughout the room. At the instant the source is cut off there is a flow of wave energy in the room along a large number of assorted paths, with numerous reflections at the wall surfaces. This flow continues after the source is stopped, the energy density in the room diminishing rapidly as each reflection extracts a fraction of the energy in the incident wave. With such a process, one would expect some kind of exponential decay of the acoustic energy within the room. When the energy has reached a low level, the sound becomes inaudible. This whole phenomenon of decay is called *reverberation.*

The reflecting properties of a room are both advantageous and detrimental. With a steady source of sound, the extraction of energy at the walls is proceeding at a steady rate. Indeed, if the observed intensity of the sound remains constant, the total energy leaving per second through wall absorption must then equal the energy fed into the room per second by the source. This intensity, when measured at any one point, will be found to be much greater than would be expected at that distance from the source in free space. The presence of reflected energy thus greatly increases the efficiency of sound transmission to the hearer. Even in large halls, a speaker can be heard from any position, even at the extreme rear, whereas communication over such distances is usually difficult in the open. Not only does the existence of reflection make possible a more effective steady state sound transmission, but from the transient point of view a certain amount of reverberation or "acoustic hangover" is apparently agreeable to the ear, particularly for music. Too abrupt a cessation of the various sounds in orchestral music is considered to detract from the smooth blending of the sound from the different instruments. Some of this effect is no doubt due to psychological conditioning on the part of the listener, since most music has in the past been played in highly reverberant rooms.

The detrimental aspect of reflection is also important, particularly if the room is to be used by a speaker. If the energies associated with consecutive sounds overlap at the ear of the listener to too great an extent, because of reverberation, speech loses clarity and the articulation score is low. Even in music there results an unpleasant blurring, often amounting to a discordant effect.

From the above discussion it is apparent that there must be an optimum degree of reverberation for which the room and its wall surfaces should be designed. To reach some definite conclusions as to the proper acoustic design of rooms in general, we shall discuss first a room having certain ideal properties.

12–2 An ideal reverberant room. Let us consider a large room whose walls absorb only a small fraction of the incident energy with each reflection. If the mean path between reflections is long and if the walls are highly reflecting, any wave motion started within the room will persist for a relatively long time after the source is stopped. The room is then called acoustically "live." Also, when the source is first started it will take a relatively long time, as we shall see, before a steady state is reached, that is, an instant when the total rate of disappearance of energy at the walls is equal to the total rate of influx of energy from the source. If the wavelengths are small compared with any room dimension (and if the room is irregular in shape), the wave energy will be distributed quite uniformly. As a result, the effect of stationary wave patterns will be small and may be ignored. It will be correct to say that in such a room *the energy associated with the sound is at any one instant distributed uniformly throughout the volume of the room.* With the situation as described, the energy density everywhere in the room will go up or down at the same rate during those transient periods of relatively slow change when the energy is either increasing towards the equilibrium steady state level, or decreasing towards the zero level.

12–3 Rate of disappearance of energy from the ideal reverberant room. The uniform distribution of energy assumed above implies that all possible directions of flow are represented at any one point, distributed in a completely random manner. This means that a velocity microphone, at any one instant, would give the same response regardless of its orientation. (This ideal state of affairs assumes no significant stationary wave patterns.) For the same reason, the total incident energy striking any given wall area per unit time will be uniform regardless of location. With this in mind, it is not difficult to compute the total rate of arrival of energy per unit area of wall surface from all directions in terms of the instantaneous energy density e_i. The Sabine absorption coefficient α_s, it will be recalled, assumes energy arriving from all possible directions, and under these conditions represents the ratio of absorbed to incident energy. Therefore a fraction α_s of the energy incident per unit time will disappear from the room. Knowing the total surface area, it is then a simple matter to determine the total time rate of disappearance of energy over the surfaces of all the walls. Details of the above analysis are given in Appendix IV; the theory is due to Sabine and Jaeger. The equation for the energy u arriving per unit area of wall surface per unit time is

$$u = \frac{e_i c}{4}, \tag{12-1}$$

where e_i is the instantaneous acoustic energy density in the room and c is the velocity of sound. The total energy U_a absorbed per unit time over the total wall surface S will be

$$U_a = \frac{e_i c}{4} \bar{\alpha}_s S, \tag{12-2}$$

where $\bar{\alpha}_s$ is the mean value of the Sabine absorption coefficient for all the surfaces. As would be expected, U_a is proportional to the energy density and to the surface area of the room. The total energy U_a is also linearly dependent upon the velocity of sound, since the higher the velocity, the more frequent will be the reflections. The factor 4 is a result of the integration necessary to obtain Eq. (12-1).

If the value of α_s varies for different portions of the wall surface, which is usually the case, Eq. (12-2) may be written

$$U_a = \frac{e_i c}{4} \sum (\alpha_s S). \tag{12-3}$$

In this case the wall surfaces are broken up into finite areas, over any one of which the value of α_s is constant. The summation in Eq. (12-3) is then performed over the whole room surface, and we obtain

$$\sum (\alpha_s S) = (\alpha_s)_1 S_1 + (\alpha_s)_2 S_2 \cdots (\alpha_s)_n S_n.$$

12-4 The steady state energy density. In the steady state, reached in actual rooms soon after the starting of a steady source of sound, the total energy U entering the room per second is equal to the total energy U_a leaving the room per second. Making use of Eq. (12-1), therefore,

$$U - \frac{e_0 c}{4} \bar{\alpha}_s S = 0, \tag{12-4}$$

where e_0 is the steady state energy density and $\bar{\alpha}_s$ is the average absorption coefficient. Solving for e_0, we obtain

$$e_0 = \frac{4U}{c\bar{\alpha}_s S}. \tag{12-5}$$

It is important not to confuse the energy density e_0 as used in discussing room properties with the energy density concept introduced in the chapters on progressive, unidirectional waves, either plane or spherical. In the case of free space waves, the energy enters and leaves any given small volume of space along one fixed direction. In the room now being considered there is a simultaneous flow of energy *in all possible directions*, and because of this random directional distribution the *intensity* of the sound is *not* the product of the energy density and the sound velocity c, as it is for unidirectional flow. The correct relationship between the energy density and the sound intensity in the reverberant room (defined, as for unidirectional waves, as the energy flow *in a given direction* through unit area per unit time) is given by Eq. (12-1). This expression, while used for energy incident at the wall, is equally valid for an imaginary surface out in the room. Therefore the steady state sound intensity I_0 anywhere in our ideal reverberant room is, in view of Eqs. (12-1) and (12-5),

$$I_0 = \frac{e_0 c}{4} = \frac{U}{\bar{\alpha}_s S}. \tag{12-6}$$

12-5 The transient equations. During the time that the acoustic energy is building up to the steady state condition, the flow of energy into the room from the source is greater than the flow out through the walls. If V is the volume of the room, the total rate of increase in the acoustic energy in the room during this transient period is $V(de_i/dt)$. Equating this rate of increase to the difference between the rate of influx and the rate of efflux, we may write, using Eq. (12-2),

$$V\frac{de_i}{dt} = U - \frac{e_i c}{4}\bar{\alpha}_s S.$$

This equation may be rearranged:

$$V\frac{de_i}{dt} + \frac{e_i c}{4}\bar{\alpha}_s S = U. \tag{12-7}$$

This is the differential equation which describes how the energy density e_i varies with the time while the source is feeding in energy at the constant rate U. It will be recognized as identical in form with the differential equation which describes the variation in the *current* in an L–R circuit upon which has been impressed a constant emf, E, that is,

$$L\frac{di}{dt} + Ri = E. \tag{12-8}$$

The integration of this electrical equation gives

$$i = \frac{E}{R}(1 - \epsilon^{-(R/L)t}).$$

By analogy, the expression for the acoustic energy density as a function of the time must be given by

$$e_i = \frac{4U}{c\bar{a}_s S}(1 - \epsilon^{-[(c\bar{a}_s S)/4V]t}). \tag{12-9}$$

We may simplify Eq. (12–9) by making use of Eq. (12–5):

$$e_i = e_0(1 - \epsilon^{-[(c\bar{a}_s S)/4V]t}). \tag{12-10}$$

The energy density e_0 is plainly the final steady state value which e_i approaches asymptotically with the time.

For the transient period after the sound source is turned off, U in Eq. (12–7) is zero. The proper differential equation is then

$$V\frac{de_i}{dt} + \frac{e_i c}{4}\bar{a}_s S = 0. \tag{12-11}$$

This equation is analogous to the differential equation for the current in an L–R circuit after the removal of the emf. The electrical equation is then similar to Eq. (12–8), but with the right-hand side equal to zero. The integration of this equation yields

$$i = \frac{E}{R}\epsilon^{-(R/L)t},$$

and the analogous equation for e_i is therefore

$$e_i = \frac{4U}{c\bar{a}_s S}\epsilon^{-[(c\bar{a}_s S)/4V]t} \tag{12-12}$$

or

$$e_i = e_0\epsilon^{-[(c\bar{a}_s S)/4V]t}. \tag{12-13}$$

Since the sound intensity in the room is equal to $e_i c/4$ and therefore proportional to e_i, Eqs. (12–10) and (12–13) may be written in terms of the intensities, where I_i replaces e_i and I_0 replaces e_0. We then have for the rise in intensity,

$$I_i = I_0(1 - \epsilon^{-[(c\bar{a}_s S)/4V]t}), \tag{12-14}$$

and for the decay,

$$I_i = I_0\epsilon^{-[(c\bar{a}_s S)/4V]t}. \tag{12-15}$$

12–6 Reverberation time. Graphs of Eqs. (12–10) and (12–13) are given in Fig. 12–1. The time required for e_i to reach the steady state value in the one case, or the zero value in the other, is obviously infinite. The ear, however, will judge these limits to have been reached in a finite time. The length of time required for the acoustic energy density to drop to some fixed fraction of its initial steady state value after the sound is turned off is a convenient measure of the importance of the transient period to the ear. This reverberation time, T_r, is defined as the length of time required for the energy density to decrease to one millionth of its initial value. Such a decrease corresponds to a drop of 60 db in intensity. For a sound whose initial intensity level is just 60 db, as referred to the standard zero level (which is near the average threshold of hearing), the reverberation time then corresponds to the actual time duration of the sound for the average ear.

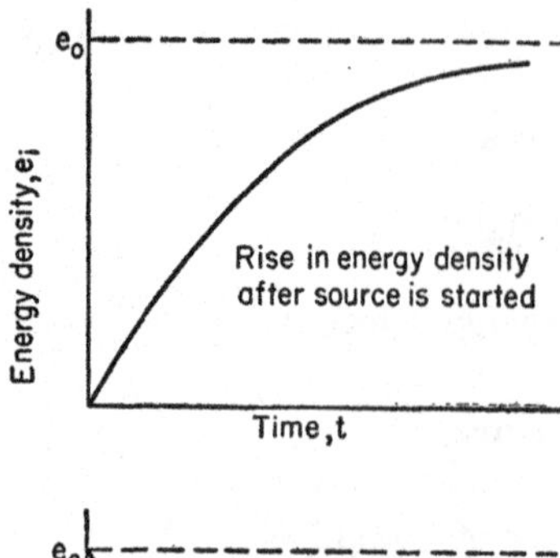

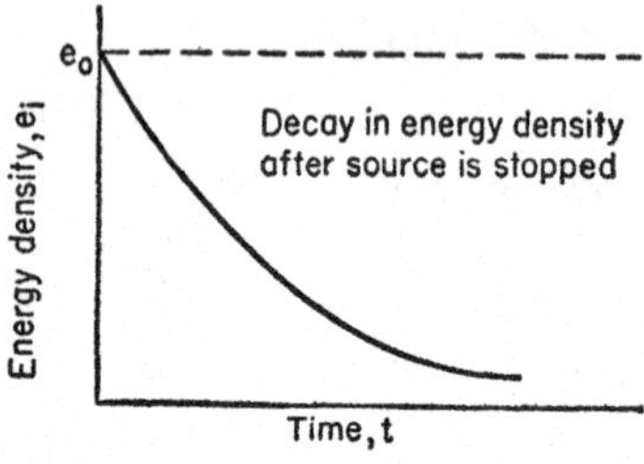

FIG. 12–1. Variation in energy density in ideal reverberant room.

If we set e_i equal to $10^{-6}e_0$ in Eq. (12–13) and solve for the corresponding time $t = T_r$, we find

$$T_r = \frac{4V}{c\bar{\alpha}_s S} \log_e 10^6. \tag{12–16}$$

Room volumes and surfaces are usually measured in cubic feet and square feet respectively. Using these units and expressing the velocity of sound in air in ft-sec^{-1}, we may write for Eq. (12–16),

$$T_r = 0.049 \frac{V}{\bar{\alpha}_s S} \text{ sec}, \tag{12–17}$$

where $\bar{\alpha}_s$ is the mean value of the Sabine absorption coefficient for all the wall surfaces. For rooms whose walls have variable absorption properties, this becomes

$$T_r = 0.049 \frac{V}{\sum (\alpha_s S)} \text{ sec}. \tag{12–18}$$

There has been considerable discussion as to the optimum value of the reverberation time. There must obviously be a compromise, because of the conflict between acoustic efficiency and clarity of speech and music. Experience indicates that a reverberation time of 1 or 2 seconds is desirable for purposes of speech, the longer time being allowable in large halls in the interests of efficiency of sound diffusion. For music more hangover time is permissible and, as mentioned earlier, may be desirable on aesthetic grounds.

12-7 Partially live rooms. Only large, highly reverberant auditoriums approach the behavior of the ideal room whose reverberation time is given by Eq. (12-18). A somewhat different point of view as to the process of sound absorption by the room walls has been presented by Norris * and Eyring.† According to this theory, there is a series of discontinuous drops in energy with each reflection, rather than the continuous disappearance visualized in the theory of Sabine and Jaeger. The equation for the decay in energy density becomes, by this theory,

$$e_i = e_0 \epsilon^{[cS \log_\epsilon (1-\bar{\alpha}_s)/4V]t}. \tag{12-19}$$

The algebraic sign of the exponent is intrinsically negative, since α_s is always less than unity and $\log_\epsilon (1 - \bar{\alpha}_s)$ is therefore always negative. The expression for the reverberation time becomes

$$T_r = 0.049 \frac{V}{-S \log_\epsilon (1 - \bar{\alpha}_s)}. \tag{12-20}$$

Equation (12-18) may lead to errors of 20-30% in rooms whose reverberation time is less than 3 seconds. In these cases Eq. (12-20) is in better agreement with experiment.

A further modification of Eq. (12-20), proposed by Millington and Sette, may be applied to rooms where one large surface is much more highly absorbent than another. This leads to somewhat better agreement with experiment than Eq. (12-20), whose derivation depends on the properties of all the walls being fairly uniform.

12-8 Determination of α_s. In Chapter 8 the Sabine absorption coefficient was defined as the ratio of absorbed to incident energy, assuming a completely random distribution of incident angles. Since this is also assumed in the theories just discussed, we may use Eq. (12-17) or (12-20) to compute the average absorption coefficient $\bar{\alpha}_s$, once T_r has been measured experimentally. Equation (12-17) is suitable for large auditoriums whose

* Norris, Appendix II, *Architectural Acoustics* by V. O. Knudsen, John Wiley & Sons (1932).

† Eyring, *Jour. Acous. Soc. Amer.* **1**, 217 (1930).

walls are highly reflecting, and Eq. (12–20) is better for smaller rooms having walls of higher absorbing power. The determination of T_r is a simple matter, once an oscillograph or similar record of the sound pressure as a function of decay time has been made. For a number of reasons to be discussed presently, the experimental curves are rarely the smooth ones predicted by theory. The rate of decay is usually approximately exponential, however, and the curves can be used to determine $\bar{\alpha}_s$ with fair precision.

12–9 Effect of varying frequency. As is indicated in Table 8–1, the value of α_s for any one type of surface is a function of the frequency, because of the nature of the acoustic impedance at the surface. Accordingly, as would be expected, the reverberation time obtained for a given type of wall surface will depend on the frequency used. In view of the broad band character of speech and music, it is desirable to average the reverberation time over a band of frequencies. Several methods are used, such as multitoned generators, "white noise" generators, and the modulation of the frequency of a "warbled" tone at a rate rapid compared with the decay time. For speech, the middle and higher frequencies are the important ones, as has been pointed out in Chapter 9. Hence it is beneficial to have a high value of $\bar{\alpha}_s$ for the low frequencies, whose reverberation tends to mask the higher frequencies essential to articulation. Under these conditions it is more useful to measure T_r for a series of rather narrow bands of frequencies than to obtain a single average value for the whole audible spectrum.

12–10 Absorbing surfaces of limited area. No practical room or auditorium has surfaces all of uniform absorbing properties. The usual procedure is to place material of high absorbing power on certain limited portions of the wall area in order to achieve the desired value for the over-all absorption coefficient. If the value of α_s for each portion of the wall surface can be determined, an equation in the form of Eq. (12–18) can be used to predict the reverberation time. To determine the effective value of α_s for a piece of material of restricted area, the sample is introduced into a "live" chamber whose normal reverberation time T_r is known. If the reverberation time is again measured after the introduction of the sample, it is an easy matter to compute the value of α_s for the material, in terms of the reverberation times, the area of the sample, and the total surface area of the room.* (Because of diffraction effects at the edges of an absorb-

* See Beranek, *Acoustic Measurements*, p. 863.

ing panel of limited dimensions, the absorption coëfficient of a small piece of material will be less than for a larger piece. The size of the sample should be large enough to minimize diffraction effects.)

12–11 Computation of α_s from z_n. In Chapter 8 it was pointed out that the coefficient of absorption α_n for normal incidence is directly related to the normal specific acoustic impedance z_n at the absorbing surface. There are several methods for the measurement of the absorption coefficient (or the related quantity, the specific acoustic impedance) as a function of the incident angle.* Once these measurements have been made, the Sabine absorption coefficient can be computed from the following expression:

$$\alpha_s = \frac{1}{2\pi}\int_0^{2\pi} d\phi \int_0^{\pi/2} \alpha_\theta \cos\theta \, d\theta, \qquad (12\text{–}21)$$

where α_θ is the absorption coefficient for the incident angle θ, and ϕ is the azimuth coordinate angle whose variation takes care of all possible directions of energy arrival. Because of the indirect nature of this method of computing the Sabine coefficient, the direct measurement of reverberation time is still the method most commonly used.

12–12 Effect of room resonances. Steady state. In actual rectangular rooms with walls which are good reflectors, there is rarely the uniform energy distribution so far assumed, even during the steady state phase. This nonuniformity is due at least in part to the presence of stationary waves, always a possibility in a medium of limited extent. The presence of such patterns along directions perpendicular to a pair of opposite walls is to be expected, but stationary waves may occur because of waves traveling in many other directions. Several such paths are indicated in Fig. 12–2. If a musical sound constituting a harmonic series is radiating into the room, the modes of vibration in the room that happen to coincide with the

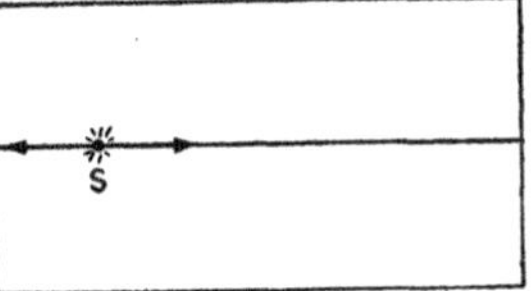

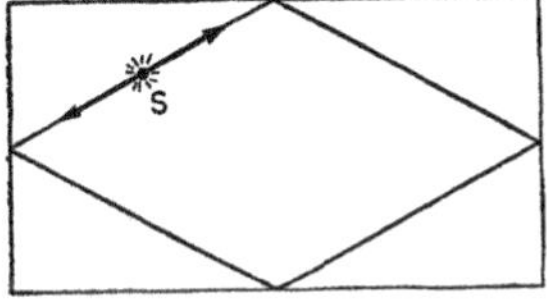

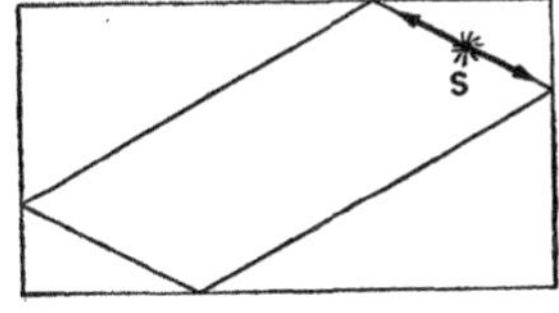

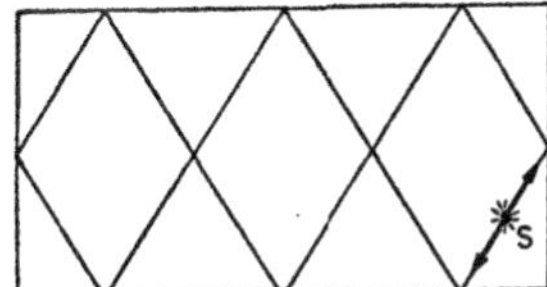

Fig. 12–2. Paths associated with four of the normal modes of vibration of a rectangular room. The exciting source is labeled *s*.

* See Beranek, *Acoustic Measurements*, pp. 864 and 867.

frequency components present in the sound source will be excited through resonance. As a consequence, there will be an uneven distribution of acoustic pressure due to the existence of numerous nodes and antinodes, which accounts for many of the irregularities observed in the radiation from a radio loudspeaker as the microphone is moved from one point in the room to another.

12–13 Normal modes of vibration. The transient period. As mentioned in Section 12–8, the experimental decay curves for room reverberation are never smooth. The irregularity may be considerable, as indicated by Fig. 12–3. Some of this irregularity is due to the different decay times associated with the nonuniform energy distribution which always exists to some extent. If there is not complete and rapid sound diffusion, highly absorbent wall surfaces may remove energy in their vicinities more rapidly than do other less absorbent surfaces. Another source of irregularity is the variation of α_s with the frequency, when the source is radiating a *mixture* of frequencies.

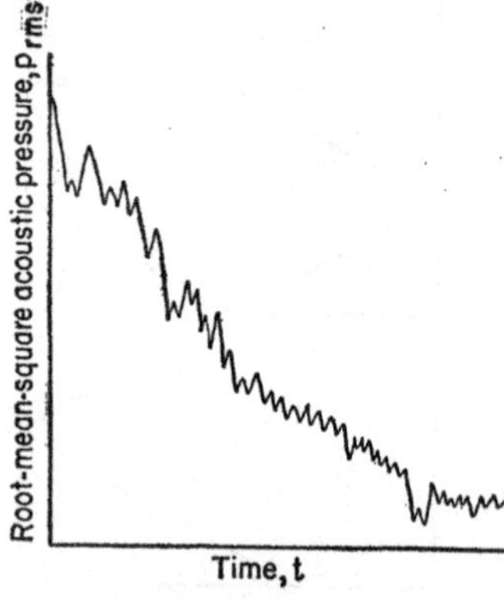

FIG. 12–3. Typical variation of acoustic pressure with the time during the decay period. Many modes of room vibration are present.

Perhaps the most important cause of the fluctuations observed in the decay curves is the existence of the room resonance referred to in the previous section. The room may be considered as a system capable of many natural frequencies of oscillation. When the sound source is turned off, the behavior of the room is similar to that of any other system after the driving force is removed: the room "oscillator" enters a period of transient motion, executing those frequencies of vibration associated with the various stationary wave patterns, just as in the case of a stretched string. As in the case of a plucked string, the modes that appear will be consistent with the particular initial conditions at the time the sound source is stopped. So numerous are the various modes of vibration associated with the wide variety of possible paths of reflection that many such frequencies are close enough together to produce beats. It is the beat effect between such pairs of frequencies that accounts for many of the fluctuations of sound pressure with time picked up by the recording microphone.

It is possible, with rooms of such simple geometrical shape as the rectangle or the cylinder, to develop a mathematical technique whereby the various normal modes of vibration of the room can be determined. *Vibration and Sound* by Morse gives full details of this analysis. One of the interesting results of a study of room reverberation is that in a room of highly irregular shape the number of normal modes of vibration that lie within any one narrow frequency range (often called "bunching") is much reduced. Hence the chance of any considerable energy being associated with the beat effect referred to above is much lessened. For the same reasons, such rooms show less annoying resonance during the steady state period. These general advantages of irregularly shaped rooms have been known empirically for many years.

12-14 Transmission of wave energy through partitions. There are many times when the amount of vibrational energy transmitted from one room to another through a partition is of more interest than the reverberation properties of either room alone. The problem is here one of wave attenuation. A commonly used measure of the effectiveness of any such sound insulating layer is the *transmission loss TL* through the structure, defined as

$$TL = 20 \log_{10} \frac{p_1}{p_2} \text{ db}, \tag{12-22}$$

where p_1 is the acoustic pressure in front of the incident face and p_2 is the acoustic pressure on the far side of the conducting layer. The ratio p_1/p_2 can be measured directly with small microphones placed on each side of the structure. A laboratory arrangement suitable for the measurement of the transmission loss through small light samples is shown in Fig. 12-4. In the case of large partitions, the panel to be studied is usually mounted between two highly reverberant rooms and, with a test source running in one room, the pressure levels are measured in both rooms. These pressure measurements need not be made near the panel, since in each room the energy is distributed uniformly (or nearly so). Measurements made in this

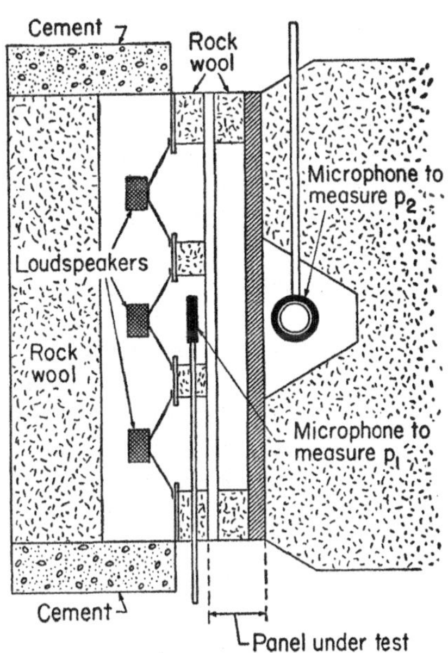

FIG. 12-4. Arrangement used to measure transmission loss for lightweight structures. (*After* Wallace, Dienel, and Beranek.)

manner must be interpreted somewhat differently than when small-scale laboratory measurements are made, because energy is incident upon the panel from all directions, and also because the values of the acoustic pressures are affected by wall absorption.*

In general, good sound insulators (having high transmission loss) are loose and porous to provide the maximum attenuation through viscosity and heat conduction losses. However, internal attenuation is actually a factor of secondary importance in the sound insulating phenomenon. Rather it is the impedance "mismatch" that accomplishes the desired end, and mounting the soft material upon a relatively hard surface is good practice, since an abrupt change in acoustic impedance at the boundary will turn back much of the energy that succeeds in penetrating the absorbing layer. Indeed, the whole process of transmission of wave energy through a laminated partition may be looked upon as a mechanical filter network whose primary purpose is to attenuate all pressures applied to the "input" surface.† For low frequencies, where the wavelength is large compared with the wall thickness, the "circuit" contains "lumped" constants. For high frequencies, the wavelength may become comparable to the dimensions of the wall, and the circuit then must be considered to contain *distributed* properties of inductance, capacitance, and resistance.

Much of the low frequency vibrational energy which penetrates partitions does so by virtue of the resonance properties of large sections of the wall acting as a unit. The wall itself then acts as a secondary radiator of sound waves into the next room (a rather efficient one, because of its large area). Hence the second wall surface should be coupled as loosely as possible to the wall surface in the room containing the source of sound. Sound transmission into an anechoic chamber is obviously undesirable. To prevent the effect just mentioned, it is usual to suspend the whole framework of the room on a system of springs or rubber pads that act as mechanical filters for vibrations of audible frequency. Such construction is expensive and, of course, not practical for ordinary buildings.

12–15 Acoustic filters. Electrical engineers often make use of electrical filters. These filter circuits are electrical networks with the peculiar property of offering selective transmission for currents of varying frequency. The action is fundamentally based on the variation in the reactance of an inductance or a capacitance with the frequency. A filter may have dis-

* See Buckingham, *Natl. Bur. Standards (U. S.) Sci. Technol. Papers* **20**, 193–219 (1925).

† See Morse, *Vibration and Sound*, p. 365, for some equivalent circuits.

tributed properties of inductance and capacitance, as in the case of a transmission line, or it may be designed with lumped circuit elements, that is, made up of separate coils and capacitors connected by wires having negligible impedance.

A simple filter network of the lumped element type is shown in Fig. 12-5. The impedances z_1 are all alike, as are the impedances z_2. The symmetry of the network structure is apparent, each impedance z_1 and its immediate neighbor, z_2, to its right being connected across the preceding impedance z_2.

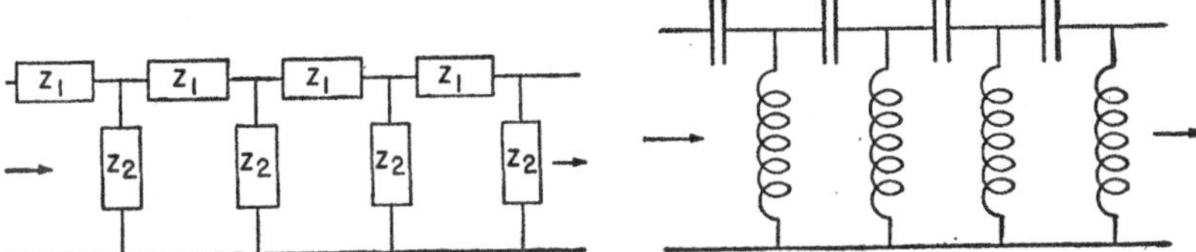

FIG. 12-5. Generalized filter circuit.

FIG. 12-6. High-pass electrical filter.

The currents in each succeeding impedance z_1 would appear to diminish from left to right, but it can be shown * that the currents in each impedance z_1 will be the same in magnitude for those electrical frequencies where the ratio z_1/z_2 lies within the range 0 to -4. There will be attenuation for frequencies where the ratio z_1/z_2 lies outside this numerical range, and so the last impedance z_1 carries a much smaller current than the first impedance z_1. If a resistor is placed at the right-hand end of the network, a small voltage will therefore appear across such a load at frequencies outside the "pass" range.

The range of frequencies which such a filter will pass with negligible attenuation can be controlled by proper selection of the impedance elements z_1 and z_2. In Fig. 12-6 is shown a high-pass filter, made up of a reiterated inductance L and capacitance C (the larger the number of sections, the greater the attenuation outside the pass band). The resistance is assumed to be negligible throughout the network. In this case,

$$\frac{z_1}{z_2} = \frac{-j(1/\omega C)}{j\omega L} = -\frac{1}{\omega^2 LC}. \qquad (12\text{-}23)$$

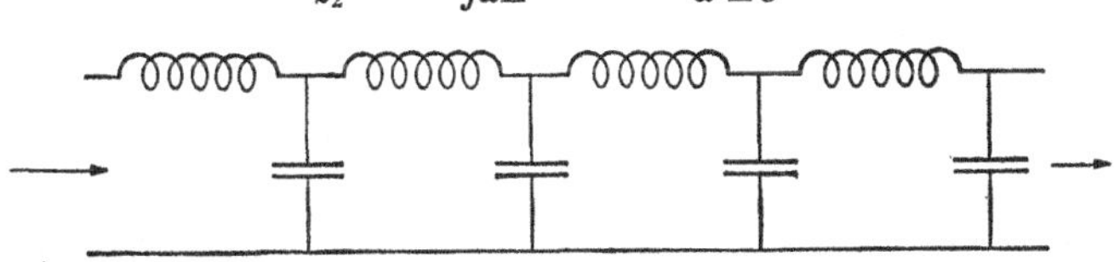

FIG. 12-7. Low-pass electrical filter.

* See A. B. Wood, *A Textbook of Sound*, pp. 498-502, G. Bell & Sons (1941).

This ratio will be zero at the frequency $f = \infty$, and equal to -4 when $f = 1/(4\pi\sqrt{LC})$. The filter is called a high-pass filter because all frequencies from infinity down to a particular frequency are passed without attenuation.

To design a low-pass filter it is necessary only to transpose the L's and C's. This circuit is shown in Fig. 12–7. In this case,

$$\frac{z_1}{z_2} = \frac{j\omega L}{-j(1/\omega C)} = -\omega^2 LC. \qquad (12\text{–}24)$$

For z_1/z_2 to lie within the range of zero to -4, the frequency must lie between zero and the value $f = 1/(\pi\sqrt{LC})$. Hence the network is a low-pass filter.

In view of the close analogy that can be drawn between the behavior of electrical circuits and certain acoustical systems, one would expect to be

FIG. 12–8. High-pass acoustic filter.

able to construct acoustical analogs to electrical filters of both the high-pass and low-pass types. G. W. Stewart has made theoretical and experimental studies of acoustical filters of all sorts. The construction of an acoustical high-pass filter is shown in Fig. 12–8. The sound waves enter a pipe having short side tubes open to the surrounding air and attached at regular intervals along the pipe. In Fig. 12–8 the acoustic elements which are analogous to the inductances are the masses or "inertances" of the volumes of air free to move back and forth in the side tubes (as in the neck of the Helmholtz resonator). Between the side tubes are sections of air whose

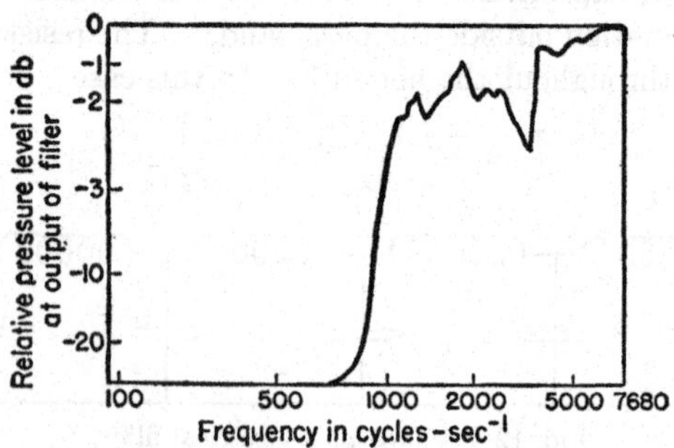

FIG. 12–9. Transmission properties of acoustic high-pass filter of type shown in Fig. 12–8. (*After* Stewart)

elastic properties furnish a compliance analogous to the capacitance of an electrical filter. A consideration of the acoustical behavior of this type of structure indicates that sound frequencies above a certain value will be transmitted down the main tube with very little attenuation, if we neglect friction and dissipation at the walls. Stewart constructed a filter of this type with six sections, and measured its transmission. The resulting graph is shown in Fig. 12-9. Remembering the compressed nature of the decibel scale, it will be seen that at the cutoff frequency (about 600 cycles-

FIG. 12-10. Low-pass acoustic filter.

sec^{-1} in this case) the drop in intensity at the output end of the filter is more than 20 db, and the intensity is therefore less than 1% of its initial value. The irregularities in the curve for the transmitted frequencies are probably due to edge effects, etc., not taken into account in the equations.

Figure 12-10 shows an acoustic low-pass filter. Here the air enclosed in the attached side chambers furnishes a compliance analogous to the capacitance of the circuit of Fig. 12-7, while the volumes of air in the main tube between successive chambers supply the inertance, analogous to the inductances. The transmission characteristics are indicated in the graph of Fig. 12-11.

Electrical filters are usually designed to be terminated at either end by resistances of specified value. If such terminating resistances are simulated in acoustical filters, even better agreement is found between the behavior of the acoustical and the electrical filter. In Fig. 12-12 are

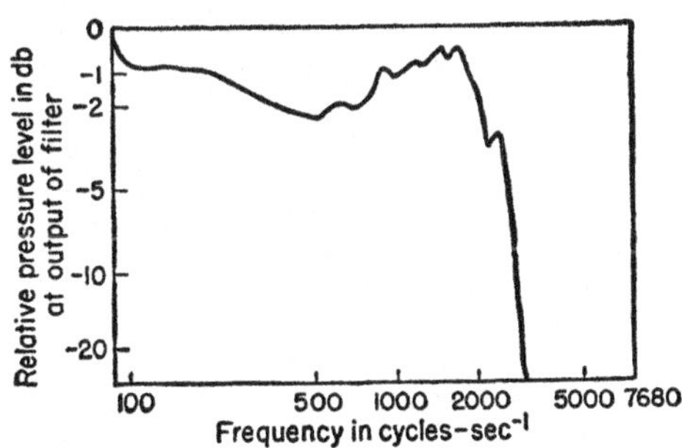

FIG. 12-11. Transmission properties of acoustic low-pass filter of type shown in Fig. 12-10. (*After* Stewart)

shown for comparison an electrical and an acoustical filter designed for the same pass characteristics. Both are simple low-pass filters.* In the equivalent acoustical circuit, the compliance of the chamber V corresponds to the capacitance C in the electrical filter. The masses of air moving in the channels t supply the inductances corresponding to L in the electrical circuit. Frictional forces along the walls of the channels furnish elements equivalent to the resistances R. To terminate the filter properly, the

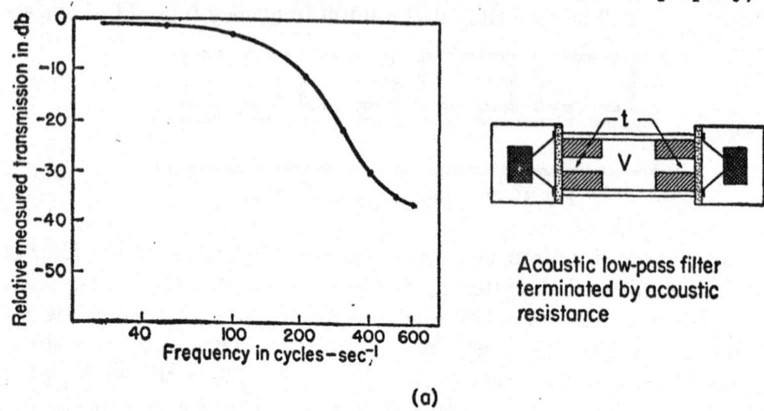

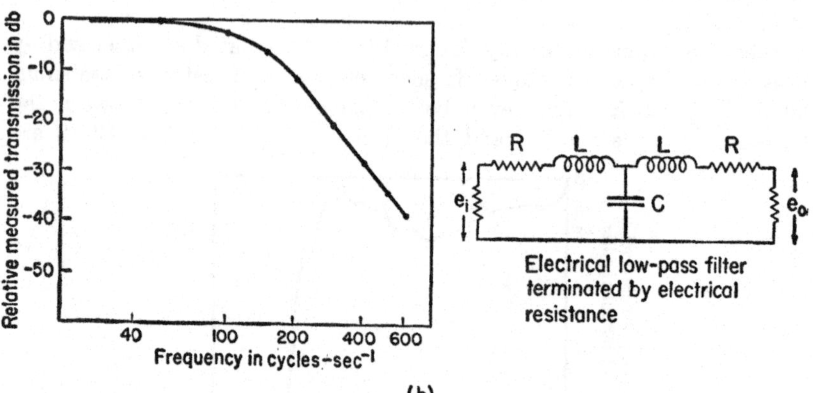

FIG. 12–12. Electrical filter and analogous acoustic filter. (*After* White and Baxter.)

---

* *Quarterly Progress Report*, Acoustic Laboratory, M.I.T., July–September 1948, pp. 7–9.

vibrating air set into motion by the loudspeaker is forced through a layer of porous ceramic supplying effective resistance. A similar layer of ceramic is placed at the output of the filter, as shown. The transmission properties of the electrical and the acoustical filter are almost identical, as will be seen from the graphs. The behavior of the acoustical filter deviates slightly from that of the electrical filter at frequencies above about 400 cycles-sec^{-1}, partly because at the higher frequencies the dimensions become comparable to the wavelength and the filter begins to have distributed rather than lumped properties.

In systems made up partly of electrical and partly of acoustical components, it is easier to perform filter operations in the electrical sections. Filter behavior, however, is an important factor in many strictly acoustical problems. From one point of view, an exponential horn is a type of high-pass filter, since frequencies below a certain critical value will not pass through. Another example of filter action is the manner in which the acoustic impedance at a partially absorbing surface controls the degree of absorption as a function of the frequency.

12-16 Ultrasonics. Frequent reference has been made throughout this book to the range of frequencies for longitudinal waves to which the human ear will respond. The upper frequency limit for an average ear is no higher than about 20,000 cycles-sec^{-1}. Frequencies higher than this are called *ultrasonic.*

The transmission of longitudinal waves of ultrasonic frequency through gases, liquids, and solids follows, for the most part, the same laws of behavior as in the case of waves within the audible frequency range. There are some anomalies characteristic of the higher frequencies, as already pointed out in Chapter 6. At very high ultrasonic frequencies, the wave velocity in gases tends towards the isothermal rather than the adiabatic. Also, there are appreciable absorptions of frequencies of the order of 10^8 cycles-sec^{-1} by certain gases and vapors having polyatomic molecules. In general, however, the behavior of ultrasonic waves constitutes simply an extension of the properties of high frequency audible waves into the super-audible region.

To take one example, the diffraction pattern of a fair-sized source radiating at frequencies above the audible is similar to that discussed in Chapter 4 for high audible frequencies, except that the various diffraction lobes are even more crowded together because of the extreme shortness of the wavelength. The wavelength in air for ultrasonic frequencies of the order of several hundred thousand cycles-sec^{-1} (about the upper limit of frequencies which it is practical to produce) is in the neighborhood of 1 mm.

Hence almost any ordinary sized generator will radiate a beam in air. In liquids and solids, where the propagation speed is higher, the wavelength will be correspondingly longer.

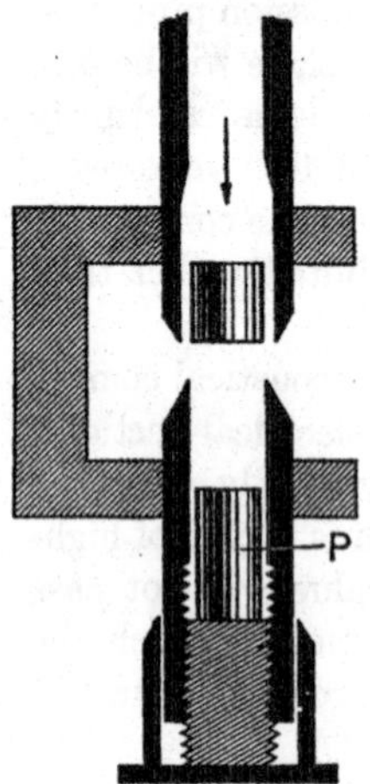

FIG. 12–13. Simplified drawing of section of Galton whistle.

While the transmission problems are similar to those for waves of audible frequency, there are special difficulties in the design of ultrasonic sources and ultrasonic receivers. We shall consider these in the following sections.

12–17 Ultrasonic sources. Generators of the whistle type have long been used as ultrasonic sources. Most familiar is the Galton whistle, designed by Edelman in 1900. In this whistle there is a short resonating cavity, the air in which is set into vibration by blowing a stream of air against the sharp edge of the opening to the cavity (Fig. 12–13). By adjusting the axial position of the plunger P with a screw, the frequency of resonance can be varied critically over a wide range. Frequencies up to about 100,000 cycles-sec^{-1} can be obtained and held fairly constant, provided the air pressure is kept steady. This whistle is capable of generating a considerable amount of power in the ultrasonic region, although its use requires some care in adjustment.

It is possible to design a transducer of the electromagnetic type which will have a fair efficiency at frequencies as high as 20,000–30,000 cycles-sec^{-1}, and is therefore capable of operating in what might be called the near-ultrasonic region. Above 30,000 cycles-sec^{-1}, the output falls off rapidly, due mainly to the difficulty of reducing the mass to the required low value while still maintaining the rigidity characteristic of an acoustic piston.

For frequencies up to about 60,000 cycles-sec^{-1} the magnetostriction generator is a most effective radiator. Its development as an ultrasonic source is largely due to Pierce. All ferromagnetic metals experience a small change in dimension when in the presence of a magnetic field. This effect is called *magnetostriction*, and may be either an increase or a decrease in length. The extent of the change depends on the material, its history, the temperature, and the strength of the field. A rod made of nickel, a material whose magnetostrictive properties are well known, will shorten if placed parallel to the field. Hence if the rod is placed inside a coil of wire carrying an alternating current, the long dimension of the rod being along the axis of the coil, longitudinal vibrations will be set up in the rod. The frequency of the rod vibration will be twice that of the alternating current, since the rod shortens regardless of the direction of the field. Just as for the ordinary telephone receiver, the addition of a constant magnetic field as a bias will ensure a rod frequency which is the same as that of the alternating current. If the frequency of the current happens to coincide with the frequency of a natural mode of the rod resonance, the amplitude of the rod motion may build up to a fairly high value. When working at the fundamental mode of vibration of the rod, with a velocity antinode at each end and a node at the center, amplitudes of the order of $10^{-4}l$ may be obtained, where l is the length of the rod.

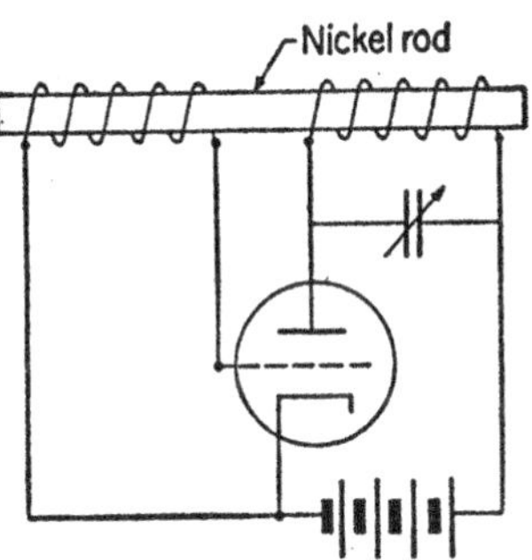

Fig. 12-14. Essential circuit for magnetostriction oscillator. (Circuit to supply steady magnetic field bias not shown.)

The process involved in magnetostriction is a reversible one, as might be expected. If the rod is compressed or stretched in the presence of an external field, the flux density within the rod is varied. This makes possible an electronic oscillator of the feed-back type, where the mechanical vibrations of the rod constitute the predominant factor determining the electrical frequency of oscillation of the circuit. The elements of this circuit are shown in Fig. 12-14. The rod is clamped at the center to encourage the fundamental mode. The L-C circuit is tuned to the natural frequency of the rod. The necessary steady field bias is not shown.

The velocity of longitudinal waves in metals is such that a nickel rod resonating at its fundamental mode with a frequency of 60,000 cycles-sec^{-1} is about 4 cm long. With shorter rods it is difficult to deliver much power. The efficiency of power delivery for modes higher than the fundamental is considerably less than for the fundamental mode. Hence 60,000 cycles-sec^{-1} is near the upper practical frequency limit for magnetostriction generators. For frequencies in the neighborhood of 20,000-30,000 cycles-sec^{-1}, however, such generators have good efficiency and are quite rugged, and hence are much used in acoustic ranging devices (Section 12-21).

12-18 Piezoelectric generators. We have already mentioned the application of the piezoelectric effect to the design of microphones, loudspeakers, and phonograph transducers. In all these applications the crystal is driven at assorted audible frequencies, all of which are well below the natural frequency of longitudinal vibration of the crystal. In the crystal microphone and phonograph pickup, particularly, the power level is low and the use of a comparatively fragile crystal like Rochelle salt is possible.

The reversible character of the piezoelectric effect suggests the use of a crystal as a direct source of ultrasonic waves; the efficiency should be high at frequencies corresponding to the natural resonances of the crystal. Quartz is a very strong material of exceptionally low internal damping. It can be cut in the form of thin slabs whose thickness, if properly oriented with respect to the crystal axes, can be made the determining factor as far as the resonant frequency is concerned. In Fig. 12–15 is shown such a crystal plate. The plate is so cut with respect to the crystal planes that the direction z is the optic axis and the direction x, normal to the flat surfaces of the plate, is one of the polar axes of the crystal. If alternating potentials are applied to electrodes clamped or cemented to the flat surfaces of the quartz slab, two types of periodic deformation will occur. One of these will be a change in thickness along the x-direction and the other will be a change in length along the y-direction. By choosing the proper frequency, longitudinal stationary waves can be set up in the crystal along either the x- or y-direction. If we consider the fundamental modes only, "thickness vibrations" will obviously make possible the higher frequencies; "length vibrations" are more suitable when lower frequencies are desired. In either case, one face of the crystal becomes the direct acoustical radiator of longitudinal waves.

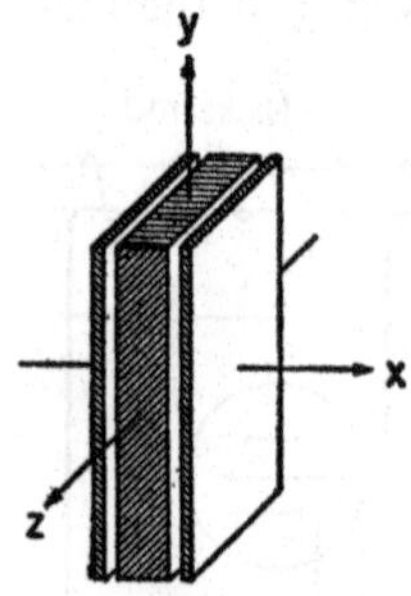

FIG. 12–15. Quartz plate between electrodes.

A simplified circuit for an electrical oscillator of the electronic type is shown in Fig. 12–16. As in the magnetostriction generator arrangement, the mechanical resonance of the radiator (in this case the crystal) essentially determines the frequency of oscillation. By tuning the L–C combination to one of the higher modes of vibration of the crystal, the latter may be made to generate waves whose frequencies are multiples of the fundamental.

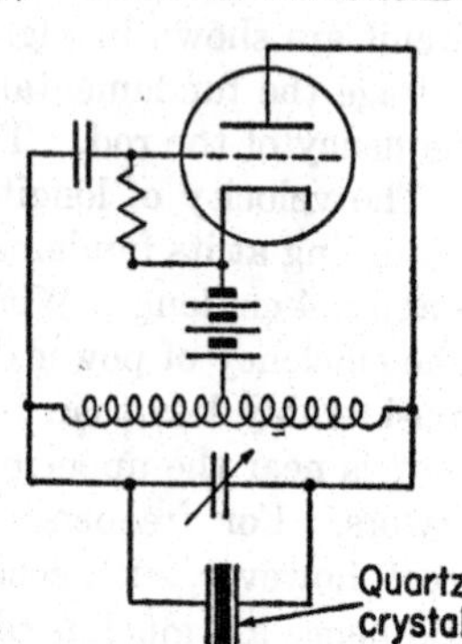

FIG. 12–16. Hartley oscillator circuit with quartz crystal.

It should be mentioned that during thickness vibrations the flat surface of the crystal does not necessarily move as a unit, like a piston. Instead, there are variations in phase and amplitude due to the simultaneous existence of longitudinal waves in the crystal parallel to the flat surfaces.

These waves are the result of shrinkages and expansions in the length. Great care must be exercised to prevent such vibrations from building up until the crystal is shattered.

12-19 Detectors of ultrasonic waves. Ordinary microphones are unsuitable for the detection of waves of ultrasonic frequency for the same reasons that ordinary types of loudspeakers make poor radiators at frequencies above the audible. Since magnetostriction generators and generators of the piezoelectric variety are essentially tuned vibrators operating at a fixed frequency, it seems quite feasible to use a receiver that is tuned to detect the one frequency radiated by the source. Piezoelectric crystals make especially good detectors, since the crystal is an efficient reversible transducer. The crystal is also a vibrator of very low damping, and its response as a detector depends critically on the frequency. The crystal is first ground to have a natural frequency as near that of the source as possible. Small differential changes in frequency of vibration may then be brought about by mounting one of the metallic plates constituting the electrodes slightly above the surface of the crystal and varying the distance between the plate and the crystal face by means of a screw adjustment. This adjustment introduces a variable capacitance in series with the effective impedance of the crystal itself, and so alters the over-all resonant frequency. The same thing may be accomplished with an external series variable capacitor. The electrical output of the crystal will, of course, be small in response to the incident wave, but it can be amplified electronically. With a liquid medium the radiation efficiency is greater than for air; the increased damping also broadens the resonance curve. As a result, the above refinements in adjustment are unnecessary.

In Chapter 10 mention was made of the possibility of making a sound wave detector whose response is due to radiation pressure. For intensities in the audible range, such a detector is not sensitive enough for any practical use. In the ultrasonic field intensities are often much greater, especially with high frequency longitudinal waves in liquids. Under these conditions the radiation pressure principle has proved useful.* "Excess" acoustic pressures of one atmosphere or more are not uncommon in liquids. The corresponding radiation pressure upon a reflecting surface in the liquid is then of the order of several hundred dynes-cm^{-2}, a force of sizable proportions which will give good deflection if allowed to actuate a torsion vane.

---

* See Bergmann, *Ultrasonics*, pp. 44–49, John Wiley & Sons (1939), for a description of radiation pressure detectors for use in liquids.

12–20 Coupling between transducer and medium. In Chapter 5 and elsewhere considerable attention was given to the problem of efficient transformation of the vibrational energy of a source of sound into radiant acoustic energy. One of the great difficulties in introducing large amounts of power into air lies in the relatively small value of the specific acoustic impedance z_s ($= 42$ gm-cm^{-2}-sec^{-1}). Even in the ideal case where the waves are plane, the average acoustic power per unit area, $(\xi_{\text{rms}})^2\rho_0 c$, will be numerically small unless ξ_{rms} is itself impossibly large. We are then driven to the use of sources having large surface areas, a solution which creates difficulties of its own. Fortunately, because of the great sensitivity of the human ear, very large amounts of acoustic power are rarely necessary or desirable.

In the ultrasonic frequency range the picture is much brighter. Due to the shorter wavelength in air, the radiation from most ultrasonic sources has less angular divergence. In liquids, where the wavelengths are several times as long as in air, the advantage in this respect is much less apparent. The important difference between a liquid and a gas, as far as the efficiency of energy transfer into the medium is concerned, lies in the greatly increased value of $\rho_0 c$ in the case of the liquid. The specific acoustic impedance for fresh water is 1.43×10^5 cgs units. For sea water it is slightly greater, 1.54×10^5 cgs units. In the commonly used ultrasonic generators, the mechanical energy imparted to the vibrating source exists within the bar or plate in the form of longitudinal waves. For the nickel rod so often used in magnetostriction generators, $\rho_0 c$ is 4.2×10^6 cgs units. For the quartz almost universally used in the piezoelectric generator, $\rho_0 c$ is 1.5×10^6 cgs units. These values are much closer to the value of $\rho_0 c$ for water, or any other ordinary liquid for that matter, than they are to the value for air. Hence, as pointed out in Chapter 8, the energy will rapidly pass into the medium, instead of remaining as useless local energy of vibration.

The transfer is still further aided because of the small amplitudes possible with waves of high frequency. Since the radiated power depends on the square of the particle velocity imparted to the medium, and since the velocity is in turn equal to $2\pi f\xi$, it should be clear that as the frequency is raised, the amplitude may be dropped with no diminution of radiated power. Therefore very small amplitudes of motion at the source may give rise to large amounts of acoustic power. This makes possible small area sources operated through conservative amplitudes, as compared with the large area, large amplitude sources necessary at the lower audible frequencies.

All of the above factors leading to efficient energy transfer are important in the problem of underwater signaling, to be discussed in the next section.

12-21 Undersea signaling and ranging. The idea of using high frequency longitudinal waves to locate underwater obstacles was suggested as long ago as 1912 by L. F. Richardson. It was the careful work of Paul Langevin during and after World War I that really laid the foundation for the principles of undersea signaling and for much of the general subject of ultrasonics as well. Despite the desperate scientific study of aqueous ultrasonics during World War I, with an eye to the detection of enemy submarines, little actual use of the technique was made before the ending of hostilities. Between the two World Wars, however, considerable progress was made in the design of suitable transducers and their associated circuits. During World War II and afterwards much research has been done, not only in the development of suitable sources and receivers, but in the careful study of the fundamental physical phenomena associated with the propagation of high frequency waves through water. A considerable portion of the results of these studies has not been published, since the findings are of military interest.

Because electromagnetic waves will not propagate through sea water, ultrasonic waves are used as a means of friendly communication between submerged submarines, or in the case of disaster, between a sunken submarine and a surface vessel. All that is necessary is to mount a suitable transducer on the hull of each vessel. By a change of connections, a transmitter can also be made to serve as a receiver. By electrically modulating the superaudible frequency (usually of the order of 30,000 cycles-sec^{-1}) at a rate determined by the speech frequencies, telephony is made possible.

The principle of undersea acoustic ranging is essentially that of the ordinary echo in air. If an acoustic pulse of short duration is radiated in the direction of an obstacle, such as the bottom of the sea or a submerged vessel, some of the energy will return in the direction of the source, arriving back after a small finite interval of time. The time required for this return depends upon the speed of the waves in water, which is known, and upon the distance from the obstacle. Hence it is possible to compute from the time delay the distance of the obstacle from the source. The operation of such an acoustic system is quite similar to that of the well-known radar ranging system, which makes use of electromagnetic waves. In the acoustic procedure, the time lag between the outgoing and returning pulse is much greater than in the radar system, since acoustic waves travel much more slowly than electromagnetic waves. Therefore simpler means, some of them mechanical in nature, can be used to record the acoustic time interval.* A record on a moving strip of paper will give good precision.

---

* See Bergmann, *op. cit.*, p. 198, for a description of a system used for depth sounding.

Practical depth sounding systems have the transmitter and the receiver mounted on the bottom of the vessel. For locating distant enemy submarines, it is advantageous (as in radar) to have the waves concentrated as far as possible in the form of a beam. This energy concentration is desirable in the interests of energy conservation and it also gives an indication of the position of the enemy vessel, as well as its distance, since the scattered energy which returns will plainly be a maximum when the beam is directed towards the obstacle.

12–22 Diffraction of light by liquids carrying ultrasonic waves. In 1932 Debye and Sears * reported the diffraction of light by liquids carrying ultrasonic waves. The experimental arrangement is shown in Fig. 12–17. A parallel beam of light, originating at the slit t, is allowed to pass through the cell C, containing a liquid. A lens on the other side of the cell focuses an image of the slit on the screen S. At the bottom of the container is an ultrasonic generator that sends longitudinal waves into the liquid in a direction transverse to the beam of light. In the presence of acoustic waves, a diffraction pattern is visible on the screen, characterized by a central maximum with symmetrical subsidiary orders on each side, much as in the usual pattern of an ordinary diffraction grating.

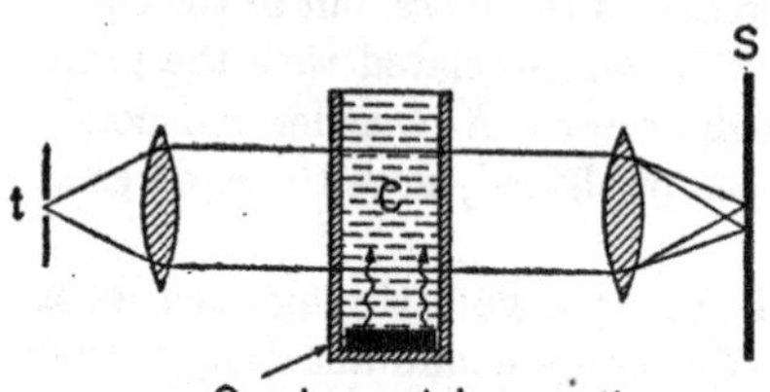

FIG. 12–17. Arrangement to show diffraction of light by liquid carrying ultrasonic waves.

The cause of the above phenomenon is the presence in the liquid of layers of variable density, periodically arranged along the direction of travel of the acoustic wave. The variations in density are associated with variations in the index of refraction. A scattering process results, similar in many respects to the Bragg scattering of x-ray energy by layers of atoms within a crystal. The periodic variation in the optical properties of the liquid are in this case analogous to the regular structure of the crystal lattice. The fact that the "lattice" in this case is traveling rapidly through the liquid at a speed equal to that of the acoustic wave is no complication, since this state of affairs will not change the essential *angular* relationship that determines the directions for reinforcement.

The simple notion that the diffraction originates with the scattering from regularly spaced layers would, for monochromatic light, lead to the

* Debye and Sears, *Proc. Nat. Acad. Sci. Wash.* **18,** 410 (1932).

formation of one order only, on each side of the central image. Actually, as many as ten orders can be seen, their angular dispersion agreeing with that given by the usual equation for a diffraction grating,

$$\sin \theta = n \frac{\lambda_l}{\lambda_s}, \tag{12-25}$$

where λ_l is the wavelength of the light, λ_s is the wavelength of the acoustic disturbance in the liquid, and n is an integer specifying the order.

There are many subtleties of acoustic diffraction patterns which become apparent in the complete theory of their formation. In the case of traveling acoustic waves, for instance, there is a small observable Doppler effect in the behavior of the light, due to the motion of the "grating." We are mainly concerned here with the possible uses of the Debye-Sears effect as a tool. It is quite obvious that the spacing of the lines in the diffraction pattern may be used to determine the acoustic wavelength, through the use of Eq. (12-25), otherwise a difficult problem at these frequencies. In the next section we shall mention a further application of the Debye-Sears effect.

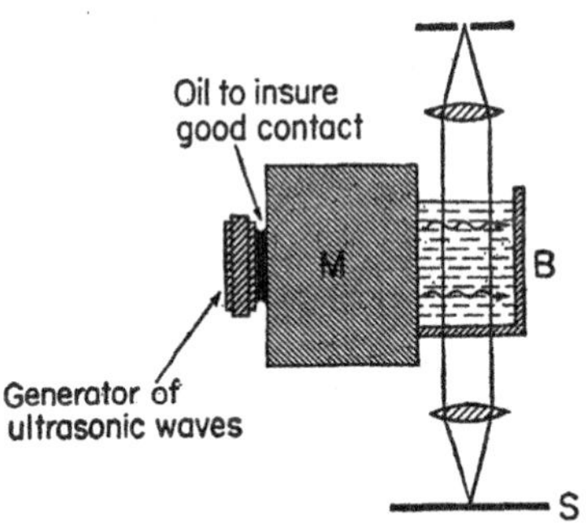

Fig. 12-18. Use of ultrasonic waves to detect inhomogeneities in solid materials. (*After* Sokoloff.)

12-23 Testing of materials with ultrasonic waves. The detection of flaws within optically opaque materials by the use of ultrasonic waves depends on the scattering effect associated with flaw regions in an otherwise homogeneous sample. To introduce appreciable amounts of energy into a sample, say, of the metal of a beam, it is common to use a liquid such as oil between the surface of the wave generator and the surface of the material. Otherwise losses due to reflection at the somewhat irregular boundaries of contact may prevent entry of sufficient acoustic energy. The same problem arises at the other side of the material, where the waves are detected.

Figure 12-18 shows the essential parts of one type of testing equipment, devised by Sokoloff.* The cross section of the bar to be tested is represented by the rectangle M. The waves are introduced at the left surface by means of a quartz transducer. In contact with the other side of the material is a liquid held in the container B. An optical system is arranged

* Sokoloff, *Phys. Z.* **36**, 142 (1935).

to send a beam of light vertically through the liquid, in the manner described in the preceding section. The acoustic waves present in the liquid because of energy transmitted through the material cause a diffraction pattern to appear on the screen S. As the sample is moved slowly along, the pattern will remain of constant intensity only if the material is perfectly homogeneous. Wherever there are internal inhomogeneities, internal scattering of energy will occur, resulting in noticeable variations in the intensity and sharpness of the interference lines. In this way internal defects are discovered.

The above testing method has been discussed largely because it is an interesting application of the Debye-Sears effect. Most of the equipment used for industrial testing today makes use of the echo principle discussed in Section 12-21. Flaws are assumed to exist wherever scattering occurs.

12-24 Other industrial applications of ultrasonic waves. Ultrasonic waves of great intensity produce marked effects upon mixtures of different liquids and upon liquids containing suspended particles. As early as 1927 Wood and Loomis performed experiments of this sort, in which they showed that a mixture of oil and water can be transformed into a very stable emulsion. As mentioned previously, it is possible to produce in a liquid variable acoustic pressures of sufficient magnitude so that peak values of several atmospheres, plus or minus, can occur. This often results in violent cavitations within the body of the liquid and consequent large mechanical dispersive forces. Too much of this effect is a detriment in some experiments. It is usually desirable to maintain a sufficiently high external pressure on the liquid so that the total internal pressure always remains positive.

Ultrasonic waves have proved of considerable value in the production of colloidal solutions of metals with particles of controllable size. The procedure consists of sending the waves through an electrolytic cell. In the presence of the wave, the minute particles of the metal are left in suspension in the liquid, and those of larger size collect at the bottom of the cell instead of adhering to the cathode. They are literally shaken loose from the cathode as soon as they touch the surface and deliver their charge.

In gases, the effect of the presence of high frequency waves is to *coagulate* small suspended particles, rather than to disperse them. This is primarily a Bernoulli effect, related to the process which causes the fine striations in the Kundt's tube experiment (Chapter 6). In this way dust, smoke particles, etc., can be removed from the air and other gases.

Many experiments are in progress to discover additional industrial applications for ultrasonic radiation. The chemical and photographic industries are particularly interested, primarily because of the effects just described.

12-25 Biological effects of ultrasonic waves. Experiments performed by Langevin, Wood and Loomis, and others indicate the great destructive power of high frequency vibrations upon living tissue. The effect seems to be largely indirect, rather than the result of a simple mechanical shaking of the structure. Some of the effect is due to the severe temperature rises which often ensue and some to the formation of bubbles of air within the tissue adjacent to regions where the pressure (due to the presence of the ultrasonic wave) has dropped well below the atmospheric level. It is the formation of these bubbles which often tears the tissue apart. Small organisms like protozoa, and even fish and frogs, can be destroyed by this method. Experimentation with ultrasonic waves as a possible means of destroying diseased tissue such as cancerous cells has been conducted. The effects of the waves upon bacteria and other minute organisms are varied and often difficult to interpret; they are sometimes killed, while at other times their virulence seems to be increased. Much work remains to be done along these lines before a consistent picture of the effect of high frequency mechanical vibrations can be obtained.

Since the introduction of jet-propelled aircraft, serious thought has been given to the effect upon an occupant (and others outside the plane) of constant exposure to vibrations in the ultrasonic region. It is quite possible that deleterious physiological effects result from prolonged exposure to waves of frequencies far above the limit of audibility.

12-26 Acoustics in relation to other branches of physics. The study of acoustics for its own sake is well worth the attention of every serious student of physics, but there are also cross-relationships with other branches of pure physics which are important both for their influence upon the development of acoustics and also for their beneficial effect upon other fields. In this sense the subject of acoustics may be "applied" to the solution of problems outside its own direct field. To mention an important historical example, the notion of quantized energy states originated in problems of vibration in the field of acoustics, and only later was this concept adopted for use in problems involving electromagnetic radiation. The theorem of Fourier originated with a study of heat flow, was taken over in vibration problems like that of the string, so important to the subject of acoustics, and finally found its way into modern atomic radiation theory and even into the discussion of meson waves.

The experimental aspects of modern acoustics are reflected in a current project to study the scattering of electromagnetic radiation by investigating acoustic scattering by an array of spheres. As an experimental tool, ultrasonic technique is proving most useful in experiments aimed at a closer determination of interatomic forces in solids at temperatures close to absolute zero. Such investigations involve a careful measurement of the speed and attenuation of high frequency longitudinal waves in superconducting solids. Doubtless many similar research projects making use of modern developments in experimental acoustics are in progress.

In short, acoustics, until recent years a rather neglected branch of physics, is about to take its rightful place as a subject of great intrinsic beauty and interest, and as a field of endeavor capable of continued growth and development.

PROBLEMS

1. (a) By what experimental means would one test, in a given room, the validity of the equations for the ideal reverberant chamber? (b) How, in general, would the energy density vary with position, under steady-state sound conditions, in a practical room whose reverberation time was very short? (c) Answer part (b) for an anechoic room.

2. (a) Using the Sabine equation (12–18), compute the reverberation time for a room which is a cube 20 ft on an edge, where the walls all have uniform surface treatment. The value of α_s is 0.1. (b) Compute the reverberation time for the same room if for two of the walls α_s is 0.1, and if for all of the remaining walls it is 0.2.

3. The cubical room of problem 2 has a uniform wall treatment where α_s is 0.1. Covering one wall with a different material lowers the reverberation time by 5%. Find the value of α_s for this material. (Make use of the Sabine equation.)

4. Compute the reverberation time for the room described in problem 2(a), using the Norris-Eyring equation.

5. A small single source of sound is placed at the center of a spherical room of radius R having walls which are nearly perfect reflectors. Find all the possible modes of vibration of the room which may be excited under these conditions.

6. Assume the wall surface of the room described in problem 5 to have an absorption coefficient α_s whose value is 0.25. (a) Find the reverberation time by both the Sabine and the Norris-Eyring equations. (b) How will the reverberation time vary with the diameter of the room?

7. The transmission loss through a certain partition is 40 db. If the intensity level on the high intensity side of the partition is 85 db, find the rms acoustic pressure on the low intensity side in dynes-cm^{-2}.

8. An acoustic piston of radius 10 cm is surrounded by a large flat baffle and is radiating acoustic energy as a single source, the frequency being 30,000 cycles-sec^{-1}. (a) Find the polar angle β, measured with respect to the normal to the piston surface, for the appearance of the first diffraction minimum, assuming the piston to be radiating into sea water. (b) Compute β if the piston is radiating into air and compare the result with the angle found in part (a).

APPENDIX I

The Introduction of the Velocity Potential, Φ, into the Differential Equation for Space Waves

We may start with the equation of continuity:

$$\frac{\partial(\rho u)}{\partial x} + \frac{\partial(\rho v)}{\partial y} + \frac{\partial(\rho w)}{\partial z} = -\frac{\partial \rho}{\partial t}. \tag{1}$$

The first term on the left may be written:

$$\frac{\partial(\rho u)}{\partial x} = \rho \frac{\partial u}{\partial x} + u \frac{\partial \rho}{\partial x},$$

which, since $\rho = \rho_0 + \rho_0 s$, becomes

$$\frac{\partial(\rho u)}{\partial x} = \rho_0 \frac{\partial u}{\partial x} + u\rho_0 \frac{\partial s}{\partial x}, \tag{2}$$

assuming small changes in ρ, so that the difference between ρ and ρ_0 may be neglected.

The second term on the right, for small amplitude disturbances, is small compared with the first term. This may be readily seen in the special case of a periodic plane wave disturbance, where both u and s are in phase and vary periodically in space with the wavelength. In this case, then, the ratio of $\partial u/\partial x$ to $\partial s/\partial x$ is of the same order of magnitude as the ratio of u to s. Therefore the second term on the right is to the first term as $u\rho_0 s$ is to $u\rho_0$, which is simply s. Since s is always a small fraction, we may therefore neglect the term $u\rho_0(\partial s/\partial x)$ in Eq. (2).

For the above reasons it is therefore valid to replace Eq. (1) by the simpler equation

$$\rho_0 \frac{\partial u}{\partial x} + \rho_0 \frac{\partial v}{\partial y} + \rho_0 \frac{\partial w}{\partial z} = -\frac{\partial \rho}{\partial t}. \tag{3}$$

Now, introducing the relationships between the velocity components and Φ, we may write Eq. (3) as

$$\frac{\partial^2 \Phi}{\partial x^2} + \frac{\partial^2 \Phi}{\partial y^2} + \frac{\partial^2 \Phi}{\partial z^2} = \frac{1}{\rho_0}\frac{\partial \rho}{\partial t}$$

or

$$\nabla^2 \Phi = \frac{1}{\rho_0}\frac{\partial \rho}{\partial t} = \frac{\partial s}{\partial t}. \tag{4}$$

We finally make use of the relationship (see Appendix 2)

$$\frac{\partial \Phi}{\partial t} = c^2 s.$$

Differentiating this equation partially with respect to the time and introducing the result into Eq. (4), we obtain the differential equation in terms of Φ:

$$c^2 \nabla^2 \Phi = \frac{\partial^2 \Phi}{\partial t^2}. \tag{5}$$

The same transformation to spherical coordinates discussed in Section 3–4 will then yield, for disturbances having spherical symmetry,

$$c^2 \frac{\partial^2 (r\Phi)}{\partial r^2} = \frac{\partial^2 (r\Phi)}{\partial t^2}. \tag{6}$$

APPENDIX II

The Relationship between the Velocity Potential Φ and the Condensation, s

We make use of Eq. (3–5a):

$$
\begin{aligned}
-\frac{\partial p}{\partial x} &= \frac{\partial(\rho u)}{\partial t} \\
&= \rho_0 \frac{\partial u}{\partial t} + u \frac{\partial \rho}{\partial t} \\
&= \rho_0 \frac{\partial u}{\partial t} + \rho_0 u \frac{\partial s}{\partial t},
\end{aligned}
\tag{1}
$$

neglecting the second order difference between ρ and ρ_0.

The second term on the right, for small amplitude disturbances, is small compared with the first term, essentially because the condensation s, under these conditions, is a very small fraction. Therefore we may write Eq. (1) in the form

$$
\rho_0 \frac{\partial u}{\partial t} = -\frac{\partial p}{\partial x}. \tag{2}
$$

Introducing the relationship $p = \mathcal{B}s$,

$$
\rho_0 \frac{\partial u}{\partial t} = -\mathcal{B}\frac{\partial s}{\partial x},
$$

or

$$
\frac{\partial u}{\partial t} = -c^2 \frac{\partial s}{\partial x}.
$$

Similar equations may be written for v and for w. We have, as a result, the three equations:

$$
\begin{aligned}
\frac{\partial u}{\partial t} &= -c^2 \frac{\partial s}{\partial x}, \\
\frac{\partial v}{\partial t} &= -c^2 \frac{\partial s}{\partial y}, \\
\frac{\partial w}{\partial t} &= -c^2 \frac{\partial s}{\partial z}.
\end{aligned}
\tag{3}
$$

Integrating each of these equations partially with respect to time, we obtain:

$$
\begin{aligned}
u &= -c^2 \frac{\partial}{\partial x} \int_0^t s\,dt, \\
v &= -c^2 \frac{\partial}{\partial y} \int_0^t s\,dt, \\
w &= -c^2 \frac{\partial}{\partial z} \int_0^t s\,dt.
\end{aligned}
\tag{4}
$$

(Note: Any constant of integration must be zero for periodic disturbances, since the average value of the wave parameters is always zero.)

Now introducing the velocity potential relationships, *i.e.*, $u = -\frac{\partial\Phi}{\partial x}$, $v = -\frac{\partial\Phi}{\partial y}$ and $w = -\frac{\partial\Phi}{\partial z}$, we see that

$$
\begin{aligned}
-\frac{\partial\Phi}{\partial x} &= -c^2\frac{\partial}{\partial x}\int_0^t s\,dt,\\
-\frac{\partial\Phi}{\partial y} &= -c^2\frac{\partial}{\partial y}\int_0^t s\,dt,\\
-\frac{\partial\Phi}{\partial z} &= -c^2\frac{\partial}{\partial z}\int_0^t s\,dt.
\end{aligned}
\tag{5}
$$

It may therefore be seen that

$$\Phi = c^2\int_0^t s\,dt,$$

or

$$\frac{\partial\Phi}{\partial t} = c^2 s. \tag{6}$$

APPENDIX III

TABLE OF FRESNEL INTEGRALS

$$x = \int_0^v \cos \frac{\pi v^2}{2}\, dv, \quad y = \int_0^v \sin \frac{\pi v^2}{2}\, dv$$

v	x	y	v	x	y
0.00	0.0000	0.0000	2.50	0.4574	0.6192
0.10	0.1000	0.0005	2.60	0.3890	0.5500
0.20	0.1999	0.0042	2.70	0.3925	0.4529
0.30	0.2994	0.0141	2.80	0.4675	0.3915
0.40	0.3975	0.0334	2.90	0.5626	0.4101
0.50	0.4923	0.0647	3.00	0.6058	0.4963
0.60	0.5811	0.1105	3.10	0.5616	0.5818
0.70	0.6597	0.1721	3.20	0.4664	0.5933
0.80	0.7230	0.2493	3.30	0.4058	0.5192
0.90	0.7648	0.3398	3.40	0.4385	0.4296
1.00	0.7799	0.4383	3.50	0.5326	0.4152
1.10	0.7638	0.5365	3.60	0.5880	0.4923
1.20	0.7154	0.6234	3.70	0.5420	0.5750
1.30	0.6386	0.6863	3.80	0.4481	0.5656
1.40	0.5431	0.7135	3.90	0.4223	0.4752
1.50	0.4453	0.6975	4.00	0.4984	0.4204
1.60	0.3655	0.6389	4.10	0.5738	0.4758
1.70	0.3238	0.5492	4.20	0.5418	0.5633
1.80	0.3336	0.4508	4.30	0.4494	0.5540
1.90	0.3944	0.3734	4.40	0.4383	0.4622
2.00	0.4882	0.3434	4.50	0.5261	0.4342
2.10	0.5815	0.3743	4.60	0.5673	0.5162
2.20	0.6363	0.4557	4.70	0.4914	0.5672
2.30	0.6266	0.5531	4.80	0.4338	0.4968
2.40	0.5550	0.6197	4.90	0.5002	0.4350

APPENDIX IV

Derivation of the Expression, $u = \frac{e_i c}{4}$ (Eq. 12-1).

In Fig. 1 we may consider dS to be an infinitesimal area of the wall surface and dV an infinitesimal volume element located at a distance r from the area dS.

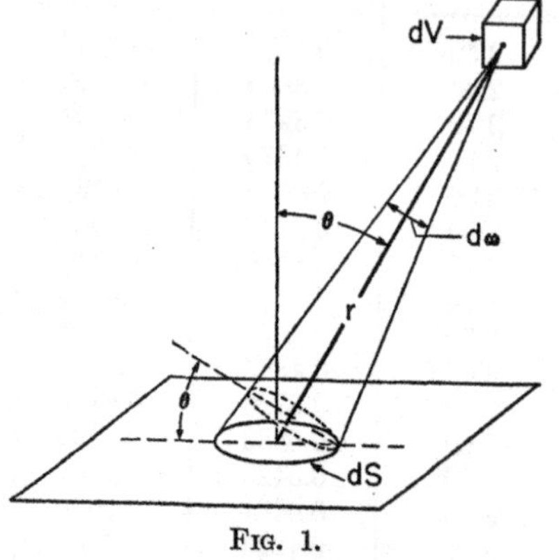

Fig. 1.

Let e_i be the instantaneous energy density in the room. The energy within the volume dV is therefore $e_i\,dV$. The fraction of this energy "directed" towards dS is $d\omega/4\pi$, where $d\omega$ is the infinitesimal solid angle subtended by dS. If we construct a ring-shaped volume element at a distance r from dS, we may express dV as

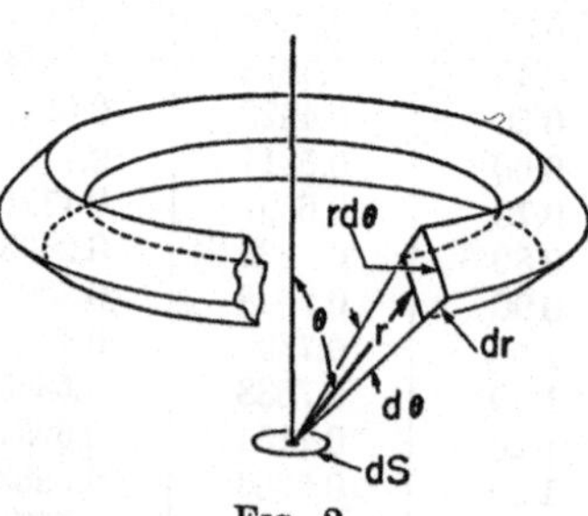

Fig. 2.

$$dV = 2\pi r^2 \sin\theta\, dr\, d\theta, \tag{1}$$

where θ is the polar angle indicated in Fig. 2. Therefore the total energy dE_r directed towards dS may be written as

$$dE_r = \frac{d\omega}{4\pi} e_i\, dV. \tag{2}$$

From Fig. 1,

$$d\omega = \frac{dS\cos\theta}{r^2}.$$

Inserting this expression in Eq. (2) and making use of the expression for dV given by Eq. (1), we have

$$dE_r = \frac{e_i\, dS}{2} \sin\theta\cos\theta\, dr\, d\theta. \tag{3}$$

The total energy incident upon the area dS in a time dt is that originating within a hemisphere of radius $c\,dt$, where c is the velocity of sound:

$$\text{Total energy incident upon the area } dS \text{ in a time } dt = \frac{e_i}{2}\,dS \int_0^{c\,dt} dr \int_0^{\pi/2} \sin\theta \cos\theta\, d\theta \quad (4)$$

$$= \frac{e_i c}{4}\,dS\,dt.$$

Therefore the energy u incident per unit area and per unit time is

$$u = \frac{e_i c}{4}. \quad (5)$$

LIST OF SYMBOLS

B' maximum strength of an acoustic source

$\mathcal{B}$ bulk modulus

$\mathcal{B}_a$ adiabatic bulk modulus

$\mathcal{B}_i$ isothermal bulk modulus

c wave velocity

C electrical capacitance

e instantaneous electrical potential

e energy density

e_i instantaneous energy density

e_0 initial or final energy density

E_m maximum value of sinusoidally varying electrical potential

E_{rms} root mean square value of sinusoidally varying electrical potential

E total energy

E_k total kinetic energy

E_p total potential energy

f frequency

F instantaneous force

F_m maximum value of sinusoidally varying force

F_{rms} root mean square value of sinusoidally varying force

i instantaneous electric current

I_m maximum value of sinusoidally varying electric current

I acoustic intensity

I_0 reference value of acoustic intensity

j $\sqrt{-1}$

k symbol abbreviation for $2\pi/\lambda$

K elastic constant (force per unit displacement)

L electrical inductance

m mass

M modulus of complex number

p instantaneous acoustic pressure

p_m maximum value of sinusoidally varying acoustic pressure

p_{rms} root mean square value of sinusoidally varying acoustic pressure

p_0 reference value of acoustic pressure

P instantaneous total pressure

P_0 average undisturbed pressure

r polar coordinate

r damping constant (force per unit velocity)

R electrical resistance

R magnitude of real part of complex impedance (electrical, mechanical, or acoustic)

s acoustic condensation

S area

t time

T period of SHM

T_r reverberation time

u energy flow per unit area and per unit time

U total energy flow per unit time

u, v, w cartesian components of instantaneous particle velocity

v variable used in Fresnel integrals

v volume change

V instantaneous volume

V_0 average undisturbed volume

W work

x instantaneous linear displacement

x_m maximum value of sinusoidally varying linear displacement

$\dot{x}$ instantaneous linear velocity

$\dot{x}_m$ maximum value of sinusoidally varying linear velocity

$\dot{x}_{rms}$ root mean square value of sinusoidally varying linear velocity

$\ddot{x}$ instantaneous linear acceleration

X magnitude of imaginary part of complex impedance (electrical, mechanical, or acoustic)

$\dot{\mathbf{X}}$ acoustic volume current

Y Young's modulus

z complex form of impedance (electrical, mechanical, or acoustic)

z_a analogous acoustic impedance (complex)

z_n normal specific acoustic impedance at a boundary (complex)

z_s specific acoustic impedance (complex)

z_m mechanical impedance (complex)

z_{em} electrical motional impedance (complex)

$(z)_R$ real part of complex impedance (electrical, mechanical, or acoustic)

$(z)_X$ imaginary part of complex impedance (electrical, mechanical, or acoustic)

Z modulus of complex impedance (electrical, mechanical, or acoustic)

List of Greek Symbols

α phase angle

α_n normal absorption coefficient

α_s Sabine absorption coefficient

$\bar{\alpha}_s$ average value of Sabine absorption coefficient

γ ratio of specific heats

δ acoustic dilatation

λ wavelength

ξ instantaneous particle displacement

ξ_m maximum value of sinusoidally varying particle displacement

$\dot{\xi}$ instantaneous particle velocity

$\dot{\xi}_m$ maximum value of sinusoidally varying particle velocity

$\dot{\xi}_{rms}$ root mean square value of sinusoidally varying particle velocity

ρ instantaneous density of medium

ρ_0 average undisturbed density of medium

Φ velocity potential

ω angular frequency

ω_u angular frequency for undamped particle vibration

REFERENCES

BOOKS

BERANEK, L. L., *Acoustic Measurements*, John Wiley & Sons, New York, 1949.
BERGMAN, L., *Ultrasonics*, John Wiley & Sons, New York, 1939.
COLBY, M. Y., *Sound Waves and Acoustics*, Henry Holt & Co., New York, 1938.
CULVER, C. A., *Musical Acoustics*, Blakiston Co., Philadelphia, 1947.
FLETCHER, H., *Speech and Hearing*, D. Van Nostrand Co., New York, 1929.
JENKINS, F. A., and WHITE, H. E., *Fundamentals of Optics*, McGraw-Hill Book Co., New York, 1950.
KNUDSEN, V. O., *Architectural Acoustics*, John Wiley & Sons, New York, 1932.
LAMB, H., *Dynamical Theory of Sound*, Edward Arnold & Co., London, 1931.
MILLER, D. C., *Anecdotal History of Sound*, Macmillan Co., New York, 1935.
MORSE, P. M., *Vibration and Sound*, McGraw-Hill Book Co., New York, 1948.
OLSON, H., *Elements of Acoustical Engineering*, D. Van Nostrand Co., New York, 1940.
STEVENS, S. S., and DAVIS, H., *Hearing, Its Psychology and Physiology*, John Wiley & Sons, New York, 1938.
WEVER, E. G., *Theory of Hearing*, John Wiley & Sons, New York, 1949.
WOOD, A. B., *A Textbook of Sound*, G. Bell and Sons, London, 1941.

PERIODICALS

ANDRADE, E. N. C., and PARKER, R. C., *Proc. Royal Soc. London*, **A159**, 507–526 (1937).
ARNOLD, H. D., and CRANDALL, I. B., *Phys. Rev.*, **10**, 22–38 (1917).
BERANEK, L. L., and SLEEPER, H. S., *Jour. Acous. Soc. Amer.*, **18**, 140 (1946).
BUCKINGHAM, E., Nat. Bur. Stand. (U. S.) *Sc. Technol. Papers*, **20**, 193–219 (1925).
CHAPIN, E. K., and FIRESTONE, F. A., *Jour. Acous. Soc. Amer.*, **5**, 173 (1934).
DEBYE, P., and SEARS, F. W., *Proc. Nat. Acad. Sci. Wash.*, **18**, 410 (1932).
DICKEY, CAULTON, and PERRY, *Radio Engineering*, **8**, No. 2, 104 (1936).
DUBOIS, R., *Revue d'acoustique*, **2**, 253–287 (1932).
DUDLEY, H., *Bell System Technical Jour.*, **19**, 496 (1940).
DUFF, A. W., *Phys. Rev.*, **11**, 64 (1900).
EYRING, C. F., *Jour. Acous. Soc. Amer.*, **1**, 217 (1930).
FLETCHER, H., *Rev. Modern Physics*, **12**, 47–65 (1940).
FLETCHER, H., and MUNSON, W. A., *Jour. Acous. Soc. Amer.*, **9**, 1 (1937).
FRENCH, N. R., and STEINBERG, J. C., *Jour. Acous. Soc. Amer.*, **19**, 90–119 (1947).
GLOVER, R., and BAUMZWEIGER, B., *Jour. Acous. Soc. Amer.*, **10**, 200–202 (1939).
HARDY, H. C., TELFAIR, D., and PIELEMEIER, W. H., *Jour. Acous. Soc. Amer.*, **13**, 226–233 (1942).
HART, M. D., *Proc. Roy. Soc.*, **A105**, 80 (1924).
HARTLEY, R. V. L., and FRY, T. C., *Phys. Rev.*, **18**, 431 (1921).

HEBB, T. C., *Phys. Rev.*, **20**, 91 (1905).
———, *Phys. Rev.*, **14**, 74 (1919).
HERZFELD, K. F., and RICE, F. O., *Phys. Rev.*, **31**, 691 (1928).
HICKMAN, C. N., *Jour. Acous. Soc. Amer.*, **6**, 108–111 (1934).
KNUDSEN, V. O., *Jour. Acous. Soc. Amer.*, **5**, 112 (1933).
KOCK, W. E., *Jour. Acous. Soc. Amer.*, **22**, 804 (1950).
KOCK, W. E., and HARVEY, F. K., *Jour. Acous. Soc. Amer.*, **21**, 471–481 (1949).
MACLEAN, W. R., *Jour. Acous. Soc. Amer.*, **12**, 140–146 (1940).
MEYER, E., *Jour. Acous. Soc. Amer.*, **7**, 88–93 (1935).
MEYER, E., and ZEITS, F., *Techn. Physik*, **7**, 253 (1930).
PIERCE, G. W., *Proc. Amer. Acad.*, **60**, 271 (1925).
POL, B. VAN DER, and MARK, J. VAN DER, *Phil. Mag.* **2**, 978 (1926).
RAMAN, C. V., *Ind. Assoc. for Cult. of Science*, **15**, 1–158 (1918).
———, *Phil. Mag.*, **38**, 573–581 (1919).
RANDALL, R. H., ROSE, F. C., and ZENER, C., *Phys. Rev.*, **53**, 343 (1939).
RANKE, O. F., *Jour. Acous. Soc. Amer.*, **22**, 772 (1950).
RUDMOSE, H. W., CLARK, K. C., CARLSON, F. D., EISENSTEIN, J. C., and WALKER, R. A., *Jour. Acous. Soc. Amer.*, **20**, 507 (1948).
SABINE, W. C., *Jour. Franklin Inst.*, **207**, 347 (1929).
SHOWER, E. G., and BIDDULPH, R., *Jour. Acous. Soc. Amer.*, **3**, 275 (1931).
SNOW, W. B., *Jour. Acous. Soc. Amer.*, **8**, 14 (1936).
SOKOLOFF, S. J. VON, *Phys. Z.*, **36**, 142 (1935).
STEVENS, S. S., and VOLKMAN, J., *Amer. Jour. Psych.*, **53**, 329 (1940).
STEWART, G. W., *Phys. Rev.*, **15**, 425 (1920).
———, *Phys. Rev.*, **28**, 1038 (1926).
WALLACE, R. L., DIENEL, H. F., and BERANEK, L. L., *Jour. Acous. Soc. Amer.*, **18**, 246 (1946) (abstract).
WIENER, F. M., *Jour. Acous. Soc. Amer.*, **19**, 446 (1947).
ZENER, C., *Phys. Rev.*, **52**, 230 (1937).
———, *Phys. Rev.*, **53**, 90 (1938).
ZWISLOCKI, J., *Jour. Acous. Soc. Amer.*, **22**, 778 (1950).

Bulletin of Acoustic Materials Assoc., **VII** (1940).
Quarterly Progress Report, Acoustic Lab. M. I. T., 7–9 (July–Sept., 1948).

INDEX

ANSWERS TO PROBLEMS

Chapter 1

1. Max displacement $= x_m$.
Max velocity $= \omega x_m$.
Max acceleration $= \omega^2 x_m$.

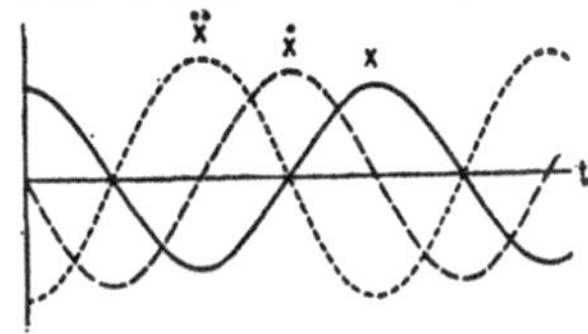

3. $x + \sqrt{x^2 + \frac{mv_0^2}{A}} = \sqrt{\frac{mv_0^2}{A}}\,\epsilon^{\sqrt{(A/m)}\,t}$.

5. (a) 4.93 cm. (b) 21°53′.

7. (a) One mathematical beat, three audible beats.
(b) Two mathematical beats and two audible beats.

9. $x = \frac{4a}{\pi}\left[\sin \omega t + \frac{1}{3}\sin 3\omega t + \frac{1}{5}\sin 5\omega t \cdots\right]$.

11. $i = \frac{I_m}{\pi}\left(1 + \frac{\pi}{2}\sin 2\pi ft - \frac{2}{3}\cos 4\pi ft - \frac{2}{15}\cos 8\pi ft - \cdots\right)$.

13. (a) Oscillatory.
(b) With damping, 390 cycles-sec^{-1}; without damping, 504 cycles-sec^{-1}.

15. 6.23 gm.

Chapter 2

1. Max particle displacement $= \xi_m$.
Max particle velocity $= \dot{\xi}_m$.
Max dilatation $= (2\pi/\lambda)\xi_m$.
Max pressure $= \rho_0 c^2 (2\pi/\lambda)\xi_m$.

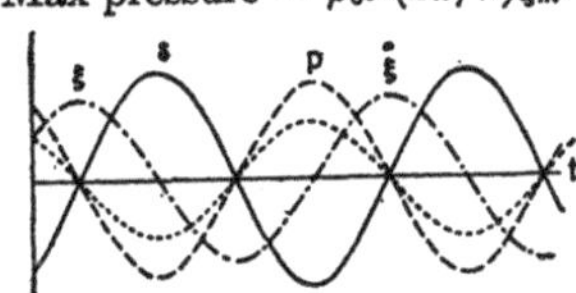

3. ξ_1 is 180° out of phase with ξ_2.
$\dot{\xi}_1$ is 180° out of phase with $\dot{\xi}_2$.
p_1 is in phase with p_2.

5. (a) $e_k = 1.8 \times 10^{-9}$ erg-cm^{-3}.
(b) $e_p = 1.8 \times 10^{-9}$ erg-cm^{-3}.
(c) $I = 1.19 \times 10^{-4}$ erg-cm^{-2}-sec^{-1}.
(d) $e_k = e_p = 1.8 \times 10^{-10}$ joule-m^{-3}.
$I = 1.19 \times 10^{-7}$ joule-m^{-2}-sec^{-1}.

7. (a) $(\xi_m)_a/(\xi_m)_h = 0.517$.
(b) $(\dot{\xi}_m)_a/(\dot{\xi}_m)_h = 0.517$.
(c) $(p_m)_a/(p_m)_h = 1.93$.

9. (a) I_c is 20 db higher than I_a.
(b) $I_b/I_a = 10/1 \cdot I_c/I_b = 10/1 \cdot I_c/I_a = 100/1$.

Chapter 3

1. (a) $\Phi_m = 10^{-3}/r$.
(b) $\Phi_m = 3.33 \times 10^{-7}$ cm^2-sec^{-1}.
(c) $\dot{\xi}_m = 6.32 \times 10^{-9}$ cm-sec^{-1}.
(d) $p_m = 2.7 \times 10^{-7}$ dyne-cm^{-2}.

3. (a) $\Phi_m = 1.5/r$.
(b) $\Phi_m = 8.5 \times 10^{-3}$ cm^2-sec^{-1}.
(c) $\dot{\xi}_m = 4.98 \times 10^{-2}$ cm-sec^{-1}.
(d) $p_m = 2.13$ dyne-cm^{-2}.

5. (a) Incorrect. (b) Correct.
   (c) Correct. (d) Incorrect.
7. (a) $B' = 39.4$ cm³-sec⁻¹.
   (b) $\Phi_m = 1.57 \times 10^{-3}$ cm²-sec⁻¹.
   (c) $p_m = 2.54 \times 10^{-3}$ dyne-cm⁻².
9. (a) $B' = 78.8$ cm³-sec⁻¹.
   (b) $\Phi_m = 3.14 \times 10^{-3}$ cm²-sec⁻¹.
   (c) $p_m = 5.08 \times 10^{-3}$ dyne-cm⁻².
11. (a) $\delta = \dfrac{1}{r^2}\dfrac{\partial(r^2\xi)}{\partial r}$.
   (b) For large values of r, $\delta \cong \dfrac{\partial \xi}{\partial r}$.

Chapter 4

1.

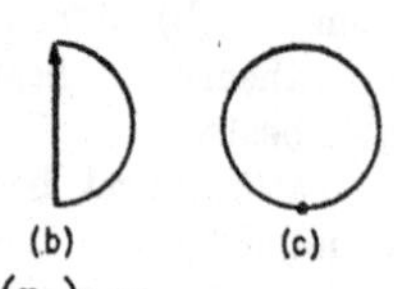

3. (a) $\dfrac{(p_m)_{\lambda/2}}{(p_m)_\lambda} = 0.707$.
   (b) $\dfrac{(p_m)_{\lambda/10}}{(p_m)_\lambda} = 0.141$.
5.

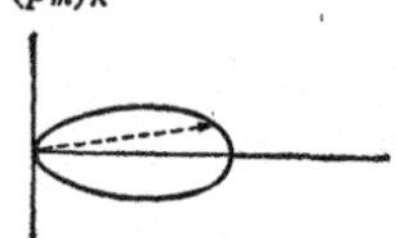

7. $R_1 = 24.6$ cm.
   $R_2 = 45.8$ cm.
   $R_3 = 68.5$ cm.
9. First order, 15°10′; second order, 31°30′.
11. (b) The slope is the tangent of the phase angle between the total instantaneous pressure (due to a given exposure of the wave front) and the pressure contribution originating at a point directly opposite a.
13. (b) $(p_m)_1/(p_m)_2 = 0.51$.

Chapter 5

1. (a) $8.66 + j5.0$.
   (b) $4.93 - j0.870$.
   (c) $14.5 + j2.58$.
   (d) $0.825 - j0.477$.
3. (a) $1.23 + j1.87$.
   (b) $1.85 - j0.326$.
   (c) $(ce + df) + j(cf - de)$.
5. (a) $\dfrac{p_m}{\dot{\xi}_m}\cos\theta - j\dfrac{p_m}{\dot{\xi}_m}\sin\theta$.
   (b) $\dfrac{p_m\dot{\xi}_m}{2}\cos\theta$.
7. (a) The frequency is reduced by the factor 0.707.
   (b) The frequency is increased by the factor 1.41.
9.

$\dot{\xi}_m$
29.5
f (cycles-sec⁻¹)

11. $Z = 80 + j19.5$ ohms.
13. Approximately 0.7.
15. At approximately 188 cycles-sec⁻¹.
17. At approximately 350 cycles-sec⁻¹.

Chapter 6

1. Only at very high ultrasonic frequencies is there any variation in wave velocity with frequency change.
3. $c \propto \sqrt{T}$. Although the density will remain constant, the velocity will change with temperature due to the variation in the pressure of the enclosed gas.

5. (a) $s_m = 0.008$.
   (b) The peak value of the condensation found in (a) will occur if the intensity level is 152 db.
7. 133 rpm.

Chapter 7

1. $f_A = f_B$.
3. (a) The harmonic frequencies are given by $f = nc/21$, where the integer n may have any value except an integral multiple of 4.
   (b) The 4th, 8th, 12th, etc., harmonics are missing, since these modes require a node at the point of plucking.
5. (a) p and ξ are out of phase by 90° everywhere along the pipe. Hence the average power flow is zero.
   (b) z_s is infinite at the pressure antinodes, zero at the pressure nodes, reactive in between.
7. For resonant frequencies = $(n)(217)$ cycles-sec^{-1}, where n is any integer. There will be little cancellation, however, since the dipole components are widely separated in space.
9. 8.26×10^6 ergs-sec^{-1}.
11. Due to the buckling of the paper under the action of the axial force, the paper along an element of the cone may execute one-half of a transverse vibration cycle during one complete cycle of the axial motion.

Chapter 8

1. (a) $\dfrac{s_r}{s_i} = \dfrac{\rho_2c_2 - \rho_1c_1}{\rho_1c_1 + \rho_2c_2}$.
   (b) $\dfrac{s_t}{s_i} = \dfrac{2\rho_1c_1^2}{c_2(\rho_1c_1 + \rho_2c_2)}$.
3. $(z_s)_2 = 7.21$ gm-cm^{-2}-sec^{-1}.
5. $p_{rms} = 6.35$ dynes-cm^{-2}.
7. (a) ξ lags p by 18°27′.
   (b) 6.8 ergs-sec^{-1}.
9. No; because the reflected wave will have spherical divergence.

Chapter 9

1. (a) No. (b) The throat and mouth constitute an inefficient radiating system at 80 cycles-sec^{-1}. (c) The observed pitch is due to the ability of the ear to supply the fundamental of a harmonic series.
3. At 40 cycles-sec^{-1}, about 70 db. At 1000 cycles-sec^{-1}, about 120 db. At 10,000 cycles-sec^{-1}, about 110 db.
5. Highly damped.
7. About 13,000 millisones.
9. (a) Percentage precision greatest at medium and high frequencies. (b) At 5 db level, 7.4 percent or cycles-sec^{-1}, at 10 db level, 5.4 percent or cycles-sec^{-1}, at 20 db level, 4 percent or cycles-sec^{-1}, at 60 db or higher level, 3 percent or cycles-sec^{-1}.

Chapter 10

1. Since the dimensions of the entrances to the air channels become less than the wavelength at the very low frequencies, some of the wave energy will be reflected due to diffraction effects.

3. (a) So that no phase differences in pressure may exist throughout the enclosure.
5. (a) A slight displacement away from this position produces a torque tending to displace the disk further in the same direction. (b) A slight displacement away from this position produces a restoring torque.
7. The spacing should be small to increase the sensitivity of the microphone. The ratio of the spacing dimension to the diameter should be small to reduce edge effects so that the change in capacitance may be nearly proportional to the diaphragm displacement.
9. Not if the amplified sound is to be received aurally.
11. (a) Possible existence of stationary waves, especially at low frequencies, where the damping by means of tufts is more difficult. Dimensions at the junction must be kept small compared with the wavelength. (b) No. (c) Best precision at the middle audio frequencies.

Chapter 11

1. (a) 58%. (b) 26.1%.
3. (a)

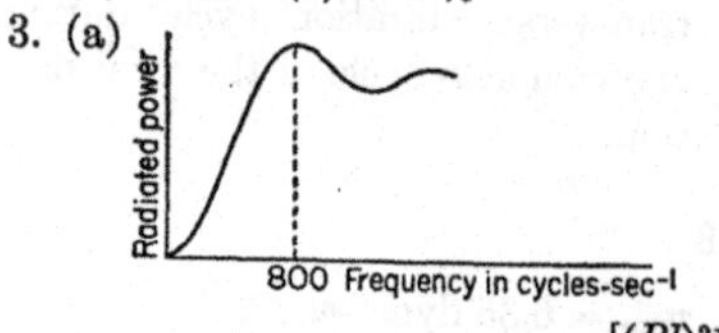

(b) No; since the radiation will fall off rapidly below about 800 cycles-sec^{-1}. Also, the efficiency will be very low.
5. 62 dynes.
7. $$z_{em} = \frac{[(Bl)^2 10^{-9}][r + S(z_s)_R]}{[r + S(z_s)_R]^2 + \left[\left(\omega m - \frac{K}{\omega}\right) + S(z_s)_X\right]^2} - j\frac{[(Bl)^2 10^{-9}]\left[\left(\omega m - \frac{K}{\omega}\right) + S(z_s)_X\right]}{[(r + S(z_s)_R)]^2 + \left[\left(\omega m - \frac{K}{\omega}\right) + S(z_s)_X\right]^2}.$$
9. (a) 1950 cycles-sec^{-1}.
(b) 0.69 ohm.
(c) It will be lowered.
11. (a) The acoustic power output will be reduced to $\frac{1}{4}$ of the original value.
(b) The acoustic power output will remain unchanged.

Chapter 12

1. (a) Set up in the room a small steady source of sound. Verify, by means of a microphone, the constancy of the acoustic pressure throughout the room for all positions and orientations of the microphone. (b) The energy density will fall off with distance from the sound source, although not according to the inverse square law. (c) The intensity will fall off inversely with the square of the distance from the source.
3. $\alpha_s = 0.13$.
5. $f = nc/4R$, where c is the velocity of sound and n is any odd integer.
7. $p_{\text{rms}} = 3.65 \times 10^{-2}$ dyne-cm^{-2}.

A CATALOG OF SELECTED

DOVER BOOKS

IN SCIENCE AND MATHEMATICS

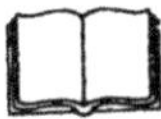

Astronomy

CHARIOTS FOR APOLLO: The NASA History of Manned Lunar Spacecraft to 1969, Courtney G. Brooks, James M. Grimwood, and Loyd S. Swenson, Jr. This illustrated history by a trio of experts is the definitive reference on the Apollo spacecraft and lunar modules. It traces the vehicles' design, development, and operation in space. More than 100 photographs and illustrations. 576pp. 6 3/4 x 9 1/4. 0-486-46756-2

EXPLORING THE MOON THROUGH BINOCULARS AND SMALL TELESCOPES, Ernest H. Cherrington, Jr. Informative, profusely illustrated guide to locating and identifying craters, rills, seas, mountains, other lunar features. Newly revised and updated with special section of new photos. Over 100 photos and diagrams. 240pp. 8 1/4 x 11. 0-486-24491-1

WHERE NO MAN HAS GONE BEFORE: A History of NASA's Apollo Lunar Expeditions, William David Compton. Introduction by Paul Dickson. This official NASA history traces behind-the-scenes conflicts and cooperation between scientists and engineers. The first half concerns preparations for the Moon landings, and the second half documents the flights that followed Apollo 11. 1989 edition. 432pp. 7 x 10. 0-486-47888-2

APOLLO EXPEDITIONS TO THE MOON: The NASA History, Edited by Edgar M. Cortright. Official NASA publication marks the 40th anniversary of the first lunar landing and features essays by project participants recalling engineering and administrative challenges. Accessible, jargon-free accounts, highlighted by numerous illustrations. 336pp. 8 3/8 x 10 7/8. 0-486-47175-6

ON MARS: Exploration of the Red Planet, 1958-1978--The NASA History, Edward Clinton Ezell and Linda Neuman Ezell. NASA's official history chronicles the start of our explorations of our planetary neighbor. It recounts cooperation among government, industry, and academia, and it features dozens of photos from Viking cameras. 560pp. 6 3/4 x 9 1/4. 0-486-46757-0

ARISTARCHUS OF SAMOS: The Ancient Copernicus, Sir Thomas Heath. Heath's history of astronomy ranges from Homer and Hesiod to Aristarchus and includes quotes from numerous thinkers, compilers, and scholasticists from Thales and Anaximander through Pythagoras, Plato, Aristotle, and Heraclides. 34 figures. 448pp. 5 3/8 x 8 1/2. 0-486-43886-4

AN INTRODUCTION TO CELESTIAL MECHANICS, Forest Ray Moulton. Classic text still unsurpassed in presentation of fundamental principles. Covers rectilinear motion, central forces, problems of two and three bodies, much more. Includes over 200 problems, some with answers. 437pp. 5 3/8 x 8 1/2. 0-486-64687-4

BEYOND THE ATMOSPHERE: Early Years of Space Science, Homer E. Newell. This exciting survey is the work of a top NASA administrator who chronicles technological advances, the relationship of space science to general science, and the space program's social, political, and economic contexts. 528pp. 6 3/4 x 9 1/4. 0-486-47464-X

STAR LORE: Myths, Legends, and Facts, William Tyler Olcott. Captivating retellings of the origins and histories of ancient star groups include Pegasus, Ursa Major, Pleiades, signs of the zodiac, and other constellations. "Classic." – *Sky & Telescope.* 58 illustrations. 544pp. 5 3/8 x 8 1/2. 0-486-43581-4

A COMPLETE MANUAL OF AMATEUR ASTRONOMY: Tools and Techniques for Astronomical Observations, P. Clay Sherrod with Thomas L. Koed. Concise, highly readable book discusses the selection, set-up, and maintenance of a telescope; amateur studies of the sun; lunar topography and occultations; and more. 124 figures. 26 halftones. 37 tables. 335pp. 6 1/2 x 9 1/4. 0-486-42820-6

Chemistry

MOLECULAR COLLISION THEORY, M. S. Child. This high-level monograph offers an analytical treatment of classical scattering by a central force, quantum scattering by a central force, elastic scattering phase shifts, and semi-classical elastic scattering. 1974 edition. 310pp. 5 3/8 x 8 1/2. 0-486-69437-2

HANDBOOK OF COMPUTATIONAL QUANTUM CHEMISTRY, David B. Cook. This comprehensive text provides upper-level undergraduates and graduate students with an accessible introduction to the implementation of quantum ideas in molecular modeling, exploring practical applications alongside theoretical explanations. 1998 edition. 832pp. 5 3/8 x 8 1/2. 0-486-44307-8

RADIOACTIVE SUBSTANCES, Marie Curie. The celebrated scientist's thesis, which directly preceded her 1903 Nobel Prize, discusses establishing atomic character of radioactivity; extraction from pitchblende of polonium and radium; isolation of pure radium chloride; more. 96pp. 5 3/8 x 8 1/2. 0-486-42550-9

CHEMICAL MAGIC, Leonard A. Ford. Classic guide provides intriguing entertainment while elucidating sound scientific principles, with more than 100 unusual stunts: cold fire, dust explosions, a nylon rope trick, a disappearing beaker, much more. 128pp. 5 3/8 x 8 1/2. 0-486-67628-5

ALCHEMY, E. J. Holmyard. Classic study by noted authority covers 2,000 years of alchemical history: religious, mystical overtones; apparatus; signs, symbols, and secret terms; advent of scientific method, much more. Illustrated. 320pp. 5 3/8 x 8 1/2. 0-486-26298-7

CHEMICAL KINETICS AND REACTION DYNAMICS, Paul L. Houston. This text teaches the principles underlying modern chemical kinetics in a clear, direct fashion, using several examples to enhance basic understanding. Solutions to selected problems. 2001 edition. 352pp. 8 3/8 x 11. 0-486-45334-0

PROBLEMS AND SOLUTIONS IN QUANTUM CHEMISTRY AND PHYSICS, Charles S. Johnson and Lee G. Pedersen. Unusually varied problems, with detailed solutions, cover of quantum mechanics, wave mechanics, angular momentum, molecular spectroscopy, scattering theory, more. 280 problems, plus 139 supplementary exercises. 430pp. 6 1/2 x 9 1/4. 0-486-65236-X

ELEMENTS OF CHEMISTRY, Antoine Lavoisier. Monumental classic by the founder of modern chemistry features first explicit statement of law of conservation of matter in chemical change, and more. Facsimile reprint of original (1790) Kerr translation. 539pp. 5 3/8 x 8 1/2. 0-486-64624-6

MAGNETISM AND TRANSITION METAL COMPLEXES, F. E. Mabbs and D. J. Machin. A detailed view of the calculation methods involved in the magnetic properties of transition metal complexes, this volume offers sufficient background for original work in the field. 1973 edition. 240pp. 5 3/8 x 8 1/2. 0-486-46284-6

GENERAL CHEMISTRY, Linus Pauling. Revised third edition of classic first-year text by Nobel laureate. Atomic and molecular structure, quantum mechanics, statistical mechanics, thermodynamics correlated with descriptive chemistry. Problems. 992pp. 5 3/8 x 8 1/2. 0-486-65622-5

ELECTROLYTE SOLUTIONS: Second Revised Edition, R. A. Robinson and R. H. Stokes. Classic text deals primarily with measurement, interpretation of conductance, chemical potential, and diffusion in electrolyte solutions. Detailed theoretical interpretations, plus extensive tables of thermodynamic and transport properties. 1970 edition. 590pp. 5 3/8 x 8 1/2. 0-486-42225-9

Engineering

FUNDAMENTALS OF ASTRODYNAMICS, Roger R. Bate, Donald D. Mueller, and Jerry E. White. Teaching text developed by U.S. Air Force Academy develops the basic two-body and n-body equations of motion; orbit determination; classical orbital elements, coordinate transformations; differential correction; more. 1971 edition. 455pp. 5 3/8 x 8 1/2. 0-486-60061-0

INTRODUCTION TO CONTINUUM MECHANICS FOR ENGINEERS: Revised Edition, Ray M. Bowen. This self-contained text introduces classical continuum models within a modern framework. Its numerous exercises illustrate the governing principles, linearizations, and other approximations that constitute classical continuum models. 2007 edition. 320pp. 6 1/8 x 9 1/4. 0-486-47460-7

ENGINEERING MECHANICS FOR STRUCTURES, Louis L. Bucciarelli. This text explores the mechanics of solids and statics as well as the strength of materials and elasticity theory. Its many design exercises encourage creative initiative and systems thinking. 2009 edition. 320pp. 6 1/8 x 9 1/4. 0-486-46855-0

FEEDBACK CONTROL THEORY, John C. Doyle, Bruce A. Francis and Allen R. Tannenbaum. This excellent introduction to feedback control system design offers a theoretical approach that captures the essential issues and can be applied to a wide range of practical problems. 1992 edition. 224pp. 6 1/2 x 9 1/4. 0-486-46933-6

THE FORCES OF MATTER, Michael Faraday. These lectures by a famous inventor offer an easy-to-understand introduction to the interactions of the universe's physical forces. Six essays explore gravitation, cohesion, chemical affinity, heat, magnetism, and electricity. 1993 edition. 96pp. 5 3/8 x 8 1/2. 0-486-47482-8

DYNAMICS, Lawrence E. Goodman and William H. Warner. Beginning engineering text introduces calculus of vectors, particle motion, dynamics of particle systems and plane rigid bodies, technical applications in plane motions, and more. Exercises and answers in every chapter. 619pp. 5 3/8 x 8 1/2. 0-486-42006-X

ADAPTIVE FILTERING PREDICTION AND CONTROL, Graham C. Goodwin and Kwai Sang Sin. This unified survey focuses on linear discrete-time systems and explores natural extensions to nonlinear systems. It emphasizes discrete-time systems, summarizing theoretical and practical aspects of a large class of adaptive algorithms. 1984 edition. 560pp. 6 1/2 x 9 1/4. 0-486-46932-8

INDUCTANCE CALCULATIONS, Frederick W. Grover. This authoritative reference enables the design of virtually every type of inductor. It features a single simple formula for each type of inductor, together with tables containing essential numerical factors. 1946 edition. 304pp. 5 3/8 x 8 1/2. 0-486-47440-2

THERMODYNAMICS: Foundations and Applications, Elias P. Gyftopoulos and Gian Paolo Beretta. Designed by two MIT professors, this authoritative text discusses basic concepts and applications in detail, emphasizing generality, definitions, and logical consistency. More than 300 solved problems cover realistic energy systems and processes. 800pp. 6 1/8 x 9 1/4. 0-486-43932-1

THE FINITE ELEMENT METHOD: Linear Static and Dynamic Finite Element Analysis, Thomas J. R. Hughes. Text for students without in-depth mathematical training, this text includes a comprehensive presentation and analysis of algorithms of time-dependent phenomena plus beam, plate, and shell theories. Solution guide available upon request. 672pp. 6 1/2 x 9 1/4. 0-486-41181-8

HELICOPTER THEORY, Wayne Johnson. Monumental engineering text covers vertical flight, forward flight, performance, mathematics of rotating systems, rotary wing dynamics and aerodynamics, aeroelasticity, stability and control, stall, noise, and more. 189 illustrations. 1980 edition. 1089pp. 5 5/8 x 8 1/4. 0-486-68230-7

MATHEMATICAL HANDBOOK FOR SCIENTISTS AND ENGINEERS: Definitions, Theorems, and Formulas for Reference and Review, Granino A. Korn and Theresa M. Korn. Convenient access to information from every area of mathematics: Fourier transforms, Z transforms, linear and nonlinear programming, calculus of variations, random-process theory, special functions, combinatorial analysis, game theory, much more. 1152pp. 5 3/8 x 8 1/2. 0-486-41147-8

A HEAT TRANSFER TEXTBOOK: Fourth Edition, John H. Lienhard V and John H. Lienhard IV. This introduction to heat and mass transfer for engineering students features worked examples and end-of-chapter exercises. Worked examples and end-of-chapter exercises appear throughout the book, along with well-drawn, illuminating figures. 768pp. 7 x 9 1/4. 0-486-47931-5

BASIC ELECTRICITY, U.S. Bureau of Naval Personnel. Originally a training course; best nontechnical coverage. Topics include batteries, circuits, conductors, AC and DC, inductance and capacitance, generators, motors, transformers, amplifiers, etc. Many questions with answers. 349 illustrations. 1969 edition. 448pp. 6 1/2 x 9 1/4. 0-486-20973-3

BASIC ELECTRONICS, U.S. Bureau of Naval Personnel. Clear, well-illustrated introduction to electronic equipment covers numerous essential topics: electron tubes, semiconductors, electronic power supplies, tuned circuits, amplifiers, receivers, ranging and navigation systems, computers, antennas, more. 560 illustrations. 567pp. 6 1/2 x 9 1/4. 0-486-21076-6

BASIC WING AND AIRFOIL THEORY, Alan Pope. This self-contained treatment by a pioneer in the study of wind effects covers flow functions, airfoil construction and pressure distribution, finite and monoplane wings, and many other subjects. 1951 edition. 320pp. 5 3/8 x 8 1/2. 0-486-47188-8

SYNTHETIC FUELS, Ronald F. Probstein and R. Edwin Hicks. This unified presentation examines the methods and processes for converting coal, oil, shale, tar sands, and various forms of biomass into liquid, gaseous, and clean solid fuels. 1982 edition. 512pp. 6 1/8 x 9 1/4. 0-486-44977-7

THEORY OF ELASTIC STABILITY, Stephen P. Timoshenko and James M. Gere. Written by world-renowned authorities on mechanics, this classic ranges from theoretical explanations of 2- and 3-D stress and strain to practical applications such as torsion, bending, and thermal stress. 1961 edition. 560pp. 5 3/8 x 8 1/2. 0-486-47207-8

PRINCIPLES OF DIGITAL COMMUNICATION AND CODING, Andrew J. Viterbi and Jim K. Omura. This classic by two digital communications experts is geared toward students of communications theory and to designers of channels, links, terminals, modems, or networks used to transmit and receive digital messages. 1979 edition. 576pp. 6 1/8 x 9 1/4. 0-486-46901-8

LINEAR SYSTEM THEORY: The State Space Approach, Lotfi A. Zadeh and Charles A. Desoer. Written by two pioneers in the field, this exploration of the state space approach focuses on problems of stability and control, plus connections between this approach and classical techniques. 1963 edition. 656pp. 6 1/8 x 9 1/4. 0-486-46663-9

Mathematics–Bestsellers

HANDBOOK OF MATHEMATICAL FUNCTIONS: with Formulas, Graphs, and Mathematical Tables, Edited by Milton Abramowitz and Irene A. Stegun. A classic resource for working with special functions, standard trig, and exponential logarithmic definitions and extensions, it features 29 sets of tables, some to as high as 20 places. 1046pp. 8 x 10 1/2. 0-486-61272-4

ABSTRACT AND CONCRETE CATEGORIES: The Joy of Cats, Jiri Adamek, Horst Herrlich, and George E. Strecker. This up-to-date introductory treatment employs category theory to explore the theory of structures. Its unique approach stresses concrete categories and presents a systematic view of factorization structures. Numerous examples. 1990 edition, updated 2004. 528pp. 6 1/8 x 9 1/4. 0-486-46934-4

MATHEMATICS: Its Content, Methods and Meaning, A. D. Aleksandrov, A. N. Kolmogorov, and M. A. Lavrent'ev. Major survey offers comprehensive, coherent discussions of analytic geometry, algebra, differential equations, calculus of variations, functions of a complex variable, prime numbers, linear and non-Euclidean geometry, topology, functional analysis, more. 1963 edition. 1120pp. 5 3/8 x 8 1/2. 0-486-40916-3

INTRODUCTION TO VECTORS AND TENSORS: Second Edition--Two Volumes Bound as One, Ray M. Bowen and C.-C. Wang. Convenient single-volume compilation of two texts offers both introduction and in-depth survey. Geared toward engineering and science students rather than mathematicians, it focuses on physics and engineering applications. 1976 edition. 560pp. 6 1/2 x 9 1/4. 0-486-46914-X

AN INTRODUCTION TO ORTHOGONAL POLYNOMIALS, Theodore S. Chihara. Concise introduction covers general elementary theory, including the representation theorem and distribution functions, continued fractions and chain sequences, the recurrence formula, special functions, and some specific systems. 1978 edition. 272pp. 5 3/8 x 8 1/2. 0-486-47929-3

ADVANCED MATHEMATICS FOR ENGINEERS AND SCIENTISTS, Paul DuChateau. This primary text and supplemental reference focuses on linear algebra, calculus, and ordinary differential equations. Additional topics include partial differential equations and approximation methods. Includes solved problems. 1992 edition. 400pp. 7 1/2 x 9 1/4. 0-486-47930-7

PARTIAL DIFFERENTIAL EQUATIONS FOR SCIENTISTS AND ENGINEERS, Stanley J. Farlow. Practical text shows how to formulate and solve partial differential equations. Coverage of diffusion-type problems, hyperbolic-type problems, elliptic-type problems, numerical and approximate methods. Solution guide available upon request. 1982 edition. 414pp. 6 1/8 x 9 1/4. 0-486-67620-X

VARIATIONAL PRINCIPLES AND FREE-BOUNDARY PROBLEMS, Avner Friedman. Advanced graduate-level text examines variational methods in partial differential equations and illustrates their applications to free-boundary problems. Features detailed statements of standard theory of elliptic and parabolic operators. 1982 edition. 720pp. 6 1/8 x 9 1/4. 0-486-47853-X

LINEAR ANALYSIS AND REPRESENTATION THEORY, Steven A. Gaal. Unified treatment covers topics from the theory of operators and operator algebras on Hilbert spaces; integration and representation theory for topological groups; and the theory of Lie algebras, Lie groups, and transform groups. 1973 edition. 704pp. 6 1/8 x 9 1/4. 0-486-47851-3

A SURVEY OF INDUSTRIAL MATHEMATICS, Charles R. MacCluer. Students learn how to solve problems they'll encounter in their professional lives with this concise single-volume treatment. It employs MATLAB and other strategies to explore typical industrial problems. 2000 edition. 384pp. 5 3/8 x 8 1/2. 0-486-47702-9

NUMBER SYSTEMS AND THE FOUNDATIONS OF ANALYSIS, Elliott Mendelson. Geared toward undergraduate and beginning graduate students, this study explores natural numbers, integers, rational numbers, real numbers, and complex numbers. Numerous exercises and appendixes supplement the text. 1973 edition. 368pp. 5 3/8 x 8 1/2. 0-486-45792-3

A FIRST LOOK AT NUMERICAL FUNCTIONAL ANALYSIS, W. W. Sawyer. Text by renowned educator shows how problems in numerical analysis lead to concepts of functional analysis. Topics include Banach and Hilbert spaces, contraction mappings, convergence, differentiation and integration, and Euclidean space. 1978 edition. 208pp. 5 3/8 x 8 1/2. 0-486-47882-3

FRACTALS, CHAOS, POWER LAWS: Minutes from an Infinite Paradise, Manfred Schroeder. A fascinating exploration of the connections between chaos theory, physics, biology, and mathematics, this book abounds in award-winning computer graphics, optical illusions, and games that clarify memorable insights into self-similarity. 1992 edition. 448pp. 6 1/8 x 9 1/4. 0-486-47204-3

SET THEORY AND THE CONTINUUM PROBLEM, Raymond M. Smullyan and Melvin Fitting. A lucid, elegant, and complete survey of set theory, this three-part treatment explores axiomatic set theory, the consistency of the continuum hypothesis, and forcing and independence results. 1996 edition. 336pp. 6 x 9. 0-486-47484-4

DYNAMICAL SYSTEMS, Shlomo Sternberg. A pioneer in the field of dynamical systems discusses one-dimensional dynamics, differential equations, random walks, iterated function systems, symbolic dynamics, and Markov chains. Supplementary materials include PowerPoint slides and MATLAB exercises. 2010 edition. 272pp. 6 1/8 x 9 1/4. 0-486-47705-3

ORDINARY DIFFERENTIAL EQUATIONS, Morris Tenenbaum and Harry Pollard. Skillfully organized introductory text examines origin of differential equations, then defines basic terms and outlines general solution of a differential equation. Explores integrating factors; dilution and accretion problems; Laplace Transforms; Newton's Interpolation Formulas, more. 818pp. 5 3/8 x 8 1/2. 0-486-64940-7

MATROID THEORY, D. J. A. Welsh. Text by a noted expert describes standard examples and investigation results, using elementary proofs to develop basic matroid properties before advancing to a more sophisticated treatment. Includes numerous exercises. 1976 edition. 448pp. 5 3/8 x 8 1/2. 0-486-47439-9

THE CONCEPT OF A RIEMANN SURFACE, Hermann Weyl. This classic on the general history of functions combines function theory and geometry, forming the basis of the modern approach to analysis, geometry, and topology. 1955 edition. 208pp. 5 3/8 x 8 1/2. 0-486-47004-0

THE LAPLACE TRANSFORM, David Vernon Widder. This volume focuses on the Laplace and Stieltjes transforms, offering a highly theoretical treatment. Topics include fundamental formulas, the moment problem, monotonic functions, and Tauberian theorems. 1941 edition. 416pp. 5 3/8 x 8 1/2. 0-486-47755-X

Mathematics–Logic and Problem Solving

PERPLEXING PUZZLES AND TANTALIZING TEASERS, Martin Gardner. Ninety-three riddles, mazes, illusions, tricky questions, word and picture puzzles, and other challenges offer hours of entertainment for youngsters. Filled with rib-tickling drawings. Solutions. 224pp. 5 3/8 x 8 1/2. 0-486-25637-5

MY BEST MATHEMATICAL AND LOGIC PUZZLES, Martin Gardner. The noted expert selects 70 of his favorite "short" puzzles. Includes The Returning Explorer, The Mutilated Chessboard, Scrambled Box Tops, and dozens more. Complete solutions included. 96pp. 5 3/8 x 8 1/2. 0-486-28152-3

THE LADY OR THE TIGER?: and Other Logic Puzzles, Raymond M. Smullyan. Created by a renowned puzzle master, these whimsically themed challenges involve paradoxes about probability, time, and change; metapuzzles; and self-referentiality. Nineteen chapters advance in difficulty from relatively simple to highly complex. 1982 edition. 240pp. 5 3/8 x 8 1/2. 0-486-47027-X

SATAN, CANTOR AND INFINITY: Mind-Boggling Puzzles, Raymond M. Smullyan. A renowned mathematician tells stories of knights and knaves in an entertaining look at the logical precepts behind infinity, probability, time, and change. Requires a strong background in mathematics. Complete solutions. 288pp. 5 3/8 x 8 1/2. 0-486-47036-9

THE RED BOOK OF MATHEMATICAL PROBLEMS, Kenneth S. Williams and Kenneth Hardy. Handy compilation of 100 practice problems, hints and solutions indispensable for students preparing for the William Lowell Putnam and other mathematical competitions. Preface to the First Edition. Sources. 1988 edition. 192pp. 5 3/8 x 8 1/2. 0-486-69415-1

KING ARTHUR IN SEARCH OF HIS DOG AND OTHER CURIOUS PUZZLES, Raymond M. Smullyan. This fanciful, original collection for readers of all ages features arithmetic puzzles, logic problems related to crime detection, and logic and arithmetic puzzles involving King Arthur and his Dogs of the Round Table. 160pp. 5 3/8 x 8 1/2. 0-486-47435-6

UNDECIDABLE THEORIES: Studies in Logic and the Foundation of Mathematics, Alfred Tarski in collaboration with Andrzej Mostowski and Raphael M. Robinson. This well-known book by the famed logician consists of three treatises: "A General Method in Proofs of Undecidability," "Undecidability and Essential Undecidability in Mathematics," and "Undecidability of the Elementary Theory of Groups." 1953 edition. 112pp. 5 3/8 x 8 1/2. 0-486-47703-7

LOGIC FOR MATHEMATICIANS, J. Barkley Rosser. Examination of essential topics and theorems assumes no background in logic. "Undoubtedly a major addition to the literature of mathematical logic." – *Bulletin of the American Mathematical Society.* 1978 edition. 592pp. 6 1/8 x 9 1/4. 0-486-46898-4

INTRODUCTION TO PROOF IN ABSTRACT MATHEMATICS, Andrew Wohlgemuth. This undergraduate text teaches students what constitutes an acceptable proof, and it develops their ability to do proofs of routine problems as well as those requiring creative insights. 1990 edition. 384pp. 6 1/2 x 9 1/4. 0-486-47854-8

FIRST COURSE IN MATHEMATICAL LOGIC, Patrick Suppes and Shirley Hill. Rigorous introduction is simple enough in presentation and context for wide range of students. Symbolizing sentences; logical inference; truth and validity; truth tables; terms, predicates, universal quantifiers; universal specification and laws of identity; more. 288pp. 5 3/8 x 8 1/2. 0-486-42259-3

Mathematics–Algebra and Calculus

VECTOR CALCULUS, Peter Baxandall and Hans Liebeck. This introductory text offers a rigorous, comprehensive treatment. Classical theorems of vector calculus are amply illustrated with figures, worked examples, physical applications, and exercises with hints and answers. 1986 edition. 560pp. 5 3/8 x 8 1/2. 0-486-46620-5

ADVANCED CALCULUS: An Introduction to Classical Analysis, Louis Brand. A course in analysis that focuses on the functions of a real variable, this text introduces the basic concepts in their simplest setting and illustrates its teachings with numerous examples, theorems, and proofs. 1955 edition. 592pp. 5 3/8 x 8 1/2. 0-486-44548-8

ADVANCED CALCULUS, Avner Friedman. Intended for students who have already completed a one-year course in elementary calculus, this two-part treatment advances from functions of one variable to those of several variables. Solutions. 1971 edition. 432pp. 5 3/8 x 8 1/2. 0-486-45795-8

METHODS OF MATHEMATICS APPLIED TO CALCULUS, PROBABILITY, AND STATISTICS, Richard W. Hamming. This 4-part treatment begins with algebra and analytic geometry and proceeds to an exploration of the calculus of algebraic functions and transcendental functions and applications. 1985 edition. Includes 310 figures and 18 tables. 880pp. 6 1/2 x 9 1/4. 0-486-43945-3

BASIC ALGEBRA I: Second Edition, Nathan Jacobson. A classic text and standard reference for a generation, this volume covers all undergraduate algebra topics, including groups, rings, modules, Galois theory, polynomials, linear algebra, and associative algebra. 1985 edition. 528pp. 6 1/8 x 9 1/4. 0-486-47189-6

BASIC ALGEBRA II: Second Edition, Nathan Jacobson. This classic text and standard reference comprises all subjects of a first-year graduate-level course, including in-depth coverage of groups and polynomials and extensive use of categories and functors. 1989 edition. 704pp. 6 1/8 x 9 1/4. 0-486-47187-X

CALCULUS: An Intuitive and Physical Approach (Second Edition), Morris Kline. Application-oriented introduction relates the subject as closely as possible to science with explorations of the derivative; differentiation and integration of the powers of x; theorems on differentiation, antidifferentiation; the chain rule; trigonometric functions; more. Examples. 1967 edition. 960pp. 6 1/2 x 9 1/4. 0-486-40453-6

ABSTRACT ALGEBRA AND SOLUTION BY RADICALS, John E. Maxfield and Margaret W. Maxfield. Accessible advanced undergraduate-level text starts with groups, rings, fields, and polynomials and advances to Galois theory, radicals and roots of unity, and solution by radicals. Numerous examples, illustrations, exercises, appendixes. 1971 edition. 224pp. 6 1/8 x 9 1/4. 0-486-47723-1

AN INTRODUCTION TO THE THEORY OF LINEAR SPACES, Georgi E. Shilov. Translated by Richard A. Silverman. Introductory treatment offers a clear exposition of algebra, geometry, and analysis as parts of an integrated whole rather than separate subjects. Numerous examples illustrate many different fields, and problems include hints or answers. 1961 edition. 320pp. 5 3/8 x 8 1/2. 0-486-63070-6

LINEAR ALGEBRA, Georgi E. Shilov. Covers determinants, linear spaces, systems of linear equations, linear functions of a vector argument, coordinate transformations, the canonical form of the matrix of a linear operator, bilinear and quadratic forms, and more. 387pp. 5 3/8 x 8 1/2. 0-486-63518-X

Mathematics–Probability and Statistics

BASIC PROBABILITY THEORY, Robert B. Ash. This text emphasizes the probabilistic way of thinking, rather than measure-theoretic concepts. Geared toward advanced undergraduates and graduate students, it features solutions to some of the problems. 1970 edition. 352pp. 5 3/8 x 8 1/2. 0-486-46628-0

PRINCIPLES OF STATISTICS, M. G. Bulmer. Concise description of classical statistics, from basic dice probabilities to modern regression analysis. Equal stress on theory and applications. Moderate difficulty; only basic calculus required. Includes problems with answers. 252pp. 5 5/8 x 8 1/4. 0-486-63760-3

OUTLINE OF BASIC STATISTICS: Dictionary and Formulas, John E. Freund and Frank J. Williams. Handy guide includes a 70-page outline of essential statistical formulas covering grouped and ungrouped data, finite populations, probability, and more, plus over 1,000 clear, concise definitions of statistical terms. 1966 edition. 208pp. 5 3/8 x 8 1/2. 0-486-47769-X

GOOD THINKING: The Foundations of Probability and Its Applications, Irving J. Good. This in-depth treatment of probability theory by a famous British statistician explores Keynesian principles and surveys such topics as Bayesian rationality, corroboration, hypothesis testing, and mathematical tools for induction and simplicity. 1983 edition. 352pp. 5 3/8 x 8 1/2. 0-486-47438-0

INTRODUCTION TO PROBABILITY THEORY WITH CONTEMPORARY APPLICATIONS, Lester L. Helms. Extensive discussions and clear examples, written in plain language, expose students to the rules and methods of probability. Exercises foster problem-solving skills, and all problems feature step-by-step solutions. 1997 edition. 368pp. 6 1/2 x 9 1/4. 0-486-47418-6

CHANCE, LUCK, AND STATISTICS, Horace C. Levinson. In simple, non-technical language, this volume explores the fundamentals governing chance and applies them to sports, government, and business. "Clear and lively ... remarkably accurate." – *Scientific Monthly*. 384pp. 5 3/8 x 8 1/2. 0-486-41997-5

FIFTY CHALLENGING PROBLEMS IN PROBABILITY WITH SOLUTIONS, Frederick Mosteller. Remarkable puzzlers, graded in difficulty, illustrate elementary and advanced aspects of probability. These problems were selected for originality, general interest, or because they demonstrate valuable techniques. Also includes detailed solutions. 88pp. 5 3/8 x 8 1/2. 0-486-65355-2

EXPERIMENTAL STATISTICS, Mary Gibbons Natrella. A handbook for those seeking engineering information and quantitative data for designing, developing, constructing, and testing equipment. Covers the planning of experiments, the analyzing of extreme-value data; and more. 1966 edition. Index. Includes 52 figures and 76 tables. 560pp. 8 3/8 x 11. 0-486-43937-2

STOCHASTIC MODELING: Analysis and Simulation, Barry L. Nelson. Coherent introduction to techniques also offers a guide to the mathematical, numerical, and simulation tools of systems analysis. Includes formulation of models, analysis, and interpretation of results. 1995 edition. 336pp. 6 1/8 x 9 1/4. 0-486-47770-3

INTRODUCTION TO BIOSTATISTICS: Second Edition, Robert R. Sokal and F. James Rohlf. Suitable for undergraduates with a minimal background in mathematics, this introduction ranges from descriptive statistics to fundamental distributions and the testing of hypotheses. Includes numerous worked-out problems and examples. 1987 edition. 384pp. 6 1/8 x 9 1/4. 0-486-46961-1

Mathematics–Geometry and Topology

PROBLEMS AND SOLUTIONS IN EUCLIDEAN GEOMETRY, M. N. Aref and William Wernick. Based on classical principles, this book is intended for a second course in Euclidean geometry and can be used as a refresher. More than 200 problems include hints and solutions. 1968 edition. 272pp. 5 3/8 x 8 1/2. 0-486-47720-7

TOPOLOGY OF 3-MANIFOLDS AND RELATED TOPICS, Edited by M. K. Fort, Jr. With a New Introduction by Daniel Silver. Summaries and full reports from a 1961 conference discuss decompositions and subsets of 3-space; n-manifolds; knot theory; the Poincaré conjecture; and periodic maps and isotopies. Familiarity with algebraic topology required. 1962 edition. 272pp. 6 1/8 x 9 1/4. 0-486-47753-3

POINT SET TOPOLOGY, Steven A. Gaal. Suitable for a complete course in topology, this text also functions as a self-contained treatment for independent study. Additional enrichment materials make it equally valuable as a reference. 1964 edition. 336pp. 5 3/8 x 8 1/2. 0-486-47222-1

INVITATION TO GEOMETRY, Z. A. Melzak. Intended for students of many different backgrounds with only a modest knowledge of mathematics, this text features self-contained chapters that can be adapted to several types of geometry courses. 1983 edition. 240pp. 5 3/8 x 8 1/2. 0-486-46626-4

TOPOLOGY AND GEOMETRY FOR PHYSICISTS, Charles Nash and Siddhartha Sen. Written by physicists for physics students, this text assumes no detailed background in topology or geometry. Topics include differential forms, homotopy, homology, cohomology, fiber bundles, connection and covariant derivatives, and Morse theory. 1983 edition. 320pp. 5 3/8 x 8 1/2. 0-486-47852-1

BEYOND GEOMETRY: Classic Papers from Riemann to Einstein, Edited with an Introduction and Notes by Peter Pesic. This is the only English-language collection of these 8 accessible essays. They trace seminal ideas about the foundations of geometry that led to Einstein's general theory of relativity. 224pp. 6 1/8 x 9 1/4. 0-486-45350-2

GEOMETRY FROM EUCLID TO KNOTS, Saul Stahl. This text provides a historical perspective on plane geometry and covers non-neutral Euclidean geometry, circles and regular polygons, projective geometry, symmetries, inversions, informal topology, and more. Includes 1,000 practice problems. Solutions available. 2003 edition. 480pp. 6 1/8 x 9 1/4. 0-486-47459-3

TOPOLOGICAL VECTOR SPACES, DISTRIBUTIONS AND KERNELS, François Trèves. Extending beyond the boundaries of Hilbert and Banach space theory, this text focuses on key aspects of functional analysis, particularly in regard to solving partial differential equations. 1967 edition. 592pp. 5 3/8 x 8 1/2. 0-486-45352-9

INTRODUCTION TO PROJECTIVE GEOMETRY, C. R. Wylie, Jr. This introductory volume offers strong reinforcement for its teachings, with detailed examples and numerous theorems, proofs, and exercises, plus complete answers to all odd-numbered end-of-chapter problems. 1970 edition. 576pp. 6 1/8 x 9 1/4. 0-486-46895-X

FOUNDATIONS OF GEOMETRY, C. R. Wylie, Jr. Geared toward students preparing to teach high school mathematics, this text explores the principles of Euclidean and non-Euclidean geometry and covers both generalities and specifics of the axiomatic method. 1964 edition. 352pp. 6 x 9. 0-486-47214-0

Mathematics–History

THE WORKS OF ARCHIMEDES, Archimedes. Translated by Sir Thomas Heath. Complete works of ancient geometer feature such topics as the famous problems of the ratio of the areas of a cylinder and an inscribed sphere; the properties of conoids, spheroids, and spirals; more. 326pp. 5 3/8 x 8 1/2. 0-486-42084-1

THE HISTORICAL ROOTS OF ELEMENTARY MATHEMATICS, Lucas N. H. Bunt, Phillip S. Jones, and Jack D. Bedient. Exciting, hands-on approach to understanding fundamental underpinnings of modern arithmetic, algebra, geometry and number systems examines their origins in early Egyptian, Babylonian, and Greek sources. 336pp. 5 3/8 x 8 1/2. 0-486-25563-8

THE THIRTEEN BOOKS OF EUCLID'S ELEMENTS, Euclid. Contains complete English text of all 13 books of the Elements plus critical apparatus analyzing each definition, postulate, and proposition in great detail. Covers textual and linguistic matters; mathematical analyses of Euclid's ideas; classical, medieval, Renaissance and modern commentators; refutations, supports, extrapolations, reinterpretations and historical notes. 995 figures. Total of 1,425pp. All books 5 3/8 x 8 1/2.
Vol. I: 443pp. 0-486-60088-2
Vol. II: 464pp. 0-486-60089-0
Vol. III: 546pp. 0-486-60090-4

A HISTORY OF GREEK MATHEMATICS, Sir Thomas Heath. This authoritative two-volume set that covers the essentials of mathematics and features every landmark innovation and every important figure, including Euclid, Apollonius, and others. 5 3/8 x 8 1/2.
Vol. I: 461pp. 0-486-24073-8
Vol. II: 597pp. 0-486-24074-6

A MANUAL OF GREEK MATHEMATICS, Sir Thomas L. Heath. This concise but thorough history encompasses the enduring contributions of the ancient Greek mathematicians whose works form the basis of most modern mathematics. Discusses Pythagorean arithmetic, Plato, Euclid, more. 1931 edition. 576pp. 5 3/8 x 8 1/2.
0-486-43231-9

CHINESE MATHEMATICS IN THE THIRTEENTH CENTURY, Ulrich Libbrecht. An exploration of the 13th-century mathematician Ch'in, this fascinating book combines what is known of the mathematician's life with a history of his only extant work, the Shu-shu chiu-chang. 1973 edition. 592pp. 5 3/8 x 8 1/2.
0-486-44619-0

PHILOSOPHY OF MATHEMATICS AND DEDUCTIVE STRUCTURE IN EUCLID'S ELEMENTS, Ian Mueller. This text provides an understanding of the classical Greek conception of mathematics as expressed in Euclid's Elements. It focuses on philosophical, foundational, and logical questions and features helpful appendixes. 400pp. 6 1/2 x 9 1/4. 0-486-45300-6

BEYOND GEOMETRY: Classic Papers from Riemann to Einstein, Edited with an Introduction and Notes by Peter Pesic. This is the only English-language collection of these 8 accessible essays. They trace seminal ideas about the foundations of geometry that led to Einstein's general theory of relativity. 224pp. 6 1/8 x 9 1/4. 0-486-45350-2

HISTORY OF MATHEMATICS, David E. Smith. Two-volume history – from Egyptian papyri and medieval maps to modern graphs and diagrams. Non-technical chronological survey with thousands of biographical notes, critical evaluations, and contemporary opinions on over 1,100 mathematicians. 5 3/8 x 8 1/2.
Vol. I: 618pp. 0-486-20429-4
Vol. II: 736pp. 0-486-20430-8

Physics

THEORETICAL NUCLEAR PHYSICS, John M. Blatt and Victor F. Weisskopf. An uncommonly clear and cogent investigation and correlation of key aspects of theoretical nuclear physics by leading experts: the nucleus, nuclear forces, nuclear spectroscopy, two-, three- and four-body problems, nuclear reactions, beta-decay and nuclear shell structure. 896pp. 5 3/8 x 8 1/2. 0-486-66827-4

QUANTUM THEORY, David Bohm. This advanced undergraduate-level text presents the quantum theory in terms of qualitative and imaginative concepts, followed by specific applications worked out in mathematical detail. 655pp. 5 3/8 x 8 1/2. 0-486-65969-0

ATOMIC PHYSICS AND HUMAN KNOWLEDGE, Niels Bohr. Articles and speeches by the Nobel Prize–winning physicist, dating from 1934 to 1958, offer philosophical explorations of the relevance of atomic physics to many areas of human endeavor. 1961 edition. 112pp. 5 3/8 x 8 1/2. 0-486-47928-5

COSMOLOGY, Hermann Bondi. A co-developer of the steady-state theory explores his conception of the expanding universe. This historic book was among the first to present cosmology as a separate branch of physics. 1961 edition. 192pp. 5 3/8 x 8 1/2. 0-486-47483-6

LECTURES ON QUANTUM MECHANICS, Paul A. M. Dirac. Four concise, brilliant lectures on mathematical methods in quantum mechanics from Nobel Prize-winning quantum pioneer build on idea of visualizing quantum theory through the use of classical mechanics. 96pp. 5 3/8 x 8 1/2. 0-486-41713-1

THE PRINCIPLE OF RELATIVITY, Albert Einstein and Frances A. Davis. Eleven papers that forged the general and special theories of relativity include seven papers by Einstein, two by Lorentz, and one each by Minkowski and Weyl. 1923 edition. 240pp. 5 3/8 x 8 1/2. 0-486-60081-5

PHYSICS OF WAVES, William C. Elmore and Mark A. Heald. Ideal as a classroom text or for individual study, this unique one-volume overview of classical wave theory covers wave phenomena of acoustics, optics, electromagnetic radiations, and more. 477pp. 5 3/8 x 8 1/2. 0-486-64926-1

THERMODYNAMICS, Enrico Fermi. In this classic of modern science, the Nobel Laureate presents a clear treatment of systems, the First and Second Laws of Thermodynamics, entropy, thermodynamic potentials, and much more. Calculus required. 160pp. 5 3/8 x 8 1/2. 0-486-60361-X

QUANTUM THEORY OF MANY-PARTICLE SYSTEMS, Alexander L. Fetter and John Dirk Walecka. Self-contained treatment of nonrelativistic many-particle systems discusses both formalism and applications in terms of ground-state (zero-temperature) formalism, finite-temperature formalism, canonical transformations, and applications to physical systems. 1971 edition. 640pp. 5 3/8 x 8 1/2. 0-486-42827-3

QUANTUM MECHANICS AND PATH INTEGRALS: Emended Edition, Richard P. Feynman and Albert R. Hibbs. Emended by Daniel F. Styer. The Nobel Prize–winning physicist presents unique insights into his theory and its applications. Feynman starts with fundamentals and advances to the perturbation method, quantum electrodynamics, and statistical mechanics. 1965 edition, emended in 2005. 384pp. 6 1/8 x 9 1/4. 0-486-47722-3

Physics

INTRODUCTION TO MODERN OPTICS, Grant R. Fowles. A complete basic undergraduate course in modern optics for students in physics, technology, and engineering. The first half deals with classical physical optics; the second, quantum nature of light. Solutions. 336pp. 5 3/8 x 8 1/2. 0-486-65957-7

THE QUANTUM THEORY OF RADIATION: Third Edition, W. Heitler. The first comprehensive treatment of quantum physics in any language, this classic introduction to basic theory remains highly recommended and widely used, both as a text and as a reference. 1954 edition. 464pp. 5 3/8 x 8 1/2. 0-486-64558-4

QUANTUM FIELD THEORY, Claude Itzykson and Jean-Bernard Zuber. This comprehensive text begins with the standard quantization of electrodynamics and perturbative renormalization, advancing to functional methods, relativistic bound states, broken symmetries, nonabelian gauge fields, and asymptotic behavior. 1980 edition. 752pp. 6 1/2 x 9 1/4. 0-486-44568-2

FOUNDATIONS OF POTENTIAL THERY, Oliver D. Kellogg. Introduction to fundamentals of potential functions covers the force of gravity, fields of force, potentials, harmonic functions, electric images and Green's function, sequences of harmonic functions, fundamental existence theorems, and much more. 400pp. 5 3/8 x 8 1/2. 0-486-60144-7

FUNDAMENTALS OF MATHEMATICAL PHYSICS, Edgar A. Kraut. Indispensable for students of modern physics, this text provides the necessary background in mathematics to study the concepts of electromagnetic theory and quantum mechanics. 1967 edition. 480pp. 6 1/2 x 9 1/4. 0-486-45809-1

GEOMETRY AND LIGHT: The Science of Invisibility, Ulf Leonhardt and Thomas Philbin. Suitable for advanced undergraduate and graduate students of engineering, physics, and mathematics and scientific researchers of all types, this is the first authoritative text on invisibility and the science behind it. More than 100 full-color illustrations, plus exercises with solutions. 2010 edition. 288pp. 7 x 9 1/4. 0-486-47693-6

QUANTUM MECHANICS: New Approaches to Selected Topics, Harry J. Lipkin. Acclaimed as "excellent" (*Nature*) and "very original and refreshing" (*Physics Today*), these studies examine the Mössbauer effect, many-body quantum mechanics, scattering theory, Feynman diagrams, and relativistic quantum mechanics. 1973 edition. 480pp. 5 3/8 x 8 1/2. 0-486-45893-8

THEORY OF HEAT, James Clerk Maxwell. This classic sets forth the fundamentals of thermodynamics and kinetic theory simply enough to be understood by beginners, yet with enough subtlety to appeal to more advanced readers, too. 352pp. 5 3/8 x 8 1/2. 0-486-41735-2

QUANTUM MECHANICS, Albert Messiah. Subjects include formalism and its interpretation, analysis of simple systems, symmetries and invariance, methods of approximation, elements of relativistic quantum mechanics, much more. "Strongly recommended." – *American Journal of Physics*. 1152pp. 5 3/8 x 8 1/2. 0-486-40924-4

RELATIVISTIC QUANTUM FIELDS, Charles Nash. This graduate-level text contains techniques for performing calculations in quantum field theory. It focuses chiefly on the dimensional method and the renormalization group methods. Additional topics include functional integration and differentiation. 1978 edition. 240pp. 5 3/8 x 8 1/2. 0-486-47752-5

Physics

MATHEMATICAL TOOLS FOR PHYSICS, James Nearing. Encouraging students' development of intuition, this original work begins with a review of basic mathematics and advances to infinite series, complex algebra, differential equations, Fourier series, and more. 2010 edition. 496pp. 6 1/8 x 9 1/4. 0-486-48212-X

TREATISE ON THERMODYNAMICS, Max Planck. Great classic, still one of the best introductions to thermodynamics. Fundamentals, first and second principles of thermodynamics, applications to special states of equilibrium, more. Numerous worked examples. 1917 edition. 297pp. 5 3/8 x 8. 0-486-66371-X

AN INTRODUCTION TO RELATIVISTIC QUANTUM FIELD THEORY, Silvan S. Schweber. Complete, systematic, and self-contained, this text introduces modern quantum field theory. "Combines thorough knowledge with a high degree of didactic ability and a delightful style." – *Mathematical Reviews*. 1961 edition. 928pp. 5 3/8 x 8 1/2. 0-486-44228-4

THE ELECTROMAGNETIC FIELD, Albert Shadowitz. Comprehensive undergraduate text covers basics of electric and magnetic fields, building up to electromagnetic theory. Related topics include relativity theory. Over 900 problems, some with solutions. 1975 edition. 768pp. 5 5/8 x 8 1/4. 0-486-65660-8

THE PRINCIPLES OF STATISTICAL MECHANICS, Richard C. Tolman. Definitive treatise offers a concise exposition of classical statistical mechanics and a thorough elucidation of quantum statistical mechanics, plus applications of statistical mechanics to thermodynamic behavior. 1930 edition. 704pp. 5 5/8 x 8 1/4. 0-486-63896-0

INTRODUCTION TO THE PHYSICS OF FLUIDS AND SOLIDS, James S. Trefil. This interesting, informative survey by a well-known science author ranges from classical physics and geophysical topics, from the rings of Saturn and the rotation of the galaxy to underground nuclear tests. 1975 edition. 320pp. 5 3/8 x 8 1/2. 0-486-47437-2

STATISTICAL PHYSICS, Gregory H. Wannier. Classic text combines thermodynamics, statistical mechanics, and kinetic theory in one unified presentation. Topics include equilibrium statistics of special systems, kinetic theory, transport coefficients, and fluctuations. Problems with solutions. 1966 edition. 532pp. 5 3/8 x 8 1/2. 0-486-65401-X

SPACE, TIME, MATTER, Hermann Weyl. Excellent introduction probes deeply into Euclidean space, Riemann's space, Einstein's general relativity, gravitational waves and energy, and laws of conservation. "A classic of physics." – *British Journal for Philosophy and Science*. 330pp. 5 3/8 x 8 1/2. 0-486-60267-2

RANDOM VIBRATIONS: Theory and Practice, Paul H. Wirsching, Thomas L. Paez and Keith Ortiz. Comprehensive text and reference covers topics in probability, statistics, and random processes, plus methods for analyzing and controlling random vibrations. Suitable for graduate students and mechanical, structural, and aerospace engineers. 1995 edition. 464pp. 5 3/8 x 8 1/2. 0-486-45015-5

PHYSICS OF SHOCK WAVES AND HIGH-TEMPERATURE HYDRO DYNAMIC PHENOMENA, Ya B. Zel'dovich and Yu P. Raizer. Physical, chemical processes in gases at high temperatures are focus of outstanding text, which combines material from gas dynamics, shock-wave theory, thermodynamics and statistical physics, other fields. 284 illustrations. 1966–1967 edition. 944pp. 6 1/8 x 9 1/4. 0-486-42002-7